W9-AAE-102

Autonomic Communication

Athanasios V. Vasilakos • Manish Parashar
Stamatis Karnouskos • Witold Pedrycz
Editors

Autonomic Communication

Editors
Athanasios V. Vasilakos
Department of Computer
and Telecommunications Engineering
University of Western Macedonia
Agios Dimitrios Park
501 00 Kozani
Greece
vasilako@ath.forthnet.gr

Manish Parashar
Department of Electrical
and Computer Engineering
Rutgers University
94 Brett Road
Piscataway, NJ 08854-8058
USA
parashar@caip.rutgers.edu

Stamatis Karnouskos
SAP AG Corporate Research Centre
Vincent-Prießnitz-Str. 1
76131 Karlsruhe
Germany
stamatis.karnouskos@sap.com

Witold Pedrycz
Department of Electrical
and Computer Engineering
University of Alberta
9107 116 Street
Edmonton, AB T6G 2V4
Canada
pedrycz@ece.ualberta.ca

ISBN 978-0-387-09752-7 e-ISBN 978-0-387-09753-4
DOI 10.1007/978-0-387-09753-4
Springer Dordrecht Heidelberg London New York

Library of Congress Control Number: 2009931557

Printed on acid-free paper

Springer is part of Springer Science+Business Media (www.springer.com)

Foreword

As information technology becomes increasingly more sophisticated and capable, repeated studies have demonstrated that this has been matched by an increasing cost and complexity in configuring, managing and servicing such systems.

Despite the obvious benefits in performance & scalability that has been provided by the ongoing evolution of faster CPUs, increased network bandwidth and storage capacity, the cost of dynamically managing the hardware, software and infrastructure has continued to rise at an alarming rate.

Autonomic Computing emerged at the dawn of the 21st century out of cross-industry and academic research and development in simplifying administration, configuration, deployment and management of IT systems.

Driven by a need to address what many had seen as a growing crisis in the IT industry for increased visibility, control and automation of complex IT systems, autonomic computing focuses upon providing embedded and integrated management systems that enable self-healing, self-configuration, self-optimization and self-protection.

As we move into the era of energy-aware IT systems and cloud-computing, the needs for autonomic systems have continue to evolve. Recently there has been an increased awareness of the need for efficient utilization of data centers associated with the changing economics of energy supply and the politics of climate change. Optimal use of communications infrastructure and the associated middleware has becoming increasingly critical. Embedded autonomic capabilities to provide self-management has become critical in delivering new dynamic infrastructure in IT.

Increased use of the Internet has led to global-scale broad-band networking. When coupled with adoption of wireless communications and mobile computing by consumers and enterprises, the need for self-managing network services and autonomic communications becomes more pressing. The same network infrastructure is being shared across more and more services. Enterprises, consumers and governments expect to have communication systems that can autonomically recover from failures or reconfigure themselves to cope with increased demand.

Those same network services are also being used to provide interfaces onto traditional physical infrastructure: Smart utility grids for efficient production and distribution of gas, water and electricity, wireless home-metering, remote wireless tracking of parcels, shipping containers and wireless monitoring of the environment and climate change. These are all expected to become the norm as we progress towards the 'Smart Planet'.

This volume is a timely overview of the evolution of autonomic computing to communications, networking and sensor systems, and provides detailed insights into the multi-disciplinary research topics that comprise the field of Autonomic Computing and Communications.

April 2009

Dr. Matt Ellis
Vice President
Autonomic Computing, IBM

Preface

The emerging world is pervasive and strives towards integrating people, technology, environment and knowledge. This emerging vision supports approaches that set the user at the center of attention, while technology becomes invisible, hidden in the natural surroundings, but still functional, autonomous, self-adaptive, available when needed, and interactive.

Achieving this vision requires innovative communication architectures and services. Communication/networking solutions should become task- and knowledge-driven, enabling a service oriented, requirement and trust based development of communication infrastructure. The growing complexity of control requires increasingly distributed and self-organizing structures, relying on simple and dependable elements that are able to collaborate to develop sophisticated behaviors, and that can adapt to an evolving situation where new resources can become available, administrative domains can change and economic models can vary.

The networking and seamless integration of concepts, technologies and devices in a dynamically changing environment poses many challenges to the research community. There remain such crucial issues as interoperability, programmability, management, openness, reliability, performance, context awareness, intelligence, autonomy, security, privacy, safety, semantics, etc. However, the overall scale, complexity, heterogeneity and dynamics of these networked environments, together, result in essential management challenges which clearly go beyond current paradigms and practices, and need a fundamentally new approach.

Autonomic Communication is such an approach. It is inspired by biological systems and envisions communication systems that are large self-managing which can organize, configure, optimize, protect and heal themselves with minimal involvement of human administrators.

This edited volume explores conceptual models and associated technologies that will help realize the vision of autonomic communication, where devices and applications seamlessly interconnect, intelligently cooperate and autonomously manage themselves, and as a result, enable the borders of virtual and real world to vanish or become significantly blurred. The chapters contributing to the edited volume are the genuine testimony to the challenges and prospects of this rapidly growing area.

The volume is composed of 14 chapters covering a wide range of issue related to autonomic communication and are organized into 3 parts as listed below.

- Autonomic Communication Infrastructure
- Autonomic Communication Services and Middleware
- Applications to Ad-Hoc (Sensor) Networks and Pervasive Systems

Part I – Autonomic Communication Infrastructure: Part I focuses on various aspects of autonomic communication infrastructure and contains 4 chapters. Chapter 1 titled "*Bio-inspired Autonomic Structures: a middleware for Telecommunications Ecosystems*" investigates a bio-inspired autonomic middleware capable of leveraging the assets of the underlying network infrastructure whilst, at the same time, supporting the development of future Telecommunications and Internet Ecosystems. Chapter 2 titled "*Social-based autonomic routing in opportunistic networks*" investigates contex-aware routing in opportunistic autonomic infrastructures under the prism of peoples' social behavior. Chapter 3 titled "*A Collaborative Knowledge Plane for Autonomic Networks*" looks at a knowledge plane architecture aiming at self-adaptation and self-organization. Chapter 4 titled "*A Rate Feedback Predictive Control Scheme Based on Neural Network and Control Theory for Autonomic Communication*" tackles the difficulty in designing an efficient congestion control scheme by using back propagation neural networks.

Part II – Autonomic Communication Services and Middleware: Part II focuses on specific autonomic communication services and middleware frameworks, and contains 5 chapters. Chapter 5 titled "*Hovering Information – Self-Organising Information that Finds its Own Storage*" investigates a hovering information model and reports on simulations performed using replication and caching algorithms. Chapter 6 titled "*The CASCADAS Framework for Autonomic Communications*" presents a prototype distributed component-ware framework for autonomic and situation-aware communication and demonstrates it via a Pervasive Behavioural Advertisement scenario. Chapter 7 titled "*Autonomic Middleware for Automotive Embedded Systems*" describes an advanced autonomic platform-independent middleware framework focused on automotive embedded systems where high flexibility and automatic run-time reconfiguration is needed. Chapter 8 titled "*Social Opportunistic Computing: Design for Autonomic User-Centric Systems*" focuses on the diffusion of data in autonomic computing environments and the way the social attitudes of mobile users impact their design. Chapter 9 titled "*Programming and Validation Techniques for Reliable Goal-driven Autonomic Software*" investigates time and concurrency which are the most critical notions of complex software interactions in autonomous flight systems.

Part III – Applications to Ad-Hoc (Sensor) Networks and Pervasive Systems: Part III focuses on the applications of autonomic communication to ad-hoc (sensor) networks and pervasive systems, and contains 5 chapters. Chapter 10 titled "*Autonomic Communication in Pervasive Multimodal Multimedia Computing System*" focuses on autonomic communication protocols involved in the detection of interaction context and the multimodal computing system's corresponding adaptation. Chapter 11 titled "*Self-healing for Autonomic Pervasive Computing*" focuses on the self-healing aspects of autonomic pervasive computing and demonstrates it as a service in the Middleware Adaptability for Resource discovery, Knowledge usability, and Self-healing platform. Chapter 12 titled "*Map-based Design for Autonomic Wireless Sensor Networks*" presents an approach that exploits the spatial correlation of sensor readings and builds a model that abstracts from low-level communication issues and supports general applications by allowing for efficient event detection, prediction and queries. Chapter 13 titled "*An Efficient, Scalable and Robust P2P Overlay for Autonomic Communication*" focuses on the applicability of P2P systems in autonomic communication domain. Chapter 14 titled "*Autonomic and Co-evolutionary Sensor Networking with BiSNET/e*" proposes a biologically inspired architecture that allows wireless sensor network applications to simultaneously satisfy conflicting operational objectives by adapting to dynamic network conditions (e.g., network traffic and node/link failures) through evolution.

Our sincere thanks go to the authors who have contributed to this edited volume by sharing their most recent research findings and expertise. The anonymous reviewers deserve our thanks for their constructive criticism, helpful comments and in-depth insights. Finally, we would like to express our gratitude to Springer for agreeing to publish this volume and especially to Valerie Schofield and her editorial team that we interacted with on a continuous basis. We do hope the community of researchers and practitioners will find the content of this volume inspiring, insightful and enjoyable.

April 2009

Athanasios Vasilakos
Manish Parashar
Stamatis Karnouskos
Witold Pedrycz

Contents

Part I Autonomic Communication Infrastructure

Bio-inspired Autonomic Structures: a middleware for Telecommunications Ecosystems 3
Antonio Manzalini, Roberto Minerva and Corrado Moiso
1 Introduction 4
2 State of Art 6
2.1 Autonomic Frameworks 6
2.2 Interaction Algorithms 9
3 Bio-inpired Autonomic Structures 10
3.1 Concept of Autonomic Structures 11
3.2 BAS middleware 12
3.3 Data Components interactions: primitives 14
3.4 Components interactions: mechanisms and algorithms 15
4 Engineer self-organization 18
4.1 Game Theory for cross-layer design 20
4.2 Auctions for optimized resource allocation 22
5 Application scenarios 23
5.1 Self-Management for Telecommunications Networks 23
5.2 Cloud Computing 24
5.3 Home Networking 25
6 Conclusions 27
References 28

Social-based autonomic routing in opportunistic networks 31
Chiara Boldrini, Marco Conti, Andrea Passarella
1 Introduction 32
2 The opportunistic networking concept and its applications 33
2.1 Opportunistic networking case studies and applications 35
3 Social-based mobility 36
3.1 CMM and HCMM: functional description 37

3.2 HCMM vs. CMM: Controlling Node Positions 40
4 Routing in opportunistic networks 43
4.1 Context-oblivious routing 43
4.2 Partially context-aware routing 44
4.3 Fully context-aware routing 46
4.4 The History-based Opportunistic Routing protocol 47
5 Performance of opportunistic routing approaches under social mobility patterns 48
5.1 Performance evaluation strategy 48
5.2 Impact of collective groups' movements (reconfigurations) 50
5.3 Impact of User Sociability 54
5.4 Breaking Closed Groups 59
6 Conclusions 62
References 65

A Collaborative Knowledge Plane for Autonomic Networks 69
Maïssa Mbaye and Francine Krief
1 Introduction 69
2 Autonomic Networking 71
2.1 Basic concepts 71
2.2 Related Work 71
3 Collaborative knowledge plane architecture 73
3.1 Architecture overview 73
3.2 Basic Concepts 74
3.3 Knowledge plane building blocks 76
4 Self-adaptation loop 78
4.1 Machine learning algorithm for self-adaptation 78
4.2 Study Case: self-adaptation of a DiffServ router 79
5 Collaborative loop 83
5.1 Situated View and Basic concepts 83
5.2 Situated Knowledge sharing algorithm 86
5.3 Performance and guarantees 88
6 Conclusion 90
References 90

A Rate Feedback Predictive Control Scheme Based on Neural Network and Control Theory for Autonomic Communication 93
Naixue Xiong, Athanasios V. Vasilakos, Laurence T. Yang, Fei Long, Lei Shu, and Yingshu Li
1 Introduction 94
2 Congestion Control Model 95
2.1 The Predictive Control Model of a Bottleneck Buffer 95
3 The Predictive Control Technique 98
3.1 The BP Neural Network Architecture 98
3.2 Multi-step Neural Predictive Technique 98
4 The Simulation Results 99

5 Conclusion 105
References 106

Part II Autonomic Communication Services and Middleware

Hovering Information – Self-Organizing Information that Finds its Own Storage 111
Alfredo A. Villalba Castro, Giovanna Di Marzo Serugendo, and Dimitri Konstantas
1 Introduction 111
2 Applications 113
3 Hovering Information Concept 116
3.1 Coordinates, Distances and Areas 116
3.2 Mobile Nodes 116
3.3 Hovering Information 118
3.4 Notations 120
3.5 Properties - Requirements 121
4 Algorithms for Hovering Information 123
4.1 Assumptions 124
4.2 Safe, Risk and Relevant Areas 125
4.3 Replication 127
4.4 Caching 129
4.5 Cleaning 131
5 Evaluation 133
5.1 Simulation Settings and Scenarios 133
5.2 Metrics 134
5.3 Results 135
6 Related Works 141
7 Conclusion 143
7.1 Future Works 144
References 145

The CASCADAS Framework for Autonomic Communications 147
Luciano Baresi, Antonio Di Ferdinando, Antonio Manzalini, and Franco Zambonelli
1 Introduction 148
2 Autonomic Communication Frameworks 149
3 CASCADAS Framework 151
3.1 ACE Component Model 153
4 Semantic Self-Organization 156
5 Situation-Awareness 158
6 Pervasive Supervision 160
7 Security and Self-Preservation 161
8 Pervasive Behavioral Advertisement Scenario 163
9 Conclusions 165
References 166

Autonomic Middleware for Automotive Embedded Systems 169
Richard Anthony, DeJiu Chen, Martin Törngren, Detlef Scholle, Martin Sanfridson, Achim Rettberg, Tahir Naseer, Magnus Persson, and Lei Feng
1 Introduction 169
2 Automotive challenges and DySCAS 170
3 Background and related work 173
3.1 Middleware for distributed computer systems 173
3.2 Policy-based configuration 174
4 The DYSCAS Middleware Architecture 175
5 The Component Model for DySCAS Middleware Services 178
5.1 Policy-based configuration in the DySCAS component model 181
6 Autonomic reconfiguration 185
6.1 Task migration as an actuation mechanism 186
6.2 Using policies for flexible reconfiguration mechanisms . . 186
6.3 Algorithms and an approach for Dependability and Quality Management and Autonomic Configuration Management 186
6.4 An Approach for Load Balancing 189
7 A reference implementation of DySCAS 194
7.1 Implementation of the DySCAS architecture 194
8 A framework for modelling, designing and analysing dynamically configurable systems 197
8.1 Simulation 199
8.2 Safety analysis and formal verification 200
9 Open issues and ongoing work 205
9.1 Integration with a legacy statically reconfigurable platform 205
9.2 Implementation on a resource-constrained platform 205
10 Conclusions 206
References 207

Social Opportunistic Computing: Design for Autonomic User-Centric Systems 211
Iacopo Carreras, David Tacconi, and Arianna Bassoli
1 Introduction 211
2 The Study 213
3 First Phase: Understanding The Technological and User constraints 215
3.1 Assessing contact opportunities of an office environment 215
3.2 Assessing users expectations 217
4 Opportunistic Content Distribution Application 218
4.1 The Technological Dimension 219
4.2 Evaluating User Preferences 222
5 Phase 3: combining users and technological constraints 224

6 Discussion . . . 227
7 Closing Remarks . . . 228
References . . . 228

Programming and Validation Techniques for Reliable Goal-driven Autonomic Software . . . 231
Damian Dechev, Nicolas Rouquette, Peter Pirkelbauer and Bjarne Stroustrup
1 Introduction . . . 231
2 Challenges for Mission Critical Autonomous Software . . . 232
2.1 Parallelism and Complexity . . . 233
2.2 Motivation and Contributions . . . 233
3 Temporal Constraint Networks . . . 234
4 Verification and Automatic Parallelization Framework . . . 235
4.1 The Problem of TCN Constraint Propagation . . . 235
4.2 Modeling, Formal Verification, and Automatic Parallelization . . . 238
5 Nonblocking Synchronization . . . 241
5.1 Practical Lock-Free Programming Techniques . . . 242
5.2 Overview of the Lock-free Operations . . . 242
6 Framework Application for Accelerated Testing . . . 244
7 Conclusion . . . 245
References . . . 246

Part III Applications to Ad-Hoc (Sensor) Networks and Pervasive Systems

Autonomic Communication in Pervasive Multimodal Multimedia Computing System . . . 251
Manolo Dulva Hina, Chakib Tadj, Amar Ramdane-Cherif, Nicole Lévy
1 Introduction . . . 252
2 Related Works . . . 253
3 Contribution and Novel Approaches . . . 254
4 The Interaction Context . . . 255
4.1 Context Definition and Representation . . . 255
4.2 The Virtual Machine and the Incremental Interaction Context . . . 256
4.3 Context Storage and Dissemination . . . 262
5 Modalities, Media Devices and Context Suitability . . . 263
5.1 Classification of Modalities . . . 263
5.2 Classification of Media Devices . . . 263
5.3 Relationship between Modalities and Media Devices . . . 264
5.4 Measuring the Context Suitability of a Modality . . . 264
5.5 Optimal Modalities and Media Devices' Priority Rankings . . . 265
5.6 Rules for Priority Ranking of Media Devices . . . 267
6 Context Learning and Adaptation . . . 268
6.1 Specimen Interaction Context . . . 268

6.2 Scenarios and Case-Based Reasoning with Supervised Learning . . . 271
6.3 Assigning a Scenario's MDPT . . . 276
6.4 Finding Replacement to a Missing or Failed Device . . . 277
6.5 Media Devices' Priority Re-ranking due to a Newly-Installed Device . . . 278
6.6 Our Pervasive Multimodal Multimedia Computing System . . . 279
7 Conclusion . . . 280
References . . . 281

Self-healing for Autonomic Pervasive Computing . . . 285
Shameem Ahmed, Sheikh I. Ahamed, Moushumi Sharmin, and Chowdhury S. Hasan
1 Introduction . . . 285
2 Motivation . . . 287
3 Characteristics of Self-healing Model . . . 288
4 Design Overview . . . 288
4.1 Self-healing System of Autonomic Pervasive Computing 288
4.2 Classification of Fault . . . 290
4.3 Fault Detection . . . 291
4.4 Fault Notification . . . 292
4.5 Faulty Device Isolation . . . 292
5 Self Healing in Autonomic Pervasive Computing . . . 292
5.1 Fault Detection . . . 292
5.2 Fault Notification . . . 294
5.3 Faulty Device Isolation . . . 295
5.4 An Illustrative Example . . . 295
6 Attributes of Our Proposed Model . . . 297
6.1 Efficiency . . . 297
6.2 Transparency . . . 298
6.3 Infrastructure less . . . 298
6.4 Non degradable performance . . . 298
7 Related Work . . . 298
8 Evaluation . . . 300
8.1 Prototype Implementation . . . 301
8.2 Performance Measurement . . . 303
8.3 Application that Uses Self-healing Model . . . 304
9 Conclusion and Future Work . . . 304
References . . . 305

Map-based Design for Autonomic Wireless Sensor Networks . . . 309
Abdelmajid Khelil, Faisal Karim Shaikh, Piotr Szczytowski, Brahim Ayari and Neeraj Suri
1 Introduction and Chapter Structure . . . 309
2 Models and Requirements . . . 311

2.1 Models for Sensing the Real World . . . 311
2.2 System Model . . . 312
2.3 Requirements on the MWM . . . 312
3 The Map-based World Model . . . 313
3.1 MWM Definition . . . 313
3.2 MWM Architecture . . . 314
3.3 MWM Management . . . 315
3.4 Region and Map Construction Techniques . . . 316
4 MWM-based WSN Design . . . 317
4.1 Enhancement of WSN Autonomicity . . . 318
4.2 Design Methodology . . . 319
4.3 Case Study: Designing a Network Partitioning Prediction Technique . . . 319
5 MWM Implementation in OMNeT++ . . . 321
5.1 MWM Implementation Architecture . . . 321
5.2 Uses of Simulator Extension . . . 321
6 Related Work . . . 323
7 Conclusions . . . 324
References . . . 324

An Efficient, Scalable and Robust P2P Overlay for Autonomic Communication . . . 327
Deng Li and Hui Liu and Athanasios Vasilakos
1 Introduction . . . 328
2 Background on P2P Overlay Networks . . . 328
3 Challenges and Requirements in Supporting P2P for AC . . . 329
3.1 Information reflection and collection . . . 329
3.2 Lack of Centralized Control . . . 330
3.3 Non-Cooperation . . . 330
4 The Description of ESR . . . 331
4.1 The Formation of ESR . . . 331
4.2 The Source Ranking . . . 333
4.3 The selection and performance of ICs . . . 334
5 The maintenance of ESR . . . 336
5.1 Two rules for maintenance . . . 337
5.2 Node joining . . . 338
5.3 Node leaving . . . 339
6 Evaluation and experimental results . . . 340
6.1 Modeling and methodology . . . 340
6.2 Scalability . . . 342
6.3 Query success rate . . . 343
6.4 Query messages and hops . . . 343
6.5 Cost and load balancing . . . 344
6.6 Fault-tolerance and robustness . . . 346
7 Conclusion and future directions . . . 348

References ... 348

Autonomic and Coevolutionary Sensor Networking ... 351
Pruet Boonma and Junichi Suzuki
1 Introduction ... 351
2 BiSNET/e Agents ... 353
2.1 Agent Structure and Behaviors ... 353
2.2 Behavior Sequence for DAs ... 355
2.3 Behavior Sequence for EAs ... 356
2.4 Agent Behavior Policy ... 357
3 MONSOON ... 357
3.1 Operational Objectives ... 358
3.2 Elite Selection ... 359
3.3 Genetic Operations ... 360
4 Simulation Results ... 361
4.1 Data Collection Application ... 363
4.2 Event Detection Application ... 364
4.3 Hybrid Application ... 366
4.4 Adaptive Mutation ... 366
4.5 Power Consumption ... 367
4.6 Memory Footprint ... 368
5 Related Work ... 368
6 Conclusion ... 369
References ... 370

Index ... 373

Part I
Autonomic Communication Infrastructure

Bio-inspired Autonomic Structures: a middleware for Telecommunications Ecosystems

Antonio Manzalini, Roberto Minerva and Corrado Moiso

Abstract Today, people are making use of several devices for communications, for accessing multi-media content services, for data/information retrieving, for processing, computing, etc.: examples are laptops, PDAs, mobile phones, digital cameras, mp3 players, smart cards and smart appliances. One of the most attracting service scenarios for future Telecommunications and Internet is the one where people will be able to browse any object in the environment they live: communications, sensing and processing of data and services will be highly pervasive. In this vision, people, machines, artifacts and the surrounding space will create a kind of computational environment and, at the same time, the interfaces to the network resources. A challenging technological issue will be interconnection and management of heterogeneous systems and a huge amount of small devices tied together in networks of networks. Moreover, future network and service infrastructures should be able to provide Users and Application Developers (at different levels, e.g., residential Users but also SMEs, LEs, ASPs/Web2.0 Service Providers, ISPs, Content Providers, etc.) with the most appropriate "environment" according to their context and specific needs. Operators must be ready to manage such level of complication enabling their platforms with technological advanced allowing network and services self-supervision and self-adaptation capabilities. Autonomic software solutions, enhanced with innovative bio-inspired mechanisms and algorithms, are promising areas of long term research to face such challenges. This chapter proposes a bio-inspired autonomic middleware capable of leveraging the assets of the underlying network infrastructure

Antonio Manzalini
Telecom Italia Future Centre, Via Reiss Romoli 274, Torino, Italy, e-mail: Antonio.Manzalini@telecomitalia.it

Roberto Minerva
Telecom Italia Future Centre, Via Reiss Romoli 274, Torino, Italy, e-mail: Roberto.Minerva@telecomitalia.it

Corrado Moiso
Telecom Italia Future Centre, Via Reiss Romoli 274, Torino, Italy, e-mail: Corrado.Moiso@telecomitalia.it

A.V. Vasilakos et al. (eds.), *Autonomic Communication*, DOI: 10.1007/978-0-387-09753-4_1,

whilst, at the same time, supporting the development of future Telecommunications and Internet Ecosystems.

1 Introduction

Recent emergence of Web2.0 has introduced the innovative paradigm of using the "web-as-a-platform" for mashing up and offering dynamic services. As a matter of fact, the so-called "Architecture of participation" refers not only the engagement of Customers in continuous improvement of service releases but also the adoption of new innovative business models (e.g., value-chains also based on Advertisement). On the other hand, Telecommunications industry has made efforts for defining and building new services, for example on top of basic voice call. Today, the Telco2.0 paradigm is aiming at applying more or less the same Web2.0 paradigms to Telecommunications, (e.g., through SDK) for exposing and mashing up Telecommunications, ICT network enablers, capabilities, service components, data, etc. Also new business models and value-chains, beyond the walled-garden, (e.g., broader federations of Players) are under evaluation by Telecommunication Players.

Another important trend is represented by the decreasing of both size and cost of physical systems and devices. On one side this is opening new opportunities, paving the way to a pervasive use of digital devices for information and communication services, on the other side it creates some serious technological challenges for future networks, such as interconnection and management of heterogeneous systems and a huge amount of small devices tied together in networks of networks.

One of the most attracting service scenarios for future Telecommunications and Internet is the one where people will be able to browse any object in the environment they live: communications, sensing and processing of data and services will be highly pervasive. In this vision, people, machines, artifacts and the surrounding space will create a kind of computational environment and, at the same time, the interfaces to the network resources.

Current Telecommunication service and network architectures do not support this vision; moreover, they cannot provide people with personalized services whilst, at the same time optimizing the use of network and data resources.

This chapter is elaborating innovative middleware architecture capable of exploiting a kind of pervasive self-adaptive computing environment optimizing, at the same time, the use of the underneath network and the service infrastructures. Middleware is autonomic, highly decentralized and composed by lightweight components interacting with each other with bio-inspired mechanisms and algorithms. Design philosophy has been considered a key factor: this should not be limited to solving interoperability problems between different domains, but should cover designing the rules with which different services are defined and composed thus allowing the construction of a global environment (a kind of ecosystem) from the composition of a set of simple unitary services.

Chapter starts with a brief state of art on autonomic frameworks (based on distributed components) and the related algorithms for focus limitations and required advances. Then the concept of Bio-inspired Autonomic Structures (BAS) is presented as a middleware in which with multiple levels, combining hierarchies, self-organization, and emergence. Component is the basic element of the BAS: specifically coalesce of components are capable of abstracting specific functions and data whist performing an autonomic behavior. Particular attention has been given to identify potential bio-inspired alternative for supporting the interactions of the components. BAS middleware is a kind of ecology of components where novelty stands in adopting communication primitives for the self-organization of pieces of data to aggregate data patterns for distributed applications (see Fig. 1). This challenging task can be approached viewing the interactions in terms of algorithms combining primitive mechanisms available for organizing cooperation among the components. On the other hand, coalesce of components abstracting functions and more complex patterns of data should adopt more articulated communication algorithm and mechanism.

The attention is then moved to identify some guidelines for design rules.

Chapter closes with some application scenarios: future exploitations are based on the idea of "emergence" as defined by mathematician Nils Baas for complex natural systems and by sociologist Margaret Archer for social systems. These authors have proposed similar frameworks which are unifying the essential requirements for ecological economics: clear definition of sustainability; ways to link ecological and economic trends; ways to understand stability and instability in dynamic social-ecological systems, and ways to include human self-observation within dynamic models of social-ecological systems.

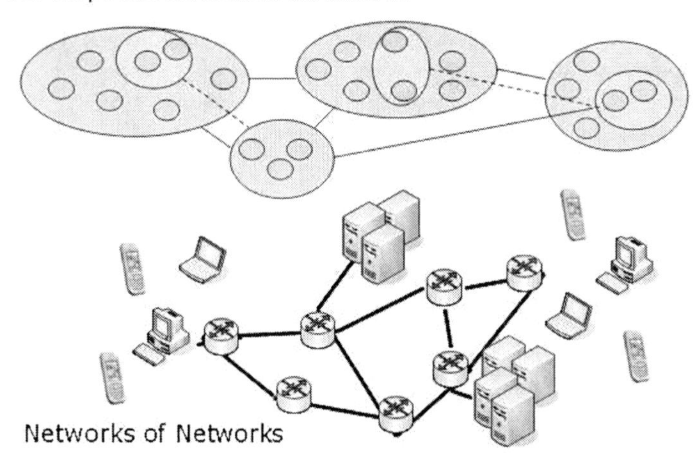

Fig. 1 Bio-inspired Autonomic Structures (BAS) abstracting Networks of Networks

2 State of Art

This section is presenting a brief state of art analysis: scope is elaborating about the technological maturity of some of the key architectural elements for developing the BAS middleware. The first element is the conceptual framework of a distributed middleware of autonomic components. The second element concerns the bio-inspired algorithms that can be adopted to allow components to interact with each other. Specifically the bio-inspired metaphor lies in the envisaging of mature capabilities as the cumulative outcome of a large number of small advances in capability. Each such advance is assumed to be associated with a particular generative mechanism, so that the nervous system as a whole can be viewed as an organized collection of such generative mechanisms, each specialized to some particular generative process.

2.1 Autonomic Frameworks

Increasing complexity of large-scale computing systems, computers and applications require learning how to manage themselves in accordance with high-level policies from human operators. This vision, which has been referred to as autonomic computing, is taking inspiration from the biological characteristics of the human Autonomic Nervous Systems. In other words, autonomic computing refers to the self-managing characteristics of distributed computing resources, adapting to unpredictable changes whilst hiding intrinsic complexity to operators and users. An autonomic system makes decisions on its own, using high-level policies; it will constantly check and optimize its status and automatically adapt itself to changing conditions.

As reported in literature (see Fig. 2), an autonomic computing framework might be seen composed by Autonomic Components (AC) interacting with each other. An AC can be modeled in terms of two main control loops with sensors (for self-monitoring), effectors (for self-adjustment), knowledge and planer/adapter for exploiting the policies:

- Local (L) control loop for self-awareness, for internal management and recovery from faults;
- Global (G) control loop for environment awareness, allowing changing behavior and even environment (through communication with other elements).

In other words, an AC is an entity capable sensing and adapting to environment changes whilst performing self-* capabilities (e.g., self-CHOP: configuration, healing, optimization, protection) through interaction with other similar components.

Fig. 3 shows an example of Autonomic Component as developed by the IST Project CASCADAS [26]. This represents an example of autonomic component as CASCADAS goal is to develop and demonstrate a prototype of Service Ecosystem for dynamic composition and execution of autonomic services.

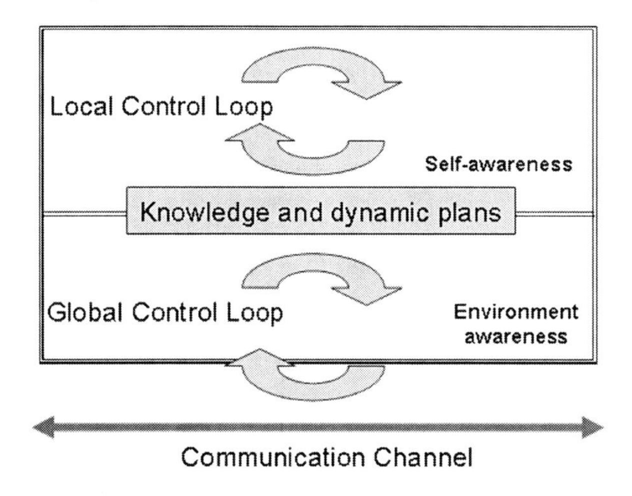

Fig. 2 Logical view of an Autonomic Component

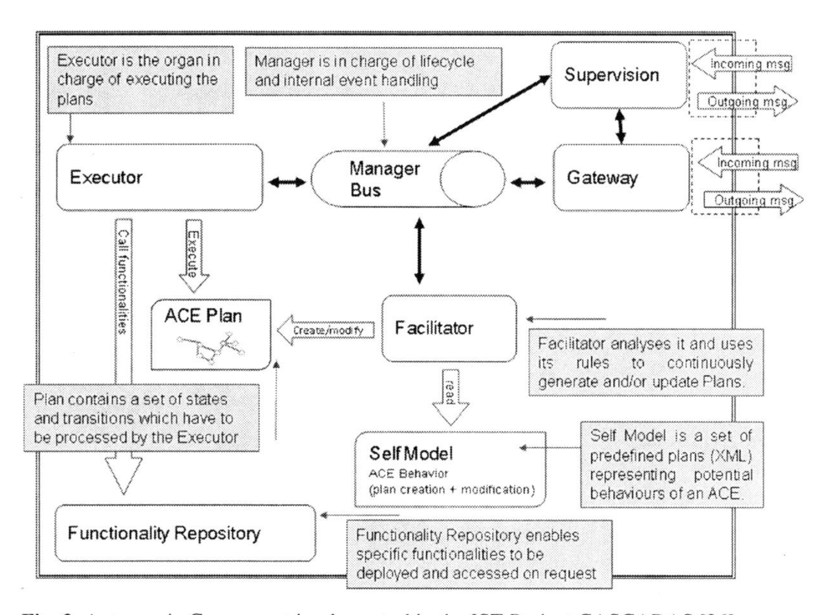

Fig. 3 Autonomic Component implemented in the IST Project CASCADAS [26]

IBM, as part of its autonomic computing initiative [21], has outlined the need for current service providers, to enforce adaptability and properties of self-configuration, self-optimization, and self-healing, via service (and server) architectures revolving around feedback loops and advanced adaptation/optimization techniques. Driven by such vision, a variety of architectural frameworks based on "self-regulating" autonomic components has been recently proposed, also with reference to the management of large data centres. A very similar trend has recently characterized significant research work in the area of multi-agent systems [5], [39]. However, most of these approaches are typically conceived with centralized or cluster-based server architectures in mind and mostly address the need of reducing management costs rather than the need of enabling complex software systems or providing innovative services.

In recent years, a variety of diverse algorithms and approaches have proved the potential of bio-inspired distributed solutions to enforce purposeful functionalities in a fully distributed, self-organizing and adaptive way [25], [11], [17], [12]. Examples are studies on: ant-inspired algorithms [33], distributed coordination based on virtual force fields, socially-inspired communication mechanisms [33].

Most of these studies (with a few exceptions detailed below) are limited to simulations and experiments, and do not propose a framework that is both theoretical and practical (e.g., a toolkit). Furthermore, each of these proposals suggests specific solutions for specific problems or classes of services, without adopting a more holistic approach.

Concerning approaches to model and build self-organizing and self-adaptive frameworks, a variety of heterogeneous proposals exists for both the basic components (e.g., reactive agents rather than proactive and goal-oriented ones) and their interactions (e.g., pheromones [33], or gossip [20]).

Beside more theoretical studies, some practical proposals exist for distributed self-organized frameworks. Service Clouds [34] is a research distributed prototype designed to facilitate rapid prototyping and deployment of autonomic self-organizing communication services. The Service Clouds infrastructure combines dynamic software configuration methods with self-organizing algorithms for the establishment of communication link in order to support both cross-layer and cross-platform cooperation. SwarmingNets [16] is a research framework for the management of complex ubiquitous services implemented by groups of autonomic objects, called TeleService Solons, which have the capabilities of fulfilling the complex tasks relating to service discovery and service activation.

A specific trend of research in the area of self-organizing service frameworks concerns those approaches that attempt to enforce self-organization and self-healing features in services and data access by relying on a Peer-to-Peer (P2P) overlay network substrate, with the goal of exploiting the self-organizing features of overlay networks [2]. The Service Oriented Peer-to-Peer System (SOPPS) [19], defined in the EU-funded Project MMAPPS [31] (Market Management of Peer-to-Peer Services) investigates how to combine P2P with SOA, in order to allow the definition of component-based P2P applications.

A similar approach was adopted in SESAM Project [36] aimed at the development of a highly flexible, scalable and technology-independent architecture for distributed service provisioning. In contrast to typical SOA, the architecture proposed by SESAM offers mechanisms that allow multiple peers to cooperate in order to provide a service, so having a positive impact on non-functional aspects, like robustness, or availability.

The proposed framework is aiming at overcoming the need for a supporting P2P layer, by trying to enforce its features via a much simpler architecture, in which data and components will play an active role in self-building, via self-organization.

Also in the area of Grid computing, scale, heterogeneity and dynamism of applications call for self-managing and autonomic properties for Grid services. The Accord service architecture [24] addresses this by enabling service and application behaviour and their interactions to be dynamically specified and adapted using high-level rules, based on current application requirements, state and execution context. Similar goals are shared by both NextGRID [32] and ASG [6] Projects, which eventually aims at developing an architecture for Next Generation Grid services for widespread use by research, industry and the ordinary citizen.

However, these approaches mostly focus on dynamic allocation, composition and management at resource level, while current proposal is also addressing these aspects at service, data, and content levels. Moreover, current proposal emphasizes the complete distributed nature of the algorithms and control solutions.

The related research work that more directly relates to current CASCADAS and BIONETS Projects [13], [26], [15] funded under the FET Situated and Autonomic Communication Initiatives. BIONETS explores innovative biologically-inspired algorithms for autonomic communication services in challenged scenarios (e.g., Digital City [17]). CASCADAS aims to deliver a research demonstrator of an innovative component-based framework for situated and autonomic communication framework, integrating advance autonomic governance tools and advanced methods for knowledge and data management. CASCADAS is providing a sort of foundational work for developing highly distributed autonomic frameworks for Telco applications.

As a final note, we want to emphasize that several approaches for automatic service composition based on semantic, goal-oriented, and pattern-matching, have been recently proposed [29]. The basic idea, in these approaches, is that semantic description can be attached to services, describing what a service can provide to services and a service needs from other services.

2.2 Interaction Algorithms

Interaction algorithms play a fundamental role for designing autonomic framework as they represent a mean through which components communicate to exploit self-organization and self-adaptation.

Each component can communicate with a few nearby neighbors. Components might interact via wired and wireless communications, whereas bio-engineered components might communicate by chemical signals (e.g., applications for amorphous computing). Although the details of the communication model can vary, the maximum distance over which two components can communicate effectively is assumed to be small compared with the size of the entire framework.

Self-adaptive and self-organizing algorithms are being extensively studied [21]. Self-adaptive systems have a sort of semantic representation of their state, and can evaluate their own behavior and change it when the evaluation indicates that they are not accomplishing what they were intended to do, or when better functionality or performance is possible. On the other hand, self-organizing systems work bottom-up without any high-level representation, based on a large number of components that interact according to simple and local rules and in which a global adaptive behavior of the system emerges from these local interactions.

Self-organization and the algorithms underlying the emergence of adaptive patterns in complex systems have been deeply investigated in communications, e.g., in P2P computing [14], social networks. Self-organization algorithms has the potential to act as enablers for service composition and aggregation, employing proven techniques to abstract from their implementation and derive design principles adapted to the requirements of artificial systems. At the same time, the presence of self-adaptive systems capable of understanding what is happening and proper reacting accordingly (as in the canonical "autonomic computing perspective" [27]) can hardly be disregarded to ensure proper reactions and adaptations to various situations.

Accordingly, a major advance with respect to most of the prior art is to provide a way to exploit self-organization approaches and enrich self-organizing components with more "semantic" and/or "cognitive" abilities, in the direction of self-adaptation. This raises the important question of evaluating the amount of information that has to be processed individually by system components, versus collectively by the self-organizing group. A key goal to be considered in the implementation of the proposed framework is to preserve the simplicity and robustness of self-organization phenomena while simultaneously bringing the benefits of semantics self-adaptation and situation-awareness, to achieve what can be defined as "semantic self-organization".

In synthesis, in order to provide self-adaptive behavior, a component-ware autonomic system must be able to reason about both its environmental data and its behavior relative to that environment. This does not necessarily imply a symbolic structure, but suggests that the system must be able to reflect on environmental data and behavior in some sense and generate feedbacks/actions as a result. Thus the autonomic system must accept goals and constraints from, and petition for the attention of, its operators using terms that are meaningful to their needs and cognitive abilities. However, this must be balanced against the need for the autonomic system to map these semantic terms deterministically to and from the self-managing capabilities of its elements. This requires a high degree of semantic interoperability between the expression of adaptive behavior and the changing of data that drives such adaptation. Semantic-based reasoning for autonomic systems requires research about models and languages for representing environmental data and components behavior.

3 Bio-inpired Autonomic Structures

In synthesis, in order to provide self-adaptive behavior, a component-ware autonomic system must be able to reason about both its environmental data and its behavior relative to that environment. This does not necessarily imply a symbolic structure, but suggests that the system must be able to reflect on environmental data and behavior in some sense and generate feedbacks/actions as a result. Thus the autonomic system must accept goals and constraints from, and petition for the attention of, its operators using terms that are meaningful to their needs and cognitive abilities. However, this must be balanced against the need for the autonomic system to map these semantic terms deterministically to and from the self-managing capabilities of its elements. This requires a high degree of semantic interoperability between the expression of adaptive behavior and the changing of data that drives such adaptation. Semantic-based reasoning for autonomic systems requires research about models and languages for representing environmental data and components behavior.

3.1 Concept of Autonomic Structures

BAS architecture take inspiration from the concept of hyper-structure [7], [8], [9], [10] as a middleware in which to study multiple levels, and in a fruitful way combine hierarchies, self-organization, and emergence. As described in literature, hyperstructure middleware has three fundamental components. The first is the primitive objects or units, which can be of physical or abstract nature. The second component is some kind of observational mechanism that observes, describes and evaluates the objects. Thirdly, there are interactions among the objects. The interactions use the properties detected by observational mechanisms.

A family of objects together with specified properties and interactions defines a process or construction. The process may generate a family of second order objects. In the reformulated version of hyper-structures [8], the interactions are bonds which directly "bind" families of primitive objects to produce second order objects. The resulting structure, consisting of objects on different levels together with their interactions and properties, is called a hyper-structure in [8]. Objects at higher levels obtain new properties, and the new interactions and observations at higher levels may or may not be meaningful at lower levels. Furthermore, hyper-structures allow for both "upwards" and "downwards" causation, in that interactions on higher levels may cause changes in interactions on lower levels.

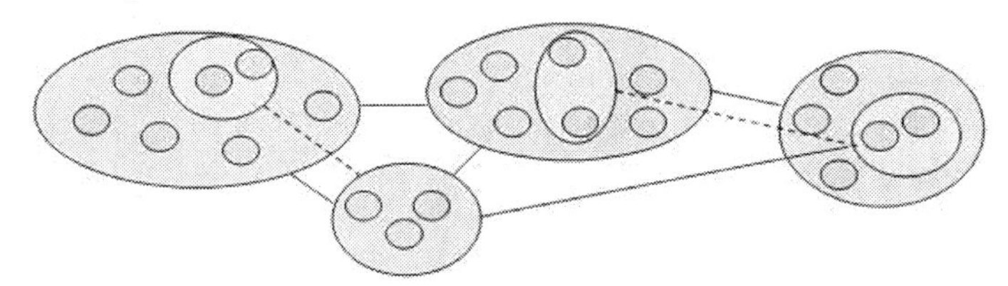

Fig. 4 BAS as an example of hyper-structures [7]

Fig. 4 represents an example of hyper-structure [7]: circles represents objects of increasing order and Figlined represent the interactions. Dashed and solid lines represent different types of interactions.

BAS are based on three fundamental ingredients: components, supervision mechanism (that observes, evaluates and manage the components), and the interactions among the components. In BAS middleware, components express some self-* autonomic features. Moreover another aspect should be considered in modeling components: middleware software architecture is assumed based on Nondeterministic Finite Automata (NFA). From a software perspective, BAS middleware is based on networks of concurrent components bound together by bio-inspired interaction protocols and decentralized algorithms.

This architectural perspective shifts focus from source code to coarse-grained components and their interconnections. Designers can concentrate on the components structure, the assignment of components to pieces of functions and data, the interactions among components and. Components are responsible for implementing self-* behavior, maintaining state information and communicating with each other. This approach tends moreover to separate (at a lower level) computing from communication allowing a system's computation and communication relationships to evolve independently of one another, including rearranging and replacing the components while the application executes as a necessary, but still insufficient, mechanism for self-adaptive software. Still all the infrastructure has be governed and orchestrated with a top-down approach in order to control the expressiveness power and the abstraction of each of its components maintaining its reflective properties. On the other hand, this is supported and integrated by a bottom-up self-organization.

3.2 BAS middleware

As mentioned, BAS middleware has three fundamental ingredients: components, supervision mechanisms, and the interactions among the components. In this subsection attention is focused on components.

BAS architecture is based on two main classes of components: High Level Components (HLC) and Low Level Components (LLC). Fig. 5 shows conceptual block diagram of a component.

HLCs (and aggregation of HLCs) include a deterministic Finite State Machine (FSM) whilst performing an autonomic behavior. HLC can be logically interconnected (with a goal-driven approach) to develop services and applications. HLC may persist or disappear in relation to the purpose for which they have been created.

It should be noted that, in principle, in order to build a flexible middleware capable of adapting to network dynamics, components with NFA should be adopted.

Motivation of the choice of NFA is that a transition from a state to another of the finite state machine is not deterministic as it depends on the conditions of the underlying networks (actually showing non deterministic behaviors). It is well-known that through a powerset construction, it is possible to convert a NFA into a DFA through

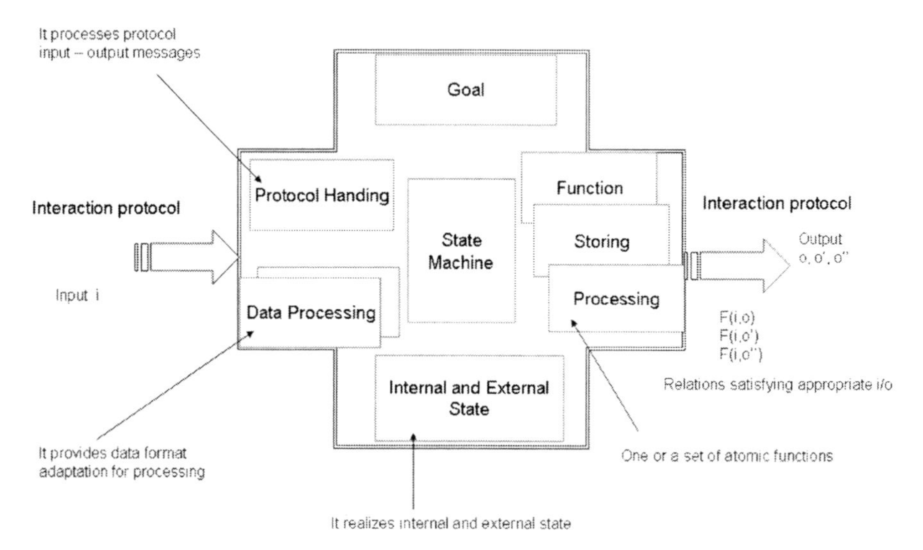

Fig. 5 Example of Component

increasing its number of states (from n states up to 2n states). Components have to designed by identifying the minimum number of states, so that to avoid excessive complications.

At the end the architecture should consist of a number of autonomic components (with a manageable number of states) able to cope with the non deterministic behavior of underlying networks, but capable of a deterministic behavior and acting as sharable resource for applications to use.

In order to self-adapt their behavior, HLCs could create/modify their internal state machine, by updating transitions and assembling simpler NFAs according to rules conditioned by external events and internal state [15].

Recursively, a HLC can be considered as a ensemble of LLCs, being latter ones simpler components that includes data structures, business logic representations, and automata; LLC can either self-organize (bottom-up aggregation) or making groups upon top-down requests from other HLCs.

In addition to a syntactic description (e.g., in terms of types, and/or interface definition), BAS components could be annotated with a kind of Semantic Description (SD). For example tuples can be used to store the semantic description of the components goal and/or behavior, for components discovery and/or goal matching. In this sense, SD "matching", in addition to "syntactic" (e.g., according to some type system) matching, could be adopted as one of the ways for composing and orchestrating composite services starting from basic service components and pieces of data and knowledge. Also approaches such as introduction of semantic-based mutation in genetic algorithms could be considered for optimizing service composition.

Service execution can be seen as a "metabolic reaction" of components consuming and producing tuples; so tuples, similarly to LINDA model [18], can also be seen as the means through which components exchange and share information, distribute tasks/goals, declare available components. The tuples are logically stored in a shared tuple space and any component can add tuples and consume tuples belonging to the types which it is able to process.

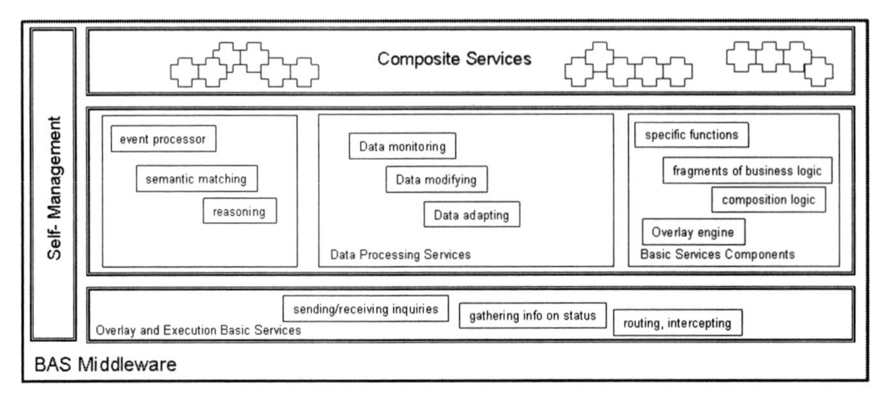

Fig. 6 BAS middleware

Fig. 6 shows conceptual view of the middleware. Computing nodes executing the middleware use adaptive component-ware to support autonomic behavior; an overlay network among these nodes serves as a vehicle to support cross-hardware adaptation, [30] and to provide communication mechanisms across heterogeneous connections. Nodes in the overlay provide an "execution canvas" on which service components can be composed, instantiated and executed as needed, and later reconfigured in response to changing conditions. The middleware is designed to be extensible in the sense that a suite of low-level services can be used to compose higher-level services.

3.3 Data Components interactions: primitives

Data gathering and analysis, monitoring and assessment looking for potentially interesting patterns in data is extremely valuable to make available the right data in the proper form at the right time for distributed applications.

Given that BAS middleware is a kind of ecology of components abstracting both functions and pieces of data, novelty of the approach is adopting communication primitives for the self-organization of pieces of data to aggregate data patterns for distributed applications. This challenging task can be approached viewing the interactions in terms of algorithms combining primitive mechanisms available for organizing cooperation among the components.

On the other hand, coalesce of components abstracting functions and more complex patterns of data should adopt more articulated communication algorithm and mechanism. For example, primitive mechanisms that are appropriate for specifying such interactions include gossip, random choice, fields, and gradients [1].

Gossiping

Gossiping, also known as epidemic communication, is aiming at obtaining an agreement about a certain value of some parameter. Each component broadcasts its belief of value of the parameter to its neighbors, and computation is performed by each component combining the values that it receives from its neighbors. a new value is re-broadcasted. The process concludes when there are no further broadcasts.

Random choice

An example of application of random choice is for establish local identity of large number of components: each component selects a random number to identify itself and communicate it to its neighbors. If the number of choices is large enough, then it is unlikely that components select the same number. Random choice can be combined with gossip to elect super-peers: every component select a value, then gossips to find the minimum. The component with the minimum value becomes the super peer.

Fields

Every component of a certain pattern is considered as a value of field over the discrete space occupied by that pattern. If the density of component is large enough this field of values may be thought of as an approximation of a field on the continuous space.

Gradients

Gradients represent an important primitive in amorphous computing: it implies the estimation of the distance from each component to the nearest component designated as a source. The gradient primitive takes inspiration from the chemical-gradient diffusion process that is crucial to biological development.

3.4 Components interactions: mechanisms and algorithms

In addition to achieving the "usual" communication patterns in distributed systems (e.g., request-response, or subscribe-notify), interactions among BAS components could have different objectives, including:

- negotiating for an optimal allocation of (local) resources to global task;
- influencing single components behavior (e.g., through tuning configuration parameters, or selecting execution plans);

- setting-up and maintaining of a self-organized overlay networks

In the followings some mechanisms and algorithms that could be adopted in BAS middleware are described and briefly elaborated.

Pull when idle

Tuples exchanges could be used for implementing an interaction model (between HLC) fully based on an innovative "pull" semantic. This approach reflects in generating an altruistic component environment with intrinsic load-balancing capabilities. In other words, each HLC is altruistic: if it is able to achieve a goal, and it is available to do it, then HLC will volunteer in helping other HLCs [30].

Novelty of the approach is enabling self-organization by propagation of information on specific capabilities (in terms of Goal Achievable) rather than what is "needed" (in terms of Goal Needed). This information could be then used for discovery and semantic self-aggregation of components.

In order to achieve this, it is necessary to ensure that the information about HLCs capability (and availability) reaches proper HLCs in a proper time: this requires the definition of a P2P overlay aimed to diffuse in a scalable way such information. This underlying middleware should be able to implement the shared tuple space in a distributed/scalable way, as a means for efficiently distributing tuples from components producing the tuples to the components able to consume them. This tuple-based approach makes it easy to associate semantic information with service execution requests in order to allow HLCs to self-organize themselves taking into account such information.

LLCs are aggregated by an executor of self-organization algorithms to let emerge basic HLCs that in turn can be seen as a blackboard that incorporates and executes a number of LLCs, i.e., a blackboard with execution (and self-adaptive) capabilities. Equivalently, an HLC can interact recursively with other HLCs. A "Poll When Idle" approach for a collaborative execution environment is adopted: each HLC executes "take" operations in order to transfer tuples on to its blackboard from overloaded HLCs.

In more detail, the HLC implements the above strategy in the following way: given a set of HLCs containing in their blackboard tuples corresponding to goals to be executed:

- each HLC calculates its own residual power (RP) or the RP of an aggregate of HLCs which represents the ability to help others;
- when a HLC realizes that it has some RP available (a few tuples in its blackboard) it offers its computational power to other HLCs, just sending them a notification of availability;
- overloaded HLCs accept one of the possible computation power offers;
- when a HLC receives acknowledgment to its computation power offers, it begins to take tuples from the HLC in trouble.

One of the main benefits of the approach described above is that when an HLC is overloaded it does not need to do anything about it. It needs simply to check if it has received some help offers from other HLCs and accept them. The collaborative behavior of the HLCs provides dynamic adaptation through self-organization, improving high availability and low response times in the executed services.

It has been mentioned that HLCs are performing autonomic capabilities. The middleware will leverage the emergence of such autonomic technologies, where systems are designed to respond dynamically to changes in the environment with only limited human guidance; in addition, the self-similar, modular, and recursive structure of the components will enable the middleware and its "components" to adapt and survive to changes in the environments (e.g., by enabling an easy replacement of modified capabilities). Moreover autonomic computing [37] can be used to support fault tolerance, enhance security and improve quality-of-service in the presence of dynamic environment conditions. Realizing autonomic behavior involves cooperation of multiple software components.

Bio-inspired mechanism

In recent years, "bio-inspired" algorithms and approaches have been investigated as potential solutions for improving the functional and non-functional characteristics of distributed component-based systems; in particular, they could enforce the implementation of fully distributed, self-organized and adaptive solutions. Several sources of inspiration from "biological systems" have been explored, ranging from swarm behaviors to genetic evolution, from social networks to epidemic spread, from embryology to cellular engineering. An overview of some of them is provided in [13].

In the following, we will sketch how some of them could be fruitfully applied.

Genetic evolution

The genetic evolution (and other forms of biological evolution, such as Embryology) can be exploited in the definition of innovative service (or components) life-cycles, by introducing steps that allow the services to evolve in order to adapt themselves to new environmental conditions or to evolution of end-users needs; for instance, "genetic" operators could be defined on component orchestration, in order to perform several transformations on composition flows [23]; fitness functions should be defined in order to select the "best" mutations, i.e., the transformed composition that optimize the new requirements of the system/environment; strategies for including mutations, either at design time or at run time, in service/component life-cycle must avoid anomalous behavior of on-line systems due to mutated components; alternatively, differentiation of components, e.g., through variation of the configuration parameters, could improve the robustness of the systems, by allowing multiple versions of a components, with slightly different behavior able to cope with possible unplanned situations, co-exist in the eco-system.

Epidemic spread

The analysis of how epidermises expand in communities, through individual-to-individual contacts, can be adapted to the autonomic components contexts, in order to pass data among them in a distributed way, without any functions implemented in a centralized way; for the mechanism ruling the epidemic diffusion can provide indication on the structure of connectivity among components (e.g., the optimal degree of connections for an overlay). Analysis of epidemic spread can be used for instance in order to provide to the system new configuration parameters to the components or to spread information computed by single components (e.g., the value computed by a network of sensors, or the load of nodes in a distributed server farm); in a similar way could be exploited the propagation of information in social networks (e.g., through the so-called "gossiping"), by adoption "socially-inspired" communication mechanisms.

Pheromone

Pheromone is one of the means used by ant swarm to achieve an intelligent behavior[1]; pheromone is a techniques that allow ants to communicate each other without a direct interaction, but leaving traces on their environments; analogous approaches can be adopted by components in order to allow an indirect communication; for instance, components can store information related to some interactions, in order to allow other components to retrieve them in the following of system evolution; as pheromone evaporates over the time, the relevance of these information decreases progressively with time, so that only the most recent information are kept; pheromone-inspired solutions can be used in order to coordinate groups of components in performing some distributed task (for instance, it could be used to coordinate components associated to rescue teams involved in an emergency [17]), or alternatively, in improving the search of information in distributed storage (e.g., in a P2P content sharing system).

4 Engineer self-organization

Design philosophy should not be limited to solving interoperability problems between different domains, but should cover designing the rules with which different services are defined and composed thus allowing the construction of a global environment (a kind of ecosystem) from the composition of a set of simple unitary services.

On one side there are traditional approaches aimed at engineering proper ISO-OSI network protocols to face such challenges, on the other side there are innovative

---

[1] When ants forage for food they release a trace of pheromone on their path on the ground: ants searching for food tend to follow the highest concentration of pheromone, so to converge towards the shortest path.

investigations, as proposed in this chapter, based on distributed middleware with autonomic self-organization features. BAS architectural model represents the tentative integration of the concepts of Complex Adaptive Systems with the principles of distributed Autonomic Computing. Any kind of complex system or structure is build up from elementary components and constructed layer by layer.

From an engineering viewpoint, it should be mentioned that a trade-off is required between Top-Down design (policies, high level rules, orchestration) and Bottom-Up self-Organization of components (based on simple local rules and distributed algorithms).

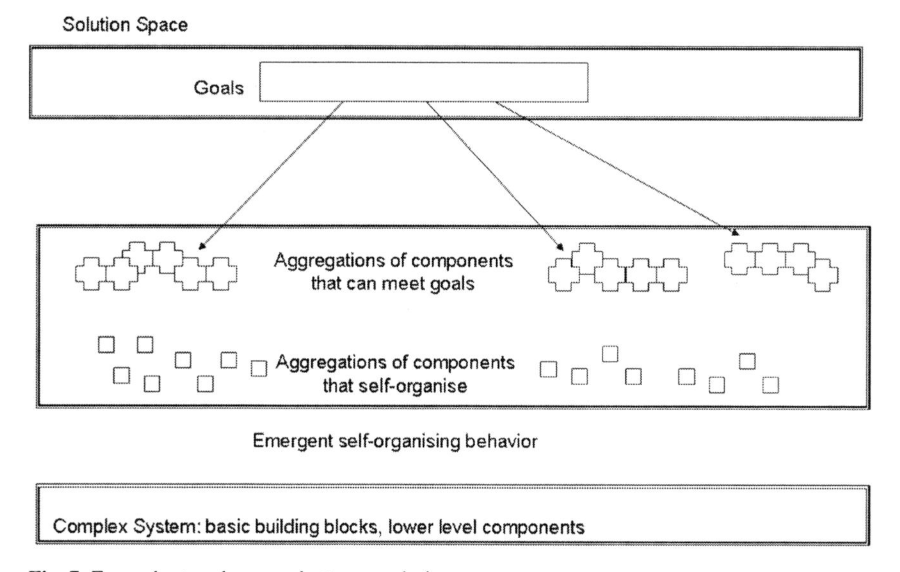

Fig. 7 Example: top-down vs. bottom-up design

Instead of making a top-down "mapping of goals into agents" (as in Multi Agent Systems approaches or in SOA orchestration with dynamic binding), engineering self-organization should be carried out in three main steps:

- applying a traditional top-down design of the overall middleware mapping from goals into (aggregation of) high level components interacting each other (according to certain algorithms);
- modeling lower level component behaviors and controlling their aggregation with self-organization algorithms (simple local rules) allowing bottom-up emergence, for example, of data patterns;
- continuous matching top-down and bottom-up patterns (in order to achieve global goals) thus "designing by variation" final local rules and component behaviors to achievable goals.

Let us consider two use cases as examples of engineering self-organization.

An attracting scenario for future Telecommunications networks and Internet is exploiting the opportunity for Users to browse any object in the environment they live. In other words, considering the evolution of RFID tagging, let us think about the possibility of using any object (we normally encounter in our day life) as a pointer to pieces of data, information and services.

This scenario will see a dramatic increase of the data clouds already overlooking our life. A tuple space of data and information might be queried to extract higher-level information, knowledge and services.

In this scenario, a bottom-up self-organization of components (wrapping recursively any pieces of data) will be able to provide a more effective way to aggregate coherent data structures to be served to services and applications. On the other hand, a top-down orchestration of service components will enable proper composition for providing services and applications (using those structured data).

Another interesting application scenario, at infrastructural level, addresses the possibility to exploit bottom-up aggregation for self-organizing "workers" offering computing and storage capabilities, such as the ones described in [34], in order to provide a scalable and robust distributed platform for execution of tasks generated by the orchestrators. Self-organization algorithm can optimize the patterns and interconnections of such workers in a way to optimize the distribution of load according to the deployment of task processing logic and data allocation.

4.1 Game Theory for cross-layer design

This sub-section is elaborating some examples about the applicability of game theory among distributed BAS components for cross-layer design.

In ISO-OSI layered architecture each layer in the protocol stack hides the complexity of the layer below and provides a service to the layer above. Different layers iterate on different subsets of the decision variables using local information to achieve individual optimality. On the other hand, vertical cross layer design implies optimizing the design of various protocol layers into a single coherent theory, thus aiming at a global optimization of the network.

An adopted strategy [43] is splitting the optimization problem into two steps: first considering each layer as a solver in isolation (iterating on a set of variables and using local information, thus implicitly assuming that also other layers are designed optimally) and then studying the cross layer interactions to perform an overall vertical optimization.

As a matter of fact, game theory has been widely applied to communications problems in the literature [35], [44]. However, existing formulations tend to focus on the physical layer exclusively.

An alternative game-theoretic approach can be played between Users and Operator(s). Idea is decomposing the overall optimization problem into two sub-problems for optimizing certain utility functions for the Users and the Operator perspectives. As an example, Users utility functions could be functions (or combinations) of bit-rate, reliability, delay, jitter, or power level; on the other hand, Operators utility functions can be functions (or combination) of congestion level, traffic engineering, network lifetime, or collective estimation error. Clearly, not every game has Nash equilibrium, neither is the equilibrium necessarily stable; as such a set of sufficient conditions should be identified for existence, uniqueness, and stability of the Nash equilibrium for both games. Finally, primal-dual algorithms should be used iteratively for executing the two games and updating shadow prices to coordinate the physical layer supply and the application services demand. This should assure that the overall optimization process reaches a right balance between the two optimization sub-problems.

An interesting example on how using the game theory for modeling an autonomic WiMax network is reported in [4]; specifically it concerns the problems of feedback suppression. Paper describes how self-configuring and self-optimizing procedures of an autonomic manager component are modeled using game theory.

Let us consider also another example focused on policy control in telecommunication network. A policy can be defined as a set of simple rules that determine which action(s) to take given a set of conditions. These set of conditions define the state. Hence, each policy rule is a simple "IF-THEN" statement which describes the action(s) that will be taken when the condition holds. Therefore a policy can be implemented as a state machine and as such in terms of autonomic components (HLC).

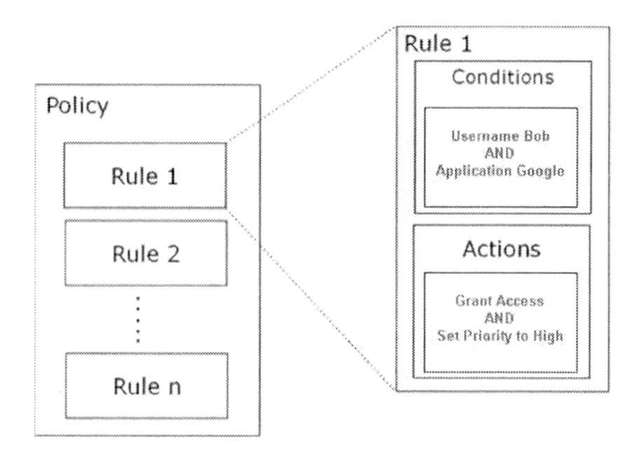

Fig. 8 Policy as a state machine

A game theoretical approach can be played among components of PDPs (Policy Decision Point, e.g., servers containing the policy decisions) in order to reach a certain equilibrium concerning policy decisions that, in turn, will be enforced to nodes via PEPs (Policy Enforcement Point, e.g., in a router). Moreover, given the enforced policies, game theory can be applied to optimize lower networks layers.

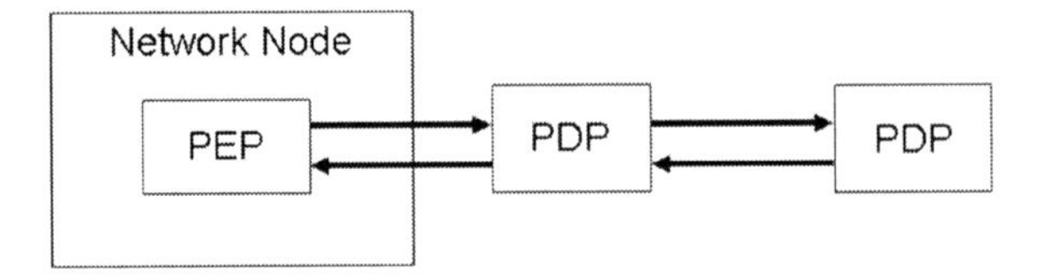

Fig. 9 PDP and PEP

4.2 Auctions for optimized resource allocation

Optimizing resource allocation in distributed network is a complex problem: a diffuse approach is enriching application services with the capabilities of negotiating for the network resources on which they are wishing to be executed [38]. A further level of complexity is introduced when considering that even the network resources may have competing goals. Auction is one of the most investigated approaches.

There are four main types of auction protocol identified by Vickrey [40]: English, Dutch, Sealed-Bid, and the Vickrey auction protocol. The English auction is the conventional open forum, ascending price, multiple bid protocol. The Dutch auction is an open forum, descending price, single bid protocol. The Sealed-Bid, or tender, is a closed forum, single bid, best price protocol in which all bids are opened simultaneously. The Vickrey auction is similar to the Sealed-Bid auction, except that the winning (highest) bid then pays the amount of the second highest bid.

Most state-of-art solutions feature a common space (playing both the role of resource discovery and allocation) in which multiple negotiations lead to contracts for the allocation of overlay resources to applications services.

Let us consider a simple example: when an application service requires some resources, it registers its requirements as an auction. This information is then distributed to the bidders (i.e., the entities able to provide the required resources) by a sort of overlay manager/agent that in turn collects any bids. Clearly applications services have the goal of obtaining maximum resources at the lowest price, whereas the overlay manager/agent has the goals of maximize the return on its overlay resources with the minimum usage. Different "pricing" policies and bidding criteria could be used to express different policies, such as optimal/balanced allocation of shared resources in a service infrastructure, or applying some power savings.

A challenge exercise might be selecting the best, w.r.t., the relevant policy, auction protocol leading to optimal equilibrium with minimal overhead messages exchange (overlay scalability requirements). Auction approach has the advantages to be fully distributed, with the possibility to dynamically and autonomously self-organize the network between resource consumers and resource providers.

5 Application scenarios

In this section some application scenarios for the Bio-inspired Autonomic Structures are briefly elaborated [28]. A first scenario concern the area of management and administration of Telecommunications Networks: growing complexities are requiring technologies and solutions for simplifying these tasks. A second scenario is considering the challenge of managing networks of networks linking together computing and storage resources (e.g., cloud computing). Third scenario is home networking environment, which is acquiring more and more relevance from the business and technical perspectives.

5.1 Self-Management for Telecommunications Networks

Telecommunications networks are becoming more and more complex an as such there is the need of technologies and solutions for simplifying these tasks. This subsection is elaborating how BAS concept can be exploited in this direction.

Local and Global control loops enable a component (or an aggregate of components) to react in an autonomous way to changes of the internal state and to events propagated by its external environment. These capabilities can be fruitfully applied to implement supervision features for controlling the behavior of a component, and for actuating corrective or optimization measures when a critical situation is detected, such as a failure state, a performance problem, or a configuration error or update. Such autonomic capabilities should be able to address several supervision areas, such as, Fault, Configuration, Accounting, Performance, and Security (FCAPS), at different levels, from single HLCs to clusters of HLCs.

Systems implemented according to BAS middleware can be supervised through complementary and co-operating approaches, at different level, such as at the level of clusters of HLCs or of single HLCs.

Clusters of HLCs implementing specific services can be supervised, according to service-specific management policies, through supervision services; as any services, also supervision services can be implemented as an aggregation of HLCs, each of them providing basic supervision functions for filtering, correlating, and elaborating events provided by the supervised HLCs, and for autonomously elaborating corrective or optimization measures. The global control loop of supervised HLCs should send events to supervision services and collect and process the suggested corrective or optimization measures. The structure of the cluster of the HLCs implementing the supervisor services can be self-adapted according to the structure of the supervised clusters.

A complementary mechanism is used to supervise the basic functions of single HLCs and the basic interaction with their neighbors. In order to supervise the huge amount of HLCs in a BAS middleware, this mechanism performs the supervision activities in a highly distributed way, by exploiting the self-adaptation characteristics of HLCs. Each HLC executes a local supervision logic, e.g., associated to the local control loop. In order to achieve a non-local supervision, the local supervision logic co-operates in a peer-to-peer way with similar logic executed by other HCLs, according to an overlay network set-up and maintained through self-organization algorithms (e.g., for aggregating HCLs providing or using similar functions/services). This approach can be used, for instance, in order to implement some FCAPS supervision functions without centralized/specialized systems.

The two mechanisms can co-operate; for instance, the supervision services must be able to react to events produced by an HCL when its local supervision logic is not able to properly react to some situation. A similar approach was elaborated in the context of the IST CASCADAS Project.

The same supervision mechanisms could be adopted in order to supervise any software system structured in a set of interacting components: each system component can be associated to an HLC in charge of performing the local supervision logic or to interact with the HLCs implementing the specific supervision services. For instance, HLCs could be associated to servers in a highly distributed server farm; by adopting the approach based on co-operating local supervision logic, it could be possible to perform, in a full decentralized way, actions to improve the performance of the system, e.g.: improving the response time through load balancing policies, for redistributing the tasks and/or redirecting the requests from the clients or reducing the energy consumption, by putting in stand-by mode some servers when there is a reduction of the service request load.

5.2 Cloud Computing

Cloud Computing vision is based on the idea of providing Users wit computing and storage services anywhere, appearing as a single point of access for all needs; most of current solutions are based on distributed resources built on servers virtualization technologies Cloud Computing encompasses many areas of tech, including software as a service, a software distribution method pioneered by `salesforce.com` about a decade ago; it also includes newer avenues such as hardware as a service, a way to order storage and server capacity on demand from Amazon and others. What all these cloud computing services have in common, though, is that they're all delivered over the Internet, on demand, from massive data centers [22].

In other words, cloud computing can be defined as infrastructures consisting of reliable services delivered through next-generation data centers that are built on compute and storage virtualization technologies [42]. Reliability and security are big challenges [22]. The access to those services is strongly based on web mechanisms (e.g., xml and http). Applications are able to use resources made available by a processing or storage "cloud" (autonomous systems) in order to processing a particular task.

In this case the Operators should compose their offering trading-off bandwidth, storage and processing power for dynamically providing the best possible services to users. Different functions can span over different clouds. It is up to the application to coordinate the different execution results into something meaningful for the application itself. Storage and processing can be "purchased" on demand from different providers at the best price. The network is becoming a sort of computer that offers capabilities on demand. For example, solutions like Amazon's "Elastic Compute Cloud" [3] or IBM's "Blue Cloud" technologies are based on open standards and open source software which link together computers that are used to deliver Web 2.0 capabilities like mash-ups [41].

Cloud Computing brings in a lot of innovation: the computing essential features are now within the network (processing and storage), the communication bus is the network itself, and the input/output devices are the end user terminals. It is possible to re-formulate the concept of the Network Computer. It should be done in a smart way and used by the Operators as a means to compete with web actors: "if the network is a commodity, then also the computer is a networked commoditized resource". The browser gets an even more prominent importance: it is really the means for interacting with services and (virtualized) PCs scattered all over the network.

Bio-inspired Autonomic Structures can be used to support load balancing, dynamic configuration, fault tolerance, enhance security and improve quality-of-service in the presence of the very dynamic conditions. Basic idea consists in adopting autonomic components to manage the high dynamicity of the cloud nodes: not only node reliability but also Users entering and exiting the cloud in an unpredictable way.

For example, each peer computing machine should be equipped with autonomic components capable of exchanging and managing events coming from other peers; an important task, for example, is informing (e.g., a super peer) periodically about execution check points in order to allow dynamic reconfiguration of the execution power. Moreover, storage strategy should face also the problems of data synchronization whilst providing the proper number of duplications for the requested persistency. Also in this case, autonomic abstractions of clusters of data can provided the required functionalities.

5.3 Home Networking

The Home Networking environment is acquiring more and more relevance from the business and technical perspectives. There are many ecosystems trying to exploit the market potential: operators bring to the home access gateways, software companies bring in the home operating systems and applications (e.g., Microsoft and Apple), and consumer electronic firms provide equipments and often entire integrated multimedia suite. In any case the user is left alone in coping with the networking and integration intricacies of putting in place a really working home network solution. Things seem to be even complicating due to the introduction in the homes of new technologies like sensor networks and the like.

There is an urgent need to relief the customers from the burden to put in place such a complex system. The needs of a typical customer in this environment are in the field of configuration, tuning, recovery of systems and the entire home network.

The ideal situation is to be able to buy new equipment and to really plug it in the home network without having to configure and tune all the system. In this context the autonomic technologies are a fundamental help for customers and even for the companies that are trying to take advantage from this potentially huge market.

Home networks could be empowered through the adoption of autonomic technologies: this means that each participating entity (e.g., the home computers, the access gateway, the printers, the TV sets, but even the fridge and the washing machine!) has an autonomic behavior. They are able to adapt their behavior to the current available context (i.e., the network capabilities, the availability of services provided by networked resources, and so on). This would improve the current situation, where, in order for them to properly interact and work, there is the need of a lot of configuration work, and after that there is the need to tune the entire network depending on particular situation that can arise during the usage of resources (e.g., watching a movie from the IPTV can be compromised by the parallel activity of file download from a computer). Also malfunctioning of one resource can have an impact on other equipment and on the whole functioning of the whole home network. Available resources need to be optimized and correctly shared and used.

An autonomic environment is put in place, for instance, by enhancing the home network elements with BAS components, collaborating for providing self-CHOP features in a distributed way. It means that, the behavior of the home network can be optimized and the human intervention can minimized. But there is more, at least two new features could improve the situation and being highly appreciated by the customers: the ability of the networked resources to accommodate for malfunctioning providing at the same time warnings about what is going on the entire system, and the ability to accommodate in advance for new resources to join the network.

In case one resource is degrading its services, it can inform the whole autonomic network and it can try to fix the problem. Error messages can be collected and used to re-balance the load of the home network and at the same time they can be passed over the network to a monitoring center that can register the events and elaborate a recovery strategy. For instance, a Network/Service Provider could use the access gateway as a sort of supervisor of the autonomic home network. It collects the events and stores a view on the topology and configuration of the whole home network. In case of malfunctions (of the network or of the single equipment) it can forward the right information to a monitoring center that could be involved in the solution of the problem. For example, if the fridge has some problems, first the heating system of the home can be asked to cooperate and then the relevant fault information can be sent to the monitoring center for being studied. The monitoring center could be able to find a solution via software. The new firmware could be downloaded in a secure way from the monitoring center down to the home network and then to the faulty equipment, or, if the fault requires human intervention, the monitoring center can provide a detailed log and possibly some hints for the solution of the problem.

On the other hand, the home network can be extended with new resources. The users can be willing to add new multimedia capabilities to the network. Let us assume that the user wants to add a new multimedia storage system. The autonomic network can help the user in many ways: first it can provide a list of needed configuration features to be available on the new resource in order to support a smooth integration in the network; secondly, it could even suggest the set of needed parameters to be set on the new resource. Moreover, the user could rely on the home networking capabilities and the monitoring center in order to get suggestion on the specific resource that better fits with the whole networked environment, i.e., a sort of suggestion about what kind of equipment to buy. If the user does not want to follow the suggestion, still the autonomic network can help: the user could choose the resources that he likes the most, and the autonomic network could provide the list of needed parameters (and their values) for a correct and easy integration of the resource. If this is not sufficient the autonomic network (together with the monitoring center) could even suggest and made available to the user (and to the vendor of the new resource) the most appropriated firmware to be used in that situation.

From this few examples it is clear the value that an autonomic approach to home networking could bring to the users, but also to the entire ecosystems engaged in supporting the market of home networking.

6 Conclusions

This chapter has presented the vision of a distributed bio-inspired autonomic middleware for the development of future Telecommunications Ecosystem. Middleware is basically an ecology of components (interacting with communication primitives for the self-adaptation and self-organization) virtualizing any devices, network resources, service components and pieces of data. The bio-inspired metaphor lies in the envisaging of mature capabilities as the cumulative outcome of a large number of small advances in capability. Each such advance is assumed to be associated with a particular generative mechanism, so that the nervous system as a whole can be viewed as an organized collection of such generative mechanisms, each specialized to some particular generative process.

It should be mentioned that the vision of developing a distributed autonomic middleware for Telecommunications Ecosystems could benefit also from the activities of the so-called artificial society discipline: for example, those results coming from understanding through simulation, emergent phenomena of large population of agents (modeling people behaviors) are valuable for trying to explain the observable macro-level characteristics of societies and building future Telecommunications Ecosystems.

Ongoing activities concern enhancements of abstractions of autonomic components and structures (with and their local proprieties), an improvement in the definition of bio-inspired primitives (fully connected, small world, random, small set of neighbors, etc.) of components interactions whilst combining top-down policies with a bottom-up self-organization emergence (based on proprieties feedbacks and environment perturbations).

References

1. Abelson, H., Beal, J., Sussman, G.J.: Amorphous Computing. MIT Technical Report CSAIL TR 2007 030 (2007)
2. Alfano, R., Manzalini, A., Moiso, C.: Distributed Service Framework: an innovative open eco-system for ICT/Telecommunications. In: Proceedings of Autonomics 2007, ACM Digital Library (2007)
3. Amazon EC2: Amazon Elastic Compute Cloud. Retrieved Feb. 10, 2009 from: http://aws.amazon.com/ec2/
4. Anastasopoulos, M. P., Vasilakos, A. V. Cottis, P.G.: An Autonomic Framework for Reliable Multicast: A Game Theoretical Approach based on Social Psychology. ACM Transactions on Autonomous and Adaptive Systems, Special Issue on Adaptive Learning in Autonomic Communication (2009)
5. Arcos, J. L., Esteva, M., Noriega, P., Rodriguez, J. A., Sierra, C.: Engineering Open Environments with Electronic Institutions. Journal on Engineering Applications of Artificial Intelligence, Vol. 18, Issue 2, (2005)
6. ASG Project: Adaptive Services Grid. www.asg-platform.org
7. Baas, N. A.: Emergence, Hierarchies, and Hyperstructures. In: Proceedings of Artificial Life III, Santa Fe Studies in the Sciences of Complexity, Vol. XVII (1994)
8. Baas, N. A.: Hyperstructures as tools in nanotechnology. Nanobiology 3(1) (1994)
9. Baas N. A.: A framework for higher order cognition and consciousness. In: Towards a science of consciousness, MIT Press/Bradford Books (1996)
10. Baas N. A.: Self-Organization and higher order structures. In: Self-organization in complex structures - from individual to collective dynamics, Gordon & Breach (1997)
11. Babaoglu, O., Canright, G., Deutsch, A., Caro, G. A. D., Ducatelle, Gambardella, F., L. M., Ganguly, N., Jelasity, M., Montemanni, R., Montresor, A., Urnes, T.: Design Patterns from Biology to Distributed Computing. ACM Transaction on Autonomous and Adaptive Systems, Vol. 1, No. 1, (2007)
12. Bernadas, A., Manzalini, A.: Demonstrating communication services based on autonomic self-organization. In: Proceedings of 2nd International Conference on Complex, Intelligent and Software Intensive Systems (CISIS2008), IEEE Computer Society CPS (2008)
13. BIONETS Project: BIOlogically inspired NETwork and Services. www.bionets.eu
14. Bouquet, P., Serafini, L., Zanobini, S.: Peer-to-Peer Semantic Coordination. Journal of Web Semantics, 2(1) (2005)
15. CASCADAS Project: Component-ware for Autonomic Situation-aware Communications, and Dynamically Adaptable Services. www.cascadas-project.org
16. Chiang, F., Braun, R.: A Nature Inspired Multi-Agent Framework for Autonomic Service Management in Pervasive Computing Environments. In: Proceedings of 10th IEEE/IFIP Network Operations and Management Symposium (NOMS2006), IEEE Computer Society (2006)
17. Fusco, A., Manzalini, A., Moiso, C., Blazquez, H., SolÃl' Pareta, J., Spadaro, S.: Autonomic wireless communications in Digital Cities: an experimental use case. In: Proceedings of Mobilware 2008, Innsbruck, ACM Digital Library (2008)
18. Gelernter D.: Generative Communication in Linda. ACM Transactions on Programming Languages, 7(1) (1985)

19. Gerke, J., Hausheer, D., Mischke, J., Stiller, B.: An Architecture for a Service Oriented Peer-to-Peer System (SOPPS). Praxis der Informationsverarbeitung und Kommunikation (PIK) 2/03, (2003)
20. Jelasity, M., Montresor, A., Babaoglu, O.: Gossip-based Aggregation in Large Dynamic Networks. ACM Transactions on Computer Systems, 23(3) (2005)
21. Kephart, J., Chess, D.: The Vision of Autonomic Computing. IEEE Computer, 36(1), (2003)
22. King, R.: How Cloud Computing Is Changing the World - Special Report. Business Week (2008)
23. Linner, D., Pfeffer, H., Steglich, S.: A genetic algorithm for the adaptation of service compositions. In: Proceedings of 2nd Conference on Bio-Inspired Models of Network, Information and Computing Systems (Bionetics 2007), IEEE Xplore (2007)
24. Liu, H., Bhat, V., Parashar, M., Klasky, S.: An Autonomic Service Architecture for Self-managing Grid Applications. In: Proceedings of 6th IEEE/ACM International Workshop on Grid Computing (Grid2005), IEEE Computer Society (2005)
25. Mamei, M., Menezes, R., Tolksdorf, R., Zambonelli, F.: Case Studies for Self-organization in Computer Science. Journal of Systems Architecture, Vol. 52, Issues 8-9, (2006)
26. Manzalini, A., Marrow, P.: CASCADAS Project: A Vision of Autonomic Self-organizing Component-ware for ICT Services. In: Proceedings of International Conference on Self-Organization and Autonomous Systems in Computing and Communications (SOAS2006), (2006)
27. Manzalini A., Zambonelli F.: Towards Autonomic and Situation-Aware Communication Services: the CASCADAS Vision. In: Proceedings of IEEE 2006 Workshop on Distributed Intelligent Systems, IEEE Computer Society (2006)
28. Manzalini, A.: Tomorrow's Open Internet for Telco and Web federations. In: Proceedings of 2nd International Conference on Complex, Intelligent and Software Intensive Systems (CISIS2008), IEEE Computer Society CPS (2008)
29. Mazzola Paluska, J., Pham H., Saif U.: Technique-based Programming. In: Proceedings of 5th IEEE International Conference on Pervasive Computing and Communications, IEEE Computer Society (2007)
30. Minerva, R.: On Some Myths about Network Intelligence. In: Proceedings of International Conference on Intelligence in Networks (ICIN2008), Neustar (2008)
31. MMAPPS Project: Market Management of Peer to Peer Services. http://www.mmapps.info/
32. NextGRID Project: Architecture for Next Generation Grids. www.nextgrid.org
33. Parunak, H. V. D.: Go to the Ant: Engineering Principles from Natural Multi-Agent Systems. Annals of Operations Research, Vol. 75 (1997)
34. Samimi, F. A., McKinley, P. K., Sadjadi, S. M., Tang, C., Shapiro, J. K., Zhou, Z.: Service Clouds: Distributed Infrastructure for Adaptive Communication Services. IEEE Transactions on Network and System Management (TNSM), Special Issue on Self-Managed Networks, Systems and Services (2007)
35. Saraydar, C., Mandayam, N. B., Goodman, D. J: Efficient Power Control via Pricing in Wireless Data Networks. IEEE Transactions on Communication, Vol. 50, No. 2 (2002)
36. ESAM Project: Self Organization and Spontaneity in Liberalized and Harmonized Markets. http://dsn.tm.uni-karlsruhe.de/english/sesam-project.php
37. Sterritta, R., Parasharb, M., Tianfieldc, H., Unland, R: A concise introduction to autonomic computing. Advanced Engineering Informatics 19, Elsevier (2005)
38. Stratford, N., Mortier, R.: An Economic Approach to Adaptive Resource Management. In: Proceedings of 7th Workshop on Hot Topics in Operating Systems (TCOS99), IEEE Computer Society (1999)
39. Valckenaers, P., Sauter, J., Sierra, C., Rodriguez-Aguilar, J. A.: Applications and Environments for Multi-Agent Systems. Journal of Autonomous Agents and Multi-Agent Systems, Vol. 14, Issue 1, (2007)
40. Vickrey, W.: Counterspeculation, Auctions, and Competitive Sealed Tenders. The Journal of Finance 16(1) (1961)
41. Webopedia.com: Cloud Computing. Retrieved Feb. 10, 2009 from: http://www.webopedia.com/TERM/C/cloud_computing.html

42. Wikipedia.org: Cloud Computing. Retrieved Feb. 10, 2009 from: http://en.wikipedia.org/wiki/Cloud_computing
43. Yates, R. D.: A Framework for Uplink Power Control in Cellular Radio Systems. IEEE Journal on Selected Areas of Communication, Vol. 13 (1995)
44. Yu, W., Ginis, G., Cioffi, J.: Distributed Multiuser Power Control for Digital Subscriber Line. IEEE Journal on Selected Areas of Communication, Vol. 20, No. 5 (2002)

Social-based autonomic routing in opportunistic networks

Chiara Boldrini, Marco Conti, Andrea Passarella

Abstract In opportunistic networks end-to-end communication between users does not require a continuous end-to-end path between source and destination. Network protocols are designed to be extremely resilient to events such as long partitions, node disconnections, etc, which are very features of this type of self-organizing ad hoc networks. This is achieved by temporarily storing messages at intermediate nodes, waiting for future opportunities to forward them towards the destination. The mobility of users plays a key role in opportunistic networks. Thus, providing accurate models of mobility patterns is one of the key research areas. In this chapter we firstly focus on this issue, with special emphasis on a class of *social-aware* models. These models are based on the observation that people move because they are attracted towards other people they have social relationships with, or towards physical places that have special meaning with respect to their social behavior. Another key research area in opportunistic networks is clearly designing routing and forwarding schemes. In this chapter we provide a survey of the main approaches to routing in purely infrastructure-less opportunistic networks, by classifying protocols based on the amount of context information they exploit. We then provide an extensive quantitative comparison between representatives of protocols that do not use any context information, and protocols that manage and exploit a rich set of context information. We mainly focus on the suitability of protocols to adapt to the dynamically changing network features, as resulting from the user movement patterns that are driven by their social behavior. Our results show that context-aware routing is extremely adaptive to dynamic networking scenarios, and, with respect to protocols that do not use any context information, is able to provide similar performance in terms of delay and loss rate, by using just a small fraction of the network resources.

Chiara Boldrini, Marco Conti, Andrea Passarella
IIT-CNR, Via G. Moruzzi, 1 - 56124 Pisa, Italy
e-mail: c.boldrini,m.conti,a.passarella@iit.cnr.it

A.V. Vasilakos et al. (eds.), *Autonomic Communication*, DOI: 10.1007/978-0-387-09753-4_2,

1 Introduction

The opportunistic networking idea stems from the critical review of the research field on Mobile Ad hoc Networks (MANET). After more than ten years of research in the MANET field, this promising technology still has not massively entered the mass market. One of the main reasons of this is nowadays seen in the lack of a *practical* approach to the design of infrastructure-less multi-hop ad hoc networks [11, 12]. One of the main approaches of conventional MANET research is to design protocols that mask the features of mobile networks via the routing (and transport) layer, so as to expose to higher layers an Internet-like network abstraction. Wireless networks' peculiarities, such as mobility of users, disconnection of nodes, network partitions, links' instability, are seen – as in the legacy Internet – as exceptions. This often results in the design of MANET network stacks that are significantly complex and unstable [6].

Opportunistic networks [33] also aim at building networks out of mobile devices carried by people, possibly without relying on any pre-existing infrastructure. However, opportunistic networks look at mobility, disconnections, partitions, etc. as *features* of the networks rather than exceptions. Actually, mobility is exploited as a way to bridge disconnected "clouds" of nodes and enable communication, rather than a drawback to be dealt with. More specifically, in opportunistic networking no assumption is made on the existence of a complete path between two nodes wishing to communicate. Source and destination nodes might never be connected to the same network, at the same time. Nevertheless, opportunistic networking techniques allow such nodes to exchange messages. By exploiting the *store-carry-and-forward* paradigm [15], intermediate nodes (between source and destination) store messages when no forwarding opportunity towards the final destination exists, and exploit any future contact opportunity with other mobile devices to bring the messages closer and closer to the destination. This approach to build self-organizing infrastructure-less wireless networks turns out to be much more practical than the conventional MANET paradigm. Indeed, despite the fact that opportunistic network research is still in its early stages, the opportunistic networking concept is nowadays exploited in a number of concrete applications (in Section 2 we provide a brief overview of them).

It is clear that understanding the real mobility patterns of users is key in this networking environment, as mobility of users is one of the enabler of end-to-end communications. To this end, after describing in more details the main concepts of opportunistic networks and their practical use cases in Section 2, the first part of this chapter is devoted to analyzing mobility models suitable for opportunistic networks (Section 3). Specifically, we consider a class of social-aware mobility models, in which users movements are driven by their social relationships and behavior. These models have shown to closely reproduce statistical features of real movement traces, and are thus very good candidate tools for designing and evaluating opportunistic networking systems. Actually, mobility modeling is one of the most active areas in the opportunistic networking field.

Another key area widely explored by researchers is clearly routing & forwarding[1], due to the inherent complexity of the problem [33, 41]. Therefore, in Section 4 we provide a brief survey of the main routing approaches available in the literature. Specifically, we categorize protocols based on the amount of *context information* they exploit, by identifying three main classes, i.e., context-oblivious, partially context-aware and fully context-aware protocols. The main idea behind using context information is to enable routing protocols to learn the network state, autonomically adapt to its dynamic evolution, and thus optimize their operations. In the final part of the chapter (Section 5) we provide performance results to evaluate the suitability of this idea in real routing protocols. To replicate realistically the users' behavior, we consider a mobility model (HCMM) that has shown to realistically reproduce real human movement patterns as driven by users' social relationships and social behavior (fully described in Section 3.1). We exploit the model's parameters to study how different routing approaches react to various levels of dynamism and users' sociability. We compare the performance of Epidemic Routing and HiBOp, which are representatives of the opposite ends of the spectrum of possible approaches, i.e. context-oblivious and fully context-aware protocols, respectively (Section 5). By analyzing their sensitiveness with respect to a number of parameters, we show that context-aware schemes are able to provide similar levels of QoS (in terms of message delay and loss rate), by spending a *small fraction* of the resources spent by context-oblivious protocols. Even more interestingly, we find that context-aware systems are much more suitable to autonomically learn the features of the network they are operating in, and the behavior of users as determined by their social relationships. We show that, unlike context-oblivious systems, context-aware protocols are able to correctly adapt their operations accordingly. This results in a much more judicious use of the available resources, also when the network scenario abruptly changes. We finally draw conclusions and identify research directions in Section 6.

This chapter blends in a unique framework both mobility modeling and social-based routing approaches that have been separately considered in [1–5]. In this chapter we provide a unique line of reasoning and a systematic presentation of these pieces of work.

2 The opportunistic networking concept and its applications

Opportunistic networks share several concepts with Delay Tolerant Networks (DTNs). The DTN architectures defined by the DTN IRTF Research Group (`http://www.dtnrg.org/docs/specs`) focus on a scenario in which independent internets, each characterized by internal Internet-like connectivity, are interconnected through a DTN *overlay*. In order to achieve end-to-end connectivity, the DTN overlay exploits

[1] As will be clear in the following, in opportunistic networks the routing and forwarding tasks are strictly intertwined and usually performed at the same time. Therefore, hereafter we use the terms routing and forwarding interchangeably.

occasional communication opportunities among the internets, which might either be scheduled over time (e.g., due to the activation of a satellite link), or completely random. In general, in conventional DTNs the points of possible disconnections are known.

Opportunistic networks can be seen as a generalization of DTNs. Specifically, in opportunistic networks no a-priori knowledge is assumed about the possible points of disconnections, nor the existence of separate Internet-like sub-networks is assumed. Opportunistic networks are formed by individual nodes, that are possibly disconnected for long time intervals, and that opportunistically exploit any contact with other nodes to forward messages. The routing approach between conventional DTNs and opportunistic networks is therefore quite different. Since in DTNs the points of disconnections (and, sometime, the duration of disconnections) are known, routing can be performed along the same lines used for conventional Internet protocols, by simply considering the duration of the disconnections as an additional cost of the links [23]. Since opportunistic networks do not assume the same knowledge about the network evolution, routes are computed dynamically while the messages are being forwarded towards the destination. Each intermediate node evaluates the suitability of encountered nodes to be a good next hop towards the destination.

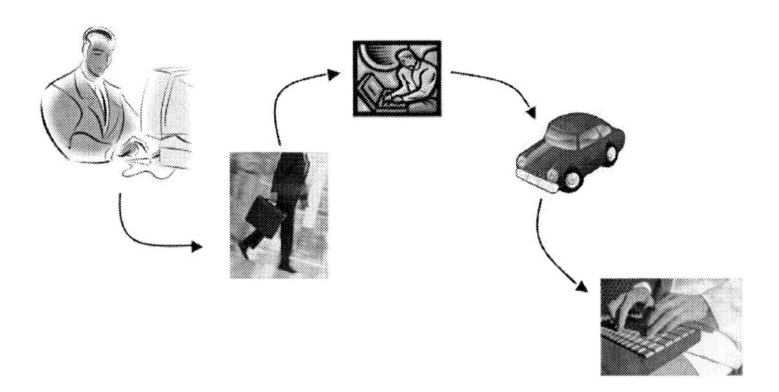

Fig. 1 The opportunistic networking concept.

For example, as shown in Figure 1, the user at the desktop opportunistically transfers, via a Wi-Fi ad hoc link, a message for a friend to a user passing nearby, "hoping" that this user will carry the information closer to the destination. This user passes close to a train station, and forwards the message to a traveler going to the same city where the destination user works. At the train station of the destination city a car driver is going in the same neighborhood of the destination's working place. The driver meets the destination user on his way, and the message is finally delivered.

2.1 Opportunistic networking case studies and applications

Despite the fact that research on opportunistic networks dates back to just a few years ago, concrete applications and real case studies are already available (for a more extensive discussion about this point please refer to [32]).

The Haggle Project (`http://www.haggleproject.org`) is a 4-year project, started in January 2006, funded by the European Commission in the framework of the FET-SAC initiative (`http://cordis.europa.eu/ist/fet/comms-sy.htm`). It targets solutions for communication in autonomic opportunistic networks. Among the various activities, the project is putting emphasis on measuring and modeling pair-wise contacts between devices. Pair-wise contacts between users/devices can be characterized by means of two parameters: contact durations and inter-contact times. The statistical properties of these parameters are used to drive the design of forwarding policies [10]. Furthermore, they are also the basis of the design of concrete applications. For example, Haggle is working with epidemiologists to experimentally study the correlation between human contact patterns and the spread of diseases such as flu. The patterns of contacts between people (measured in real experiments) are also the basis for designing "social-aware" applications. An initial example of this approach is the design of a content distribution system in urban setting [28]. Refined solutions for this type of applications are being designed in the Haggle project (e.g., [40]) thanks to the autonomic tools for detecting user social communities [22].

Opportunistic networks are also applied to interdisciplinary projects focusing on wildlife monitoring. Usually, small monitoring devices are attached to animals, and an opportunistic network is formed to gather information and carry it to a few base stations possibly connected to the Internet. Contacts among animals are exploited to aggregate data, and carry them closer and closer to the base stations. This is a reliable, cost-effective and non intrusive solution. Concrete applications implementing these ideas have been used in the ZebraNet project [25]. ZebraNet is an interdisciplinary project of the Princeton University performing novel studies of animal migrations and inter-species interactions, by deploying opportunistic networks on zebras in the vast savanna area of the central Kenya under control of the Mpala Research Centre (`http://www.mpala.org/researchctr/research/ongoing.html`).

We finally mention the use of opportunistic networks to bring Internet connectivity to rural areas. In developing countries and rural areas deploying the infrastructure required to enable conventional Internet connectivity is typically not cost-effective. However, Internet connectivity is seen as one of the main booster to bridge the digital divide. Opportunistic networks represent an easy-to-deploy and extremely cheap solution. Typically, rural villages are equipped with a few collection points that temporarily store messages addressed to the Internet. Simple devices mounted on bus, bicycles or motorbikes that periodically pass by the village collect these messages and bring them in regions where conventional Internet connectivity is available (e.g., a nearby city), where they can be delivered through the Internet. The same concept is exploited to enable communication in the opposite direction (from the Internet to villages). Projects implementing these concepts are currently ongoing. For example, the DakNet [35] and KioskNet [19] Projects focus on realising a very low-cost asynchronous ICT infrastructure to provide connectivity to rural villages in India, while the Saami Network Connectivity Project [14] provides connectivity to inhabitants of Lapland.

3 Social-based mobility

Mobility modeling for opportunistic networks is a hot topic in the research community. Opportunistic networks actually *exploit* users' mobility to bridge disconnections and partitions [17]. Therefore, it is of paramount importance to identify realistic mobility models, both to drive the protocols' design, and to provide sensible performance results. In the last few years, there has been an increasing effort aimed at reconsidering the MANET mobility models [9] for opportunistic networking scenarios. There is general agreement on the fact that popular models used in MANET research (e.g., the random waypoint model) generate quite unrealistic users' behavior (e.g., [37]). To address this issue, mobility models are reconsidered or re-designed based on real users' mobility traces available to the community (e.g., through CRAWDAD).

Several proposals [26, 29, 37] exploit WLAN association traces to derive users' association profiles and, based on these, mobility models. The resulting models are very good in capturing the fact that physical locations (WLAN hotspots in this case) exert attraction on users. The work in [20] takes this idea one step further, and provides mobility models in which general physical locations (not necessarily WLAN access points) exert attractions on users. Finally, authors of [16] explain WLAN association traces with sociological-inspired concepts, noticing that periodic association patterns follow sociological orbits, defined by the users' social behavior. Exploiting this remark, they provide a user-centric model (rather than an "AP"-centric one as in the previous works). This body of work is based on the fundamental observation that users are attracted by particular physical locations, in which they tend to preferentially spend their time. The limit we see in this approach is the fact that it does not explain the mechanisms resulting in the modelled mobility patterns. Therefore, it is not clear if the resulting models are applicable to networking scenarios other than the ones used for the initial observations (most notably, if they are applicable to opportunistic networks too).

Exploiting the social behavior of users to define the basic mechanisms of users' movements is a very interesting direction. To the best of our knowledge, the most advanced proposals of this class are the Community-based and Home-cell Community-based Mobility Models (CMM and HCMM), that have been compared in [3]. The most interesting feature of CMM is the leveraging of social network theories and models [13] to define users' movements. Besides matching well real users' mobility traces, this approach sheds light on the features of users' social behavior that result in the mobility features observed in real traces.

Despite these nice properties, in [3] we have shown that the original CMM proposal is not able to capture the attraction exerted on users by physical locations. Specifically, we have found that CMM shows a *gregarious* behavior, such that all users in a community tend to follow the first user that moves outside the physical location where the community is located. The gregarious behavior does not represent significant scenarios (e.g., working places), where users roam around preferred physical places, besides being influenced by social relationships between each other. To address this issue, we propose the *Home-cell Community-based Mobility Model (HCMM)*, which joins the concepts of CMM (for modeling social relationships between users) with the concept of defining preferential locations in which users tend to spend most of their time. Therefore, HCMM is a first step towards joining together the two promising mobility modeling approaches discussed above. HCMM still matches characteristic features of real traces (see [3]). Furthermore, we highlight that, unlike CMM, it provides very simple knobs to control the time spent by users in their preferred physical locations (Section 3.2).

3.1 CMM and HCMM: functional description

The Home-cell Community-based Mobility Model (HCMM) (fully described in [3]) is an evolution of the Community-based Mobility Model. Community-based (or group) mobility models are attracting interest of researches in the opportunistic networking area, because they are suitable to realistically model the influence of social relationships between people on the user mobility patterns.

As in CMM, in HCMM every node belongs to a social community (group). Nodes that are in the same social community are called *friends*, while nodes in different communities are called *non-friends*. Relationships between nodes are modeled through social links (each link has an associated weight). At the system start-up all friends have a link to each other. Also two nodes that are not friends can have a link, according to the *rewiring probability* (p_r) parameter. Specifically, for each node, each link towards a friend is rewired to a non-friend with p_r probability.

Social links are then used to drive node movements. Nodes move in a grid, and each community is initially randomly placed in a square of the grid. Nodes' movement is made up of two component: first, a node has to select the cell towards which to move. Node selects the target cell according to the social attraction exerted by each cell on the node. Attraction is measured as the sum of the links' weights between the node and the nodes currently moving in or towards the cell. The target cell is finally selected based on the probabilities defined by cells' attraction (i.e., if a_j is the attraction of cell j, then the probability of selecting that cell is $a_j / \sum_j a_j$). After selecting the target cell, node selects the "goal" within a cell (the precise point towards which node will be heading) according to a uniform distribution. Finally, speed is also selected accordingly to a uniform distribution within a user-specified range. HCMM (and CMM) also allows for collective group movements. Specifically, once every *reconfiguration period* nodes of each group select a (different) cell and move to that cell. Reconfigurations are synchronous across groups, i.e., all groups start moving to the new cell at the same time. Therefore, during reconfigurations nodes of different groups may get in touch.

The difference between HCMM and CMM is the way of considering the social relationships with nodes that are outside their starting cell (called "home cell" in HCMM). Let's focus on Figure 2. In CMM, when node A moves outside it's home cell, it "carries over" all its social relationships, i.e., nodes that have social relationships with A are attracted towards the same cell towards which A is moving. In [3] it is shown that this has an avalanche effect such that *all* nodes in A's home cell follow A. This behavior does not allow CMM to model relevant mobility patterns, because nodes are basically not attracted by physical locations, but only by social relationships between each other. In HCMM when A moves outside its home cell it *does not* carry over its social links. Nodes having social relationships with A are still attracted towards A's home cell. Furthermore, once A is outside its home cell, it selects its goal for the next movements outside the home cell with probability p_e, and goes back to the home cell with probability $1 - p_e$. The rationale behind these modifications is the fact that there are several scenarios in which also physical locations (besides social relationships) play a role in determining users' movements. In HCMM people wishing to meet with A (i.e., having social attraction towards A) are attracted towards A's home cell because that is the most likely *physical* place where A can be met, or because their social relationship with A is conditioned to the fact that A is in its home cell (e.g., if someone wants to meet an insurance agent, they will go to the insurance office, not to the agent's house).

In a nutshell, HCMM models the fact that humans are social (belongs to groups), move towards other people they have relationships with (most likely within their group, but also outside their group), and occasionally move collectively with their group. Furthermore, results presented in [3] show that the duration of contact and inter-contact times under HCMM are similar to those measured in real experiments, which shows that HCMM provides realistic movement traces.

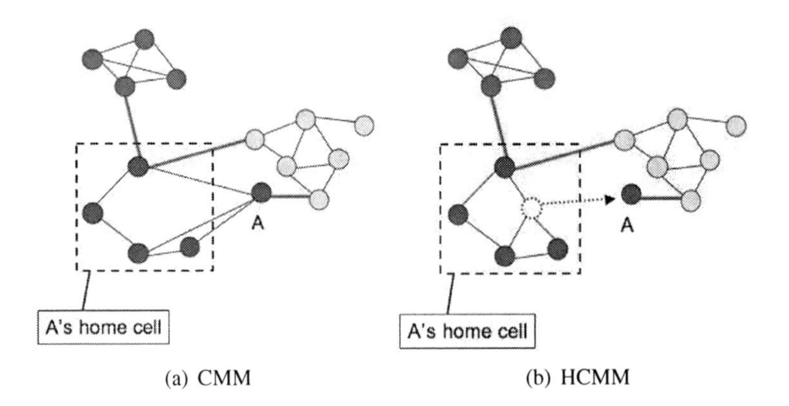

(a) CMM (b) HCMM

Fig. 2 CMM vs HCMM

In [3] we have presented an analytical model to highlight a *gregarious* behavior of CMM. To this end, we computed the *remaining probability* (P_{rem}), defined as the probability of *no other* member of node k's community to move towards the destination cell. When P_{rem} approaches 0, at least one node in the starting cell follows node k. As will be clear from the following analysis, this may generate an avalanche effect such that all nodes in node k's community follow node k in the destination cell, thus revealing the gregarious behavior. We consider the case of a single node (k) having links outside its community, because it represents the weaker condition for the gregarious behavior to take place. Therefore, the P_{rem} formula computed in [3] is actually an upper bound of the remaining probability achieved in the general case. Specifically, the final expression of P_{rem} is:

$$P_{rem} = \left[(1 - P_{out})^l\right]^{n-1} = \left[\left(1 - \frac{w_k/(fn+1)}{w_k/(fn+1)+\overline{w}}\right)^l\right]^{n-1}, \tag{1}$$

where l is the average number of times each node in the starting cell selects a new destination while node k is associated with the destination cell, $1 - P_{out}$ is the probability of each node to select the starting cell for the next step, and $n-1$ is the number of nodes in the starting cell after node k departure. Furthermore, w_k is the average weight between node k and the other nodes of its community, $fn+1$ the number of nodes in the cell towards which node k is traveling, and $\overline{w}$ is the average weight between nodes of node k community.

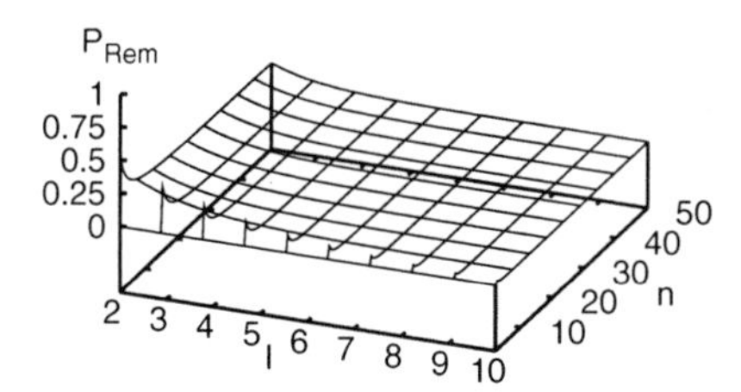

Fig. 3 P_{rem} as a function of n and l.

This model (validated in [3] against simulation results) allows us to show that the gregarious behavior occurs basically for *all* sensible ranges of the model parameters. Just to show an example, Figure 3 illustrates the P_{rem} dependence on n (the number of the nodes of k's community), and l (the ratio between the movements duration outside and within a community).

For small values of l, the grid has few cells and the duration of k's movement outside the starting cell is not so different from the duration of nodes' random movement inside a cell. Thus, a generic node i has not many opportunities of going outside the starting cell, because node k is associated with the destination cell only for a relatively small amount of time. The trend highlighted in Figure 3 generally holds true when considering the impact of l, irrespectively of the other parameters' configurations. Therefore, we will not analyze the impact of l further on.

To better understand the behavior with respect to n, let us rewrite Equation 1, by recalling that $\overline{w}=w_k$. It is easy to show that Equation 1 becomes $P_{rem}=\left[(1-(1/n+2))^l\right]^{n-1}$. The remaining probability of a *single* node $(1-(1/n+2))$ increases with n, because a large n corresponds to a "heavy" community, that exerts a strong attraction on its members. However, as the number of nodes increases, it is more and more difficult that *all* nodes remain in the starting cell. The joint effect (shown in Figure 3) is that P_{rem} is significantly greater than 0 only for small values of n.

3.2 *HCMM vs. CMM: Controlling Node Positions*

In this section we compare HCMM and CMM. Specifically, we highlight the fact that HCMM allows for a fine control of the physical locations around which users' roam, while CMM does not provide any simple control parameter on this. To this end, we generalize the analytical model presented in Section 3.1. The goal of the model we present hereafter is to provide closed formulas for the average time spent by any node inside and outside the starting cell (home cell in HCMM). For ease of presentation, we still assume to have just two cells, even though the destination cell can jointly represent all cells other than the starting cell. We assume that *all* links can be rewired at the system startup (with probability p_r). Therefore, we don't assume any difference between a tagged node (node k) and the other nodes anymore. We also don't focus anymore on the event of a particular node exiting the starting cell.

In HCMM and in CMM the status of each node can be represented with a 2-state discrete Markov chain as in Figure 4, where "IN" means the node is in the starting cell, and "OUT" means it is outside the starting cell. The difference between HCMM

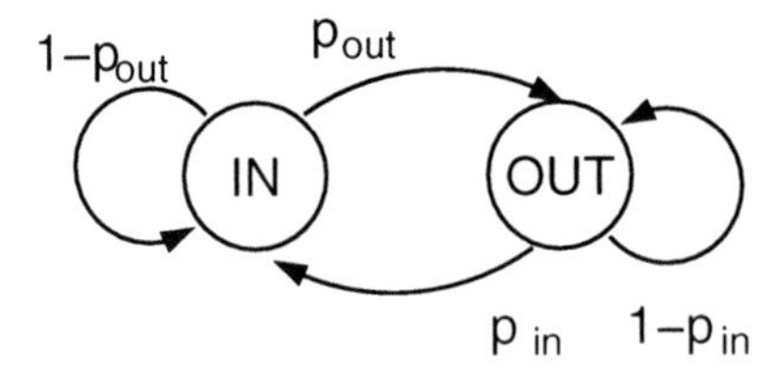

Fig. 4 Node's status in HCMM and CMM.

and CMM lies in the expressions of p_{in} and p_{out}, that we will derive at the end of this section. Otherwise, the analysis of the average time spent in the IN and OUT states is common to CMM and HCMM.

First of all, it is straightforward deriving the stationary distributions, $\pi_{in} = p_{in}/(p_{in} + p_{out})$, and $\pi_{out} = p_{out}/(p_{in} + p_{out})$. The average time spent in the IN and OUT states can be computed via the conditioned probabilities, as follows:

$$\begin{cases} E[T_{in}] &= \pi_{out} p_{in} \cdot E[T_{in}|E_{IN}] \\ E[T_{out}] &= \pi_{in} p_{out} \cdot E[T_{out}|E_{OUT}] \end{cases} \tag{2}$$

where E_{IN} and E_{OUT} denote the events "the node enters the IN state" and "the node enters the OUT state", respectively, while $\pi_{out} p_{in}$ and $\pi_{in} p_{out}$ are the probabilities of these events. By recalling that i) the number of steps spent in each state is distributed according to a geometric law, ii) the duration of each step both in the IN and OUT state can be approximated with $\overline{T}^{(in)}$, and iii) the duration of the transitions between the states can be approximated with $\overline{T}^{(out)}$, we can compute closed form expressions for $E[T_{in}]$ and $E[T_{out}]$ as follows:

$$\begin{cases} E[T_{in}] = \frac{p_{in}(1-p_{out})}{p_{in}+p_{out}} \cdot \overline{T}^{(in)} \\ E[T_{out}] = \frac{p_{out}(1-p_{in})}{p_{in}+p_{out}} \cdot \overline{T}^{(in)} + \frac{p_{in}p_{out}}{p_{in}+p_{out}} \cdot 2\overline{T}^{(out)} \end{cases} \tag{3}$$

To specialize Equation 3 to HCMM and CMM we have to compute the transition probabilities of the corresponding Markov chains, hereafter referred to as $p_{out}^{(H)}$ and $p_{in}^{(H)}$, and $p_{out}^{(C)}$ and $p_{in}^{(C)}$ respectively. By definition, $p_{in}^{(H)}$ is equal to $1 - p_e$. For the other parameters, we can use the following line of reasoning, common to HCMM and CMM. To compute p_{out}, we should focus on a node *inside* the starting cell, and compute the attractions of the starting and destination cells. To compute p_{in}, we should compute the attractions of the starting and destination cells on a node *outside* the starting cell. Then, p_{in} and p_{out} can be computed as follows:

$$\begin{cases} p_{in} &= \frac{SA_{start}^{(out)}}{SA_{dest}^{(out)}+SA_{start}^{(out)}} \\ p_{out} &= \frac{SA_{dest}^{(in)}}{SA_{dest}^{(in)}+SA_{start}^{(in)}} \end{cases} . \tag{4}$$

Clearly, the difference between HCMM and CMM turns out in different expressions for SA_{start} and SA_{dest}.

The derivation is simpler in the case of HCMM. First of all, it is easy to realize that the attractions of the starting and destination cells do not depend on the fact that the node is inside or outside the starting cell. The attraction to the starting (destination) cell depends only on the relationships with nodes having the starting (destination) cell as home, and on the number of such nodes. Thus, the attractions in HCMM are as follows:

$$\begin{cases} SA_{start}^{(H)} &= \frac{\sum_{j=1}^{n-1} w_{ij}}{n} \simeq \overline{w} \\ SA_{dest}^{(H)} &= \frac{\sum_{j=1}^{p_r(n-1)} w_{ij}}{fn} \simeq \frac{p_r(n-1)\overline{w}}{fn} \end{cases} . \tag{5}$$

Closed form expressions for the average time spent in the IN and OUT states in HCMM can be derived by replacing Equations 5 and 4 in Equation 3.

In the case of CMM computing attractions is more involved. The attraction to a cell dynamically depends on the number of nodes actually being in that cell. For the sake of simplicity, we carry on the analysis under the hypothesis that q nodes of the starting cell are roaming in the destination cell, and $q\prime$ nodes of the destination cell are roaming in the starting cell. The attraction of the destination cell on a node currently roaming in the starting cell (and belonging to the starting cell's community) are computed based on the following line of reasoning. The node is attracted to the destination cell because nodes of its community are roaming there. Since links have been rewired, the node has links just towards a fraction of these nodes, i.e., towards $(1-p_r)q$ nodes, resulting in a contribution to the attraction equal to $\sum_{j=1}^{q(1-p_r)} w_{ij} \simeq q(1-p_r)\overline{w}$. The node is also attracted by nodes of the destination's community, to which it has been rewired. The probability of the node having been rewired to a random node of the destination community is $\frac{(n-1)p_r}{fn}$, and the number of nodes exerting such attraction is $\frac{(n-1)p_r}{fn}(fn-q\prime)$. Based on the above line of reasoning (applicable also to the attraction of the starting cell) it is possible to derive the required attractions formulas for CMM, as follows:

$$\begin{cases} SA_{start}^{(C,in)} = \frac{(n-1-q)(1-p_r)+\frac{(n-1)p_r}{fn}q\prime}{n-q+q\prime} \cdot \overline{w} \\ SA_{dest}^{(C,in)} = \frac{q(1-p_r)+\frac{(n-1)p_r}{fn}(fn-q\prime)}{fn-q\prime+q} \cdot \overline{w} \\ SA_{start}^{(C,out)} = \frac{(n-q)(1-p_r)+\frac{(n-1)p_r}{fn}q\prime}{n-q+q\prime} \cdot \overline{w} \\ SA_{dest}^{(C,out)} = \frac{(q-1)(1-p_r)+\frac{(n-1)p_r}{fn}(fn-q\prime)}{fn-q\prime+q} \cdot \overline{w} \end{cases} \tag{6}$$

In the case of CMM the closed form expression of $E[T_{in}]$ and $E[T_{out}]$ is not as simple as in HCMM. The key point is the fact that in CMM these figures depend on the dynamic evolution of the users' movements. Specifically, they depend on q and $q\prime$, which are not model parameter, but change based on the nodes movements. Therefore, in CMM it is very hard to set model parameters to achieve a desired nodes' behavior as far as nodes' physical positions. On the other hand, in HCMM $E[T_{in}]$ and $E[T_{out}]$ do *not* depend on the dynamic evolution of the system, but depend only on f, n, p_r, and p_e. This means that HCMM, while retaining the social theoretical approach of CMM, also provides simple knobs to control the time spent by nodes in the preferred physical locations. These remarks are confirmed by Figure 5, which plots $E[T_{in}]$ and $E[T_{out}]$ for CMM and HCMM as functions of q (time is normalized with respect to $\overline{T}^{(in)}$).

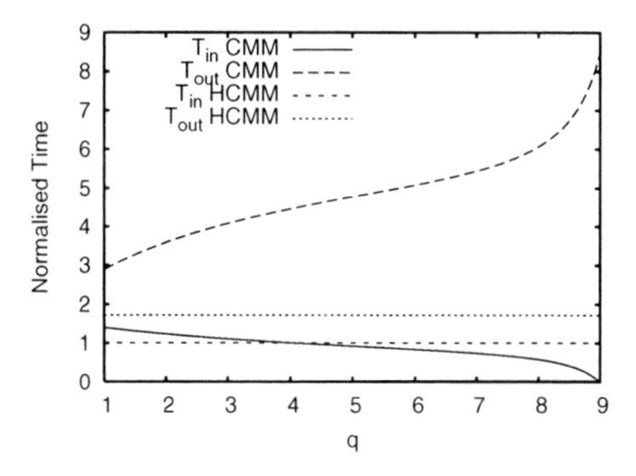

Fig. 5 Average time in the IN and OUT states as functions of q.

4 Routing in opportunistic networks

In all the case studies described in Section 2, routing is one of the most compelling challenge. The design of efficient routing strategies for opportunistic networks is generally a complicated task due to the absence of knowledge about the topological evolution of the network. Routing performance improves when more knowledge about the expected topology of the network can be exploited [23]. Unfortunately, this kind of knowledge is not easily available, and a trade-off must be met between performance and knowledge requirement. A key piece of knowledge to design efficient routing protocols is information about the *context* in which the users communicate. Context information, such as the users' working address and institution, the probability of meeting with other users or visiting particular places, can be exploited to identify suitable forwarders based on context information about the destination. In the following of this section we classify the main routing approaches proposed in the literature based on the amount of knowledge about the context of users they exploit. We specifically identify three classes, corresponding to *context-oblivious*, *partially context-aware*, and *fully context-aware* protocols.

4.1 Context-oblivious routing

Routing techniques in this class basically exploit some form of flooding. The heuristic behind this policy is that, when there is no knowledge of a possible path towards the destination nor of an appropriate next-hop node, a message should be disseminated as widely as possible. Protocols in this class might be the only solution when no context information is available. Clearly, they generate a high overhead (as we also highlight in the performance evaluation section), may suffer high contention and potentially lead to network congestion [24]. To limit this overhead, the common technique is to control flooding by either limiting the number of copies allowed to exist in the network, or by limiting the maximum number of hops a message can travel. In the latter case, when no relaying is further allowed, a node can only send directly to the destination when (in case) it is met.

The most representative protocol of this type is Epidemic Routing (Epidemic for short) [38]. Whenever two nodes come into communication range they exchange summary vectors that contain a compact unambiguous representation of the messages currently stored in the local buffers. Then, each node requests from the other the messages it is currently missing. The dissemination process is somehow bounded because each message is assigned a hop count limit giving the maximum number of hops it is allowed to traverse till the destination. When the hop count limit is set to one, the message can only be sent directly to the destination node.

Dissemination-based algorithms also include network-coding-based routing [39], which takes an original approach to limit message flooding. Just to give a classical example, let A, B, and C, be the only three nodes of a string network, such as any message traveling between A and C has to be relayed by B. Let node A generate message a addressed to node C, and node C generate the message c addressed to node A. In a conventional forwarding scheme node B has to relay message a to C and message c to A. In network coding, node B broadcasts a single packet containing $a \oplus c$. Once received $a \oplus c$, both nodes A and C can decode the messages. In general, network coding-based routing outperforms flooding, as it is able to deliver the same amount of information with fewer messages injected into the network. A more extended survey about network coding techniques can be found in [34].

An alternative, drastic way of reducing the overhead of Epidemic without relying on network coding is implemented by Spray&Wait [36]. Message delivery is subdivided in two phases: the spray phase and the wait phase. During the spray phase multiple copies of the same message are spread over the network both by the source node and those nodes that have first received the message from the source node itself. This phase ends when a given number of copies, say L, have been disseminated in the network. Then, in the wait phase each node holding a copy of the message (i.e., each relay node) stores its copy and eventually delivers it to the destination when (in case) it comes within reach. The analytical model derived in [36] shows that L can be chosen based on a target average delay. The spray phase may be performed in many ways. Under the assumption that nodes movements are i.i.d., the *Binary* Spray and Wait policy is the best one in terms of delay. Any node (including the sender) holding n copies ($n > 1$) of the message hands over $\lfloor \frac{n}{2} \rfloor$ copies to the first encountered node, and keeps the remaining copies for itself. When a node is left with only one copy of the message, it switches to direct transmission and only transmits the message to the final destination node when (if) it is met.

4.2 Partially context-aware routing

Partially context-aware protocols exploit some particular piece of context information to optimize the forwarding task. The main difference with fully context-aware protocols is the fact that the latter usually provide a full-fledged set of algorithms to gather and manage *any* type of context information, while the former are customized for a specific type of context information.

Probabilistic Routing Protocol using History of Encounters and Transitivity (PROPHET [30]) is one of the most popular examples of protocols falling in this class. PROPHET is an evolution of Epidemic that introduces the concept of delivery predictability. The delivery predictability is the probability for a node to encounter a certain destination. The delivery predictability for a destination increases when the node meets the destination, and decreases (according to an ageing function) between meetings. A transitivity law is also included in the algorithm, such that if node A frequently meets node B, and node B frequently meets node C, then nodes A and C have high delivery predictability to each other. The PROPHET forwarding algorithm is similar to Epidemic except that, during a contact, nodes also exchange their delivery predictability to the destinations of the messages they store in their buffers, and messages are requested only if the delivery predictability of the requesting node is higher than that of the node currently storing the message.

The context information used by PROPHET is the frequency of meetings between nodes. The same type of context information is also used by MV [8] and MaxProp [7], which, in addition, also exploit information about the frequency of visits to specific physical places. Other protocols use the time lag from the last meeting with a destination to estimate the probability of delivering the messages. The bottom line idea (thoroughly investigated in [18]) is that the decreasing gradient of the time lag identifies a suitable path towards the destination. Examples of protocols exploiting this piece of context information are Last Encounter Routing [18] and Spray&Focus [36].

In MobySpace Routing [27] the mobility pattern of nodes is the context information used for routing. The protocol builds up a high dimensional Euclidean space, named MobySpace, where each axis represents a possible contact between a couple of nodes and the distance along an axis measures the probability of that contact to occur. Two nodes that have similar sets of contacts, and that experience those contacts with similar frequencies, are close in the MobySpace. The best forwarding node for a message is the node that is as close as possible to the destination node in this space. Obviously, in the virtual contact space just described, the knowledge of all the axes of the space also requires the knowledge of all the nodes that are circulating in the space. This full knowledge, however, might not be required for successful routing.

The final example we mention is Bubble Rap [21] , in which the context information is the social community users belong to. In Bubble Rap communities are automatically detected via the patterns of contacts between nodes. It is assumed that communities are labeled. Messages originating in a community different from the destination's one are forwarded as follows. Assume node A is carrying a message addressed to D, and meets node B. The message is handed over to B if the community of B is the same as the community of D, or if B has a higher *ranking* with respect to node A. The ranking is measured based on the set of peers a node is usually in touch with, and is thus a measure of the "sociability" of nodes. Basically,

Bubble Rap looks for nodes belonging to the same community of the destination. If such nodes are not found, it forwards the message to increasingly sociable nodes, which have more chances to get in touch with the community of the destination. Exploiting context information related to the social behavior of people is one of the most promising research directions in the area.

4.3 Fully context-aware routing

Fully context-aware protocols not only exploit context information to optimize routing, but also provide general mechanisms to handle and use context information. The advantage of this approach is to be much more general than the approaches mentioned in Section 4.2. Indeed, these routing protocols can be used with *any* set of context information, thus allowing the system to be customized to the particular environment it has to operate in. To the best of our knowledge, two protocols only fall in this category, i.e., Context-Aware Routing (CAR [31]) and History Based Opportunistic Routing (HiBOp [1]).

CAR assumes an underlying MANET routing protocol that connects together nodes in the same MANET cloud. To reach nodes outside the cloud, a sender looks for the node in its current cloud with the highest probability of delivering the message successfully to the destination. This node temporarily stores the message, waiting either to get in touch with the destination itself, or to enter a cloud with other nodes with higher probability of meeting the destination. Therefore, nodes in CAR compute delivery probabilities proactively, and disseminate them in their ad hoc cloud. Note that context information is exploited to evaluate probabilities just for those destinations each node is aware of (i.e., that happen to have been co-located in the same cloud at some time). The main focus of CAR is on defining algorithms to combine context information (which is assumed available in some way) to compute delivery probabilities. Specifically, a multi-attribute utility-based framework is defined to this end. The framework is general enough to accommodate for different types of context information. As an example, in [31] authors use residual battery life, the rate of connectivity change and the probability of meeting between nodes as context information.

With respect to CAR, HiBOp is more general, as it does not necessarily require an underlying MANET routing protocol, and is able to exploit context information also for those nodes that have never been within the same cloud. Furthermore, the definition and management of context information is not addressed in CAR, while it is a core part of HiBOp. Finally, and most importantly, CAR does not capture, in the context definition, any information about the users social behavior, which results in [1] demonstrate being a particularly valuable piece of information to design an efficient routing scheme.

Since the performance analysis presented in this chapter focuses on the HiBOp protocol, we describe its mechanisms in more details in the following section.

4.4 The History-based Opportunistic Routing protocol

HiBOp is a fully context-aware routing protocol completely described in [1]. HiBOp includes mechanisms to handle any type of context information. As a particular instance, in [1] the context is assumed to be a collection of information that describes the community in which the user lives, and the history of social relationships among users. At each node, basic data used to build the context can be personal information about the user (e.g. name), about her residence (e.g. address), about her work (e.g. institution), etc. In HiBOp nodes share their own data during contacts, and thus learn the context they are immersed in. Messages are forwarded through nodes that share more and more context data with the message destination. Since users of HiBOp have possibly to share personal information, privacy issues should be considered. Privacy management in opportunistic networks is – in general – a topic still largely not addressed, and it is not the target of this chapter to provide complete privacy solutions for HiBOp. It should be noted that the set of information that is considered in [1] (and that we also consider hereafter) is equivalent to personal information people advertise on their public web pages (e.g., the working institution and address) which are, therefore, not perceived as sensitive information from a privacy standpoint. Designing complete privacy solutions for HiBOp is one of the main subjects of future work.

Table 1 Identity Table

Personal Information		*Residence*	
Name	John Doe	**City**	Pisa
Email	j.doe@iit.cnr.it	**Street**	Via Garibaldi, 2
	...		...

More in detail, HiBOp assumes that each node locally stores an Identity Table (IT), that contains personal information on the user that owns the device (an example is reported in Table 1). Nodes exchange ITs when getting in touch. At each node, its own IT, and the set of current neighbours' ITs, represent the *Current Context*, which provides a snapshot of the context the node is currently in.

The current context is useful in order to evaluate the *instantaneous* fitness of a node to be a forwarder. But even if a node is not a good forwarder because of its current location/neighbors, it could be a valid carrier because of its habits and past experiences. Under the assumption that humans are most of the time "predictable", it is important to collect information about the context data seen by each node in the past, and the recurrence of these data in the node's Current Context. To this end, each context attribute seen in the Current Context (i.e., each row in neighbors' ITs) is recorded in a History Table (HT), together with a Continuity Probability index, that represents the probability of encountering that attribute in the future (actually more indices are used, as described in [1]).

The main idea of HiBOp forwarding is looking for nodes that show increasing *match* with known context attributes of the destination. High match means high similarity between node's and destination's contexts and, therefore, high probability for the node to bring the message in the destination's community (possibly, to the destination). Therefore, a node wishing to send a message through HiBOp specifies (any subset of) the destination's Identity Table in the message header. Any node in the path between the sender and the destination asks encountered nodes for their match with the destination attributes, and hands over the message if an encountered node shows a greater match than its own. The detailed algorithms to evaluate matches are described in [1]. It is worth recalling here that matches are evaluated as delivery probabilities, and distinct probabilities are computed based on the Current Context (P_{CC}) only, and on the History (P_H) only. The final probability is evaluated via standard smoothed average, as $P = \alpha \cdot P_H + (1-\alpha) \cdot P_{CC}, 0 \leq \alpha \leq 1$. The α parameter allows HiBOp to tune the relative importance of the Current Context and History.

In HiBOp just the source node is allowed to replicate the message, in order to tightly control the trade-off between reliability and message spread. Specifically, the source node replicates the message until the joint loss probability of nodes used for replication is below a system-defined threshold (p_l^{max}). Specifically, if $p_{(i)}$ is the delivery probability of the i-th node used for replicating the message, and k is the number of nodes used for replication, the following equation holds: $k = \min\left\{ j | \prod_{i=0}^{j} (1 - p_{(i)}) \leq p_l^{max} \right\}$.

5 Performance of opportunistic routing approaches under social mobility patterns

The goal of this section is to compare the different opportunistic routing approaches in realistic human mobility scenarios. Specifically, we investigate the protocols' behavior with respect to a number of parameters that describe user movement patterns. The performance evaluation is carried out by considering the two opposite ends of the spectrum presented in Section 4. Specifically, we compare a context-oblivious routing protocol (Epidemic) with a fully context-aware routing protocol (HiBOp).

5.1 Performance evaluation strategy

In the following of the chapter we highlight how the different routing approaches are able to autonomically react and adapt to the dynamically evolving conditions of the operating scenario. To this end, we exploit several control knobs provided by HCMM to highlight the different autonomic properties of Epidemic and HiBOp. Specifically, we identify three main reference cases for our study. In the first one (Section 5.2), we analyze the reactivity of routing protocols to sudden contacts among groups. Specifically, we focus on closed groups (i.e., $p_r = 0$), and then we force groups to collectively move with varying frequency. Messages addressed to nodes outside the group can be delivered only during contacts between different group members during collective movements[2]. This analysis allows us to understand if routing protocols are able to exploit even those few chances to find good routes. We analyze this aspect by varying the reconfiguration interval parameter.

In the second scenario, (Section 5.3), we analyze the effect of social relationships between users. We want to understand how routing protocols react to different levels of users' sociability, measured as the probability of users having relationships outside their reference group. We clearly achieve this by varying the rewiring parameter (p_r). The higher p_r, the more nodes are "social", the lesser groups are closed communities.

In the third scenario, we look at how protocols work in completely closed groups. In this case no rewiring nor reconfigurations are allowed, and we place a different group in each cell of the grid. Therefore, the only chance of delivering messages between groups is by exploiting contacts between nodes at the borders of the cells. We study the routing protocols' performance as a function of the nodes' transmission range. Basically, this scenario allows us to understand how protocols can exploit contacts that are not related to social relationships, but just happen because of physical co-location (e.g., contacts between people working for different companies in the same floor of a building).

We test routing performance in terms of *QoS perceived by users*, and *resource consumption*. The user QoS is evaluated in terms of message delay and packet loss. Message delay is evaluated based on the first replica reaching the destination, while we count a packet loss if all replicas get lost. To highlight some specific different behavior between Epidemic and HiBOp, in some cases we also show the average number of hops required by messages to be delivered, and we separate the delay for messages addressed to friend and non-friend nodes. Resource consumption is evaluated in terms of buffer occupation and bandwidth overhead. Specifically, the bandwidth overhead is computed as the ratio between the number of bytes generated in the whole network during a simulation run, and the number of bytes generated by the senders. Note that we count in all overheads related to routing and forwarding, such as the exchanges of Identity Tables, requests for delivery probabilities, etc. To highlight specific differences, in a few cases we also show the number of copies spread in the network, and we separately highlight the bandwidth overhead related to data and non-data messages.

To highlight the effect of human mobility patterns only, we assume i) infinite buffers, ii) an ideal MAC level that completely avoids congestion impairments, iii) an ideal physical channel where nodes experience 0% packet loss within a circular transmission range and 100% packet loss outside; and iv) "infinite" bandwidth (in the sense that messages can be always exchanged when nodes get in touch). As thoroughly discussed in [1], this setup tends to favour dissemination-based schemes such as Epidemic. More specifically, in this configuration HiBOp best results would

[2] The probability of contacts due to groups choosing adjacent cells is typically low due to the high number of cells with respect to the number of groups.

be to approach the delay and packet loss achieved by Epidemic, while significantly reducing the resource consumption. Finally, unless otherwise stated, our setup consists of 30 nodes evenly divided in three groups. We assume a square simulation area 1250mx1250m large, divided in a 5x5 grid. The default transmission range is 125m. Unless otherwise stated 2 nodes in each group generate messages, with an inter-generation time exponentially distributed (with average 300s). Each message is addressed to a friend or to a non-friend node with 50% probability. Messages expire after 18000s. Each simulation run for 90000s (of simulated time). For particular setups we increased the run lenghts so as to achieve a minimum amount of characteristic events in each run (e.g. reconfiguration runs with reconfiguration interval equal to 36000s last for 397000s). To make sure that messages still not delivered at the end of a run will never be delivered (so as to achieve a correct measure of the packet loss index), during the last 18000s senders do not generate any new message. Furthermore, statistics are collected eliminating the initial and final transitory regimes, i.e., using the steady-state phase of simulation runs only. Each setup was replicated 50 times: statistics presented hereafter are averaged over the 50 replicas, with confidence interval at 95% confidence level.

5.2 Impact of collective groups' movements (reconfigurations)

It is worth recalling that in this scenario the rewiring probability is 0, and thus, except for reconfigurations, nodes do not have chances to meet. The reconfiguration interval varies between 2250s, 9000s, and 36000s. Table 2 shows the QoS performance as a function of the reconfiguration interval. As expected, both packet loss and delay increase with this parameter, because messages addressed outside the group of the sender are forced to wait for a reconfiguration. The performance in terms of delay can be better highlighted by separately focusing on delay towards friend and non-friend nodes. Specifically, Figures 6, 7, and 8 show the delay distribution towards friend nodes (left-hand-side plots) and non-friend nodes (right-hand-side plots) for the three reconfiguration periods. First of all, delays towards friends basically do not depend on the reconfiguration interval, since friends are always co-located in the same group. While only a small amount of messages destined to friend nodes experiences a delay greater than 10s, most (between 60% and 70% depending on the reconfiguration interval) of the messages addressed to non-friend nodes experience a delay greater than 10^3. Furthermore, note that depending on the frequency of reconfigurations, distributions' tails are more or less "heavy". The worst case is clearly for a reconfiguration interval equal to 36000s, where about 50% of messages towards non-friend destinations expire. Also note that, even though HiBOp provides higher packet loss and delay, the difference with Epidemic is quite thin. Note that, as buffers and bandwidth are not limited, Epidemic gives a reference upper bound on the performance achievable by any routing protocol. These results clearly show that HiBOp is able to identify very good paths even during sporadic, sudden contacts during reconfigurations among nodes belonging to different groups.

Table 2 Users QoS (focus on the reconfiguration parameter)

	reconf (s)	HiBOp	Epidemic
ploss (%)	2250	0 ± 0	0 ± 0
	9000	8.16 ± 1.68	5.52 ± 1.46
	36000	25.64 ± 1.30	24.12 ± 1.31
delay (s)	2250	1202.52 ± 91.09	907.10 ± 67.08
	9000	3651.68 ± 295.05	3204.58 ± 278.70
	36000	5615.43 ± 225.93	5445.11 ± 161.53

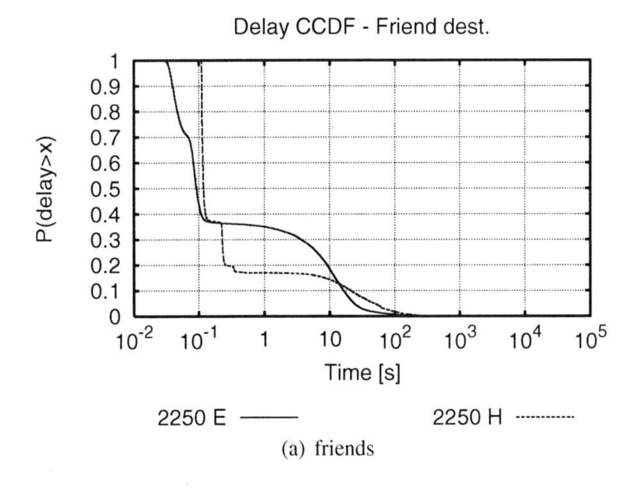

(a) friends

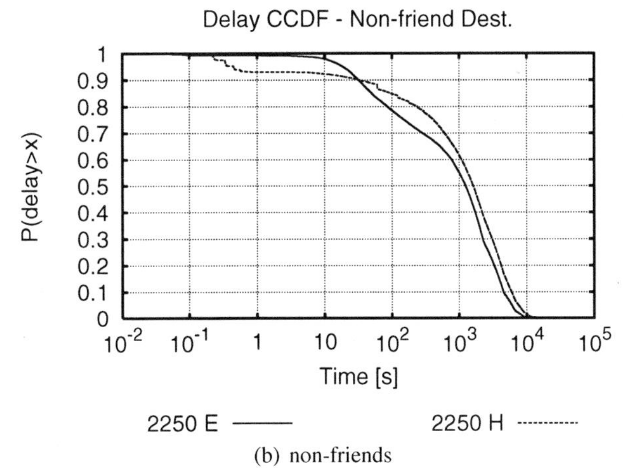

(b) non-friends

Fig. 6 Delay distributions with reconfigurations every 2250s

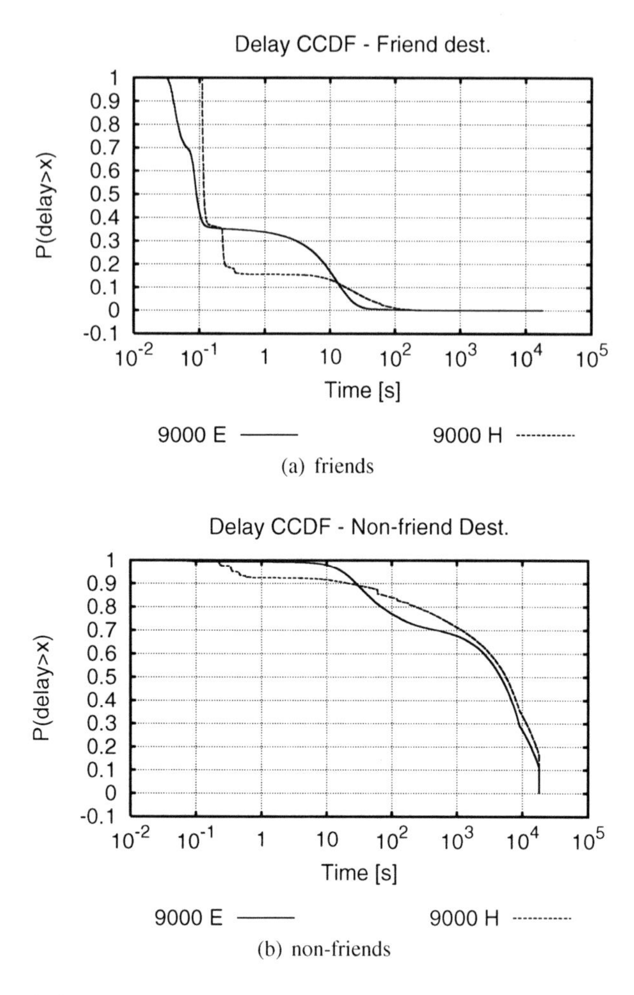

Fig. 7 Delay distributions with reconfigurations every 9000s

The good performance in terms of user QoS shown by HiBOp comes along with a drastic reduction in resource usage. Figure 9 shows the buffer occupation over time shown as a percentage of duration of a simulation run (points are average values over the replicas). HiBOp is much less greedy in spreading messages, and therefore the buffer occupation is drastically reduced. This is a general difference between Epidemic and HiBOp, which is confirmed in all scenarios we have tested. The extent of this reduction depends on the scenario, and can be as high as an order of magnitude.

Figure 10 compares Epidemic and HiBOp with respect to the number of copies generated (recall that the number of nodes in the network is 30, thus the maximum number of copies is 29). High resource consumption for Epidemic is due to the fact that each node copies all its messages to all nodes it encounters. Therefore, the

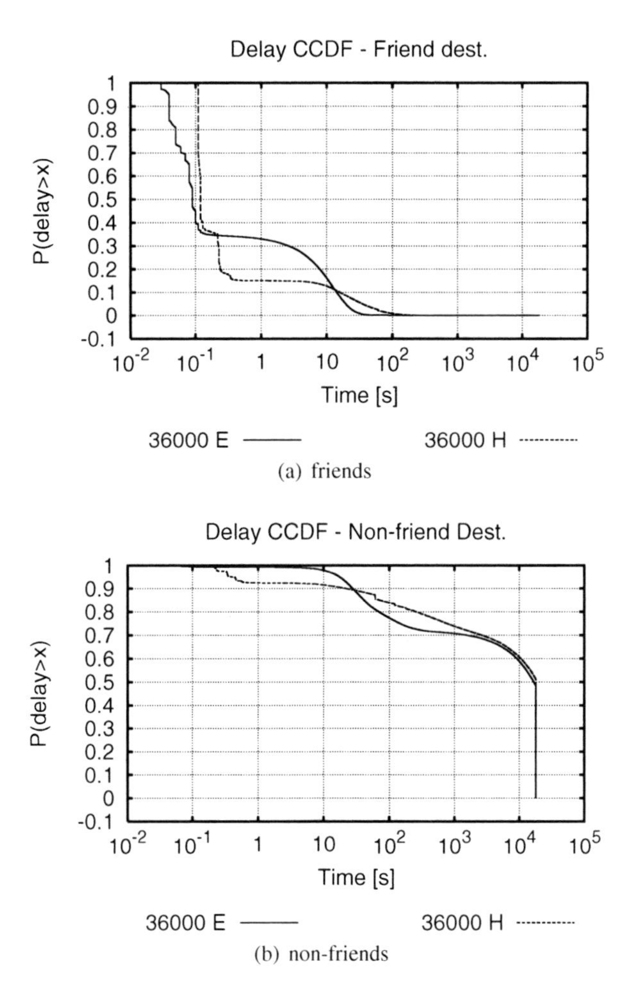

(a) friends

(b) non-friends

Fig. 8 Delay distributions with reconfigurations every 36000s

more the contacts between nodes, the more the spreading of messages. Figure 10 shows that approximately 50% of messages (corresponding to the messages with a non-friend destination) are spread by Epidemic across the *whole* network, when the reconfiguration interval is equal to 2250s and 9000s. The performance in terms of delay and packet loss shows that in this particular scenario flooding yields no significant advantages. As contacts during reconfigurations involve entire groups, a fully replication inside each group is not more convenient than replicating the message on a single node of each group. HiBOp, due to its reliability rule, tends to replicate the message inside the sender's group, but does not flood the other groups upon reconfigurations, thus resulting in lower number of copies.

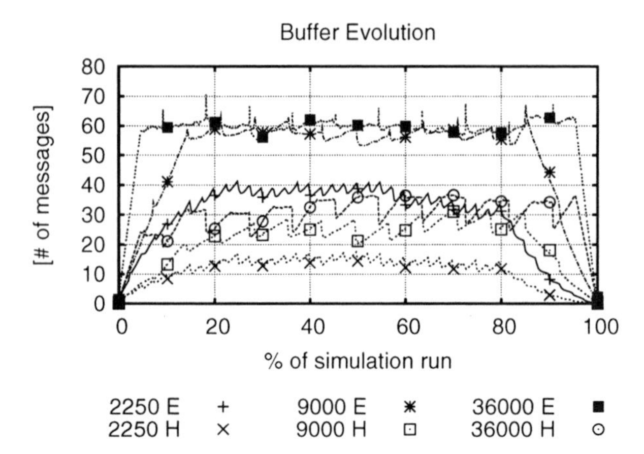

Fig. 9 Buffer occupation (focus on the reconfiguration parameter)

Finally, Figure 11 shows the bandwidth overhead of the two protocols. It allows us to highlight a main difference between HiBOp and Epidemic, related to how they react to movement patterns. Reducing the reconfiguration interval (from 36000s down to 2250s) means increasing the forwarding opportunities, because nodes get in touch with more peers more frequently. Epidemic does not use these additional "connectivity resources" wisely, as it is based on flooding. Therefore, the bandwidth overhead greatly increases. HiBOp behaves in a different way. When groups do not mix (reconfiguration interval equal to 36000s) paths for messages going outside the sender's group are seldom available. HiBOp realizes this, because context information about nodes outside the group is rarely available, and avoids consuming resources uselessly. As nodes mix more and more (reconfiguration intervals equal to 9000s and 2250s), also HiBOp (as Epidemic) generates more overhead, because more contacts become available, which may possibly lead to paths towards the destination. However, the rate of increase of the HiBOp's overhead is significantly lower than the one of Epidemic, thus showing a much more judicious use of the available network resources. These results indicate that exploiting context information makes HiBOp much more efficient than flooding-based protocols, despite the additional resources needed for context management purposes.

5.3 Impact of User Sociability

To understand the impact of user sociability on routing performance we vary the rewiring parameter (p_r). When a node goes to a cell different from its home it shows to nodes in the "foreign" cell context information related to its home cell, thus becoming a good next hop for messages destined to its friends. On the other hand,

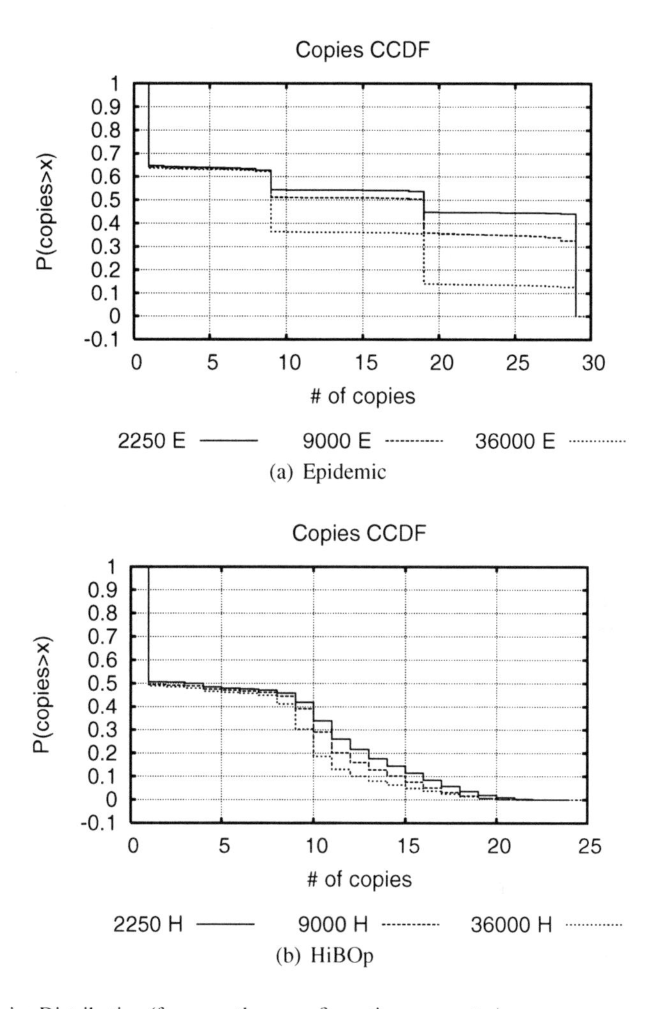

(a) Epidemic

(b) HiBOp

Fig. 10 Copies Distribution (focus on the reconfiguration parameter)

it roams in the foreign cell for a number of rounds and collects context data about nodes in that cell. When it then comes back to the home cell, this knowledge can effectively be used for sending messages to that particular foreign cell. Indeed, that node is likely to go back to the *same* foreign cell after a while, because the social links towards nodes in that cell are still active. Clearly, the routing performance is sensitive to the user sociability, because users having social relationships with other groups are the only possible way of getting messages out of the originating group. This sensitiveness impacts differently on the resource usage of HiBOp and Epidemic, as shown by Figure 12. Similar remarks drawn with respect to reconfiguration intervals apply also here. The higher the users sociability (high p_r), the higher the mix between nodes and the forwarding opportunities. While Epidemic naively

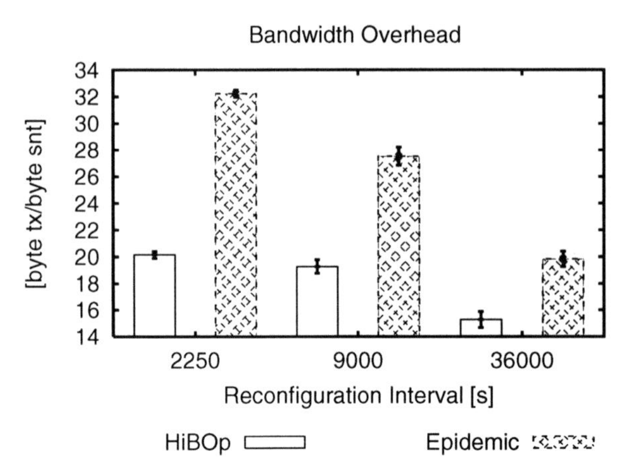

Fig. 11 Bandwidth overhead (focus on the reconfiguration parameter)

uses all these resources spreading messages, HiBOp leverages nodes' mixing (and the resulting spread of context information) to identify good paths more and more accurately.

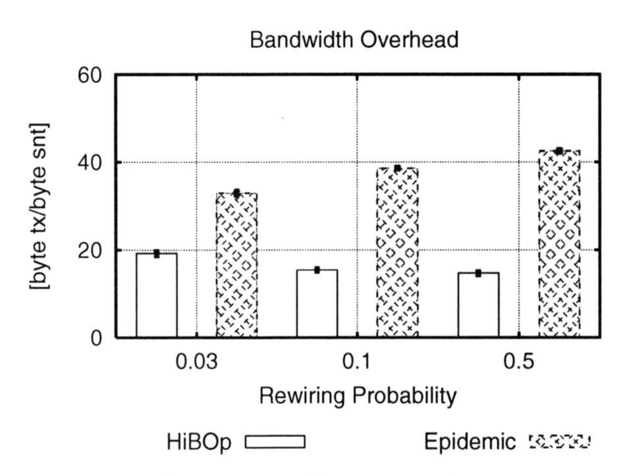

Fig. 12 Bandwidth overhead (focus on the rewiring parameter)

Figure 13 shows how data and non-data traffic contributes to the bandwidth overhead. As already said, Epidemic exploits all the possibilities of reaching the destination by copying the messages on nodes as much as possible. This results in a high overhead, which is useless particularly for highly connected scenarios where there are a lot of forwarding opportunities. Note that the high Epidemic's overhead essentially comes from the aggressive replication of messages (i.e., from data traffic). Indeed, Figure 13(b) shows that the traffic related to forwarding (i.e., the traffic related to the exchange of summary vectors) actually decreases when more connectivity opportunities are available. The buffer occupation curves (Figure 14) indicate that for higher rewiring, the buffers under Epidemic are less full, because messages can be delivered more quickly to the destinations. Therefore, the size of summary vectors decreases, and this explains the trend of Figure 13(b). However, the reduction in terms of forwarding traffic is overwhelmed by the aggressive spread of message, which results in an increase of the overhead related to the data traffic (Figure 13(a)) and, ultimately, to the overall overhead increase Figure 12. Unlike Epidemic, HiBOp "learns" the degree of connectivity of the network and uses this knowledge for adjusting the load. More specifically, HiBOp learns the current state of the network through the exchange of context messages. As context information is spread more and more widely (rewiring equal to 0.1 and 0.5) paths become more and more known, and HiBOp reduces the exchanges of both data and non-data messages.

Epidemic's high resources consumption is confirmed by Figure 15. With Epidemic, between 50% and 70% of messages are spread through the whole network. Epidemic tends to exploit all opportunities, regardless of the sociality of users. Therefore, when nodes are more mixed (higher rewiring), Epidemic floods the network more aggressively. As we will show when presenting the QoS performance figures, this is basically useless and thus results in wasting memory and bandwidth resources. HiBOp, instead, is aware of the current state of the network and adjusts the number of replicas of each packet based on the sociality of the network. Note that, even with the lowest sociality ($rewiring = 0.03$), only about 30% of messages are copied to more than ten nodes. Note also that, unlike Epidemic, this percentage decreases to zero with higher levels of sociality.

As far as the QoS performance figures (Table 3), again the packet loss is negligible (so we do not show it), while – as expected – the average delay decreases as users become more social. The performance of HiBOp are still not far from the bound represented by Epidemic. It is also interesting to note (Figure 16) that the delay of messages towards friend nodes tends to slightly *increase* as users become more social, because they spend (on average) more time outside their home group. However, as shown by Table 3, the advantage of connecting more efficiently users between groups as users become more social overwhelms the slight performance reduction experienced by friends.

Table 3 Average delay (focus on the rewiring parameter)

	p_r	HiBOp	Epidemic
	0.03	170.86 ± 25.86	130.28 ± 20.59
delay (s)	0.1	129.42 ± 12.51	83.20 ± 8.57
	0.5	104.91 ± 8.87	73.69 ± 7.16

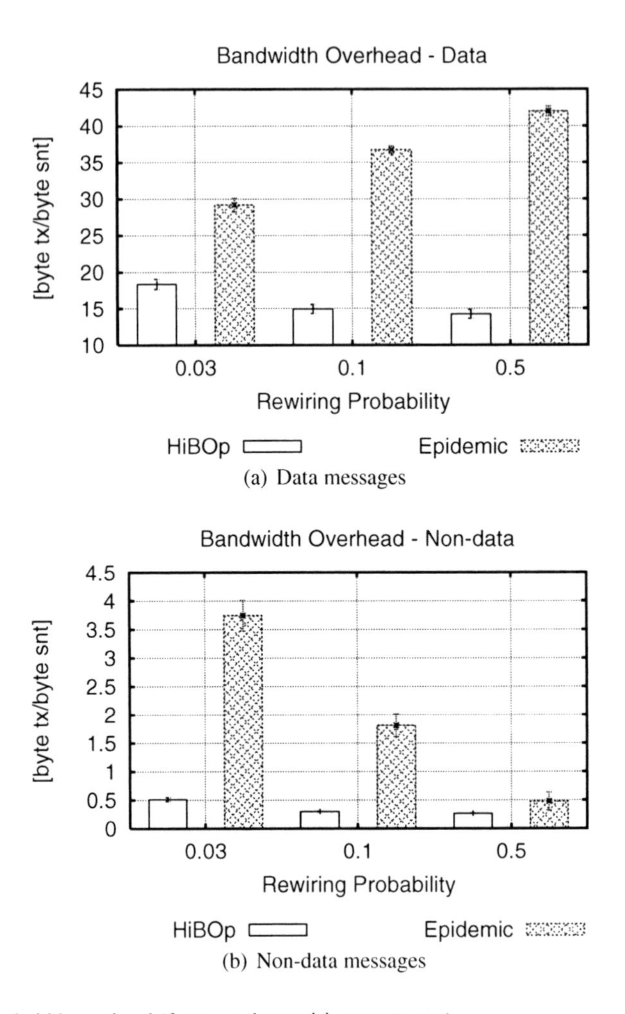

(a) Data messages

(b) Non-data messages

Fig. 13 Bandwidth overhead (focus on the rewiring parameter)

Mobility affects also the number of hops a message passes through before reaching its final destination (see Figure 17). As our setup simulates a *social* network, nodes belonging to the same community are expected to meet more frequently and for a longer time. This results in better QoS performances for messages destined to friends. As the network becomes more mixed, nodes tend to spend more time outside their community, thus becoming good forwarders for messages destined outside. The proximity between friends reduces as rewiring increases and more forwarding hops are needed in order to reach the destination (Figure 17(a)). On the other hand, the proximity between non-friend nodes increases and the number of hops a message passes through decreases (Figure 17(b)).

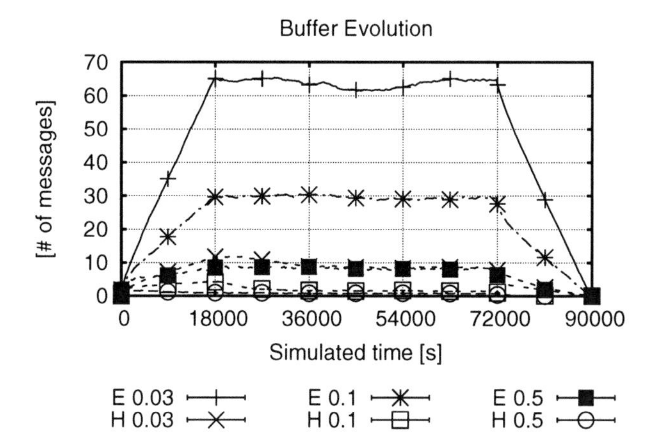

Fig. 14 Buffer occupation (focus on the rewiring parameter)

5.4 Breaking Closed Groups

In this set of simulations we use a 3x3 grid with 9 groups of 5 nodes each. Just one node, located in the upper left cell sends messages, destined to a node in the lower right cell. Recall that the only way a message can reach its final destination is through edge contacts with nodes between which no social relation exists. By varying nodes' transmission range we can analyse how this edge effect impacts on forwarding. We use three values for the transmission range, i.e. 62.5m, 125m and 250m. Therefore, nodes cover – on average – less than half a cell, slightly less than a cell, and one and a half cell.

The bottom line of the results is that HiBOp is not suitable for networks with no sociability. At very small transmission ranges (62.5m) HiBOp is not able to deliver acceptable QoS (Table 4). HiBOp needs a minimum number of contacts between users to spread context information around. Indeed, at 125m HiBOp restores acceptable QoS at least in terms of packet loss, and is fully effective at 250m. Also in this case Epidemic and HiBOp behave differently with respect to the bandwidth overhead (Figure 18). At 62.5m HiBOp seldom forwards messages. As context data is not circulating, nodes in the sender's group are almost all equally fit to carry the messages closer to the destination. At a high transmission range the context data is circulating effectively, and therefore good paths can be identified soon. In the intermediate cases (e.g., transmission range equal to 125m) HiBOp is not (yet) able to correctly learn the status of the network, and this results in a higher overhead with respect to Epidemic. However, note that these results confirm that Epidemic is not able to exploit rich connectivity scenarios without flooding the network, since it increases its overhead at high transmission ranges.

Figure 19 shows the average number of hops (recall that in this configuration statistics are related to non-friend nodes only). We can see that Epidemic generates

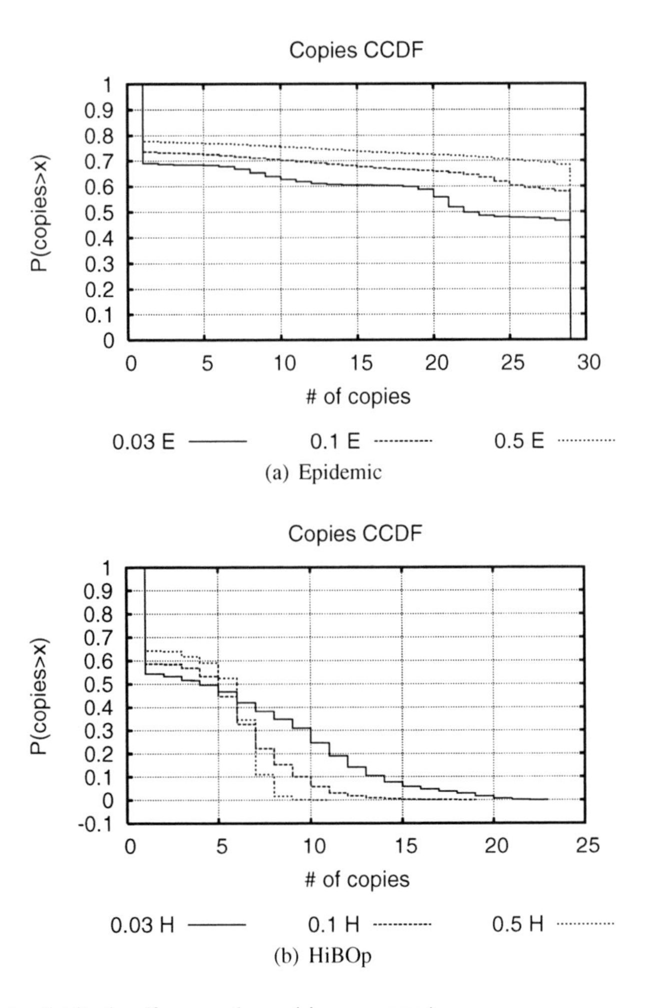

(a) Epidemic

(b) HiBOp

Fig. 15 Copies distribution (focus on the rewiring parameter)

44 copies of each message, i.e. it replicates messages on all nodes, as it is not aware of the current state of the network. In HiBOp, the number of copies increases as context information spreads, i.e., for increasing transmission ranges. This is because when the transmission range is low there is no reason to replicate messages, since no good path can be found in a context-aware scheme if context information cannot spread. As soon as context information can be exploited, paths can be found and HiBOp starts replicating messages. Finally, Figure 20 shows the average number of hops. In both cases this figure decreases with higher transmission ranges, as more contact opportunities become available, and a single hop is able to bring messages closer to the destination.

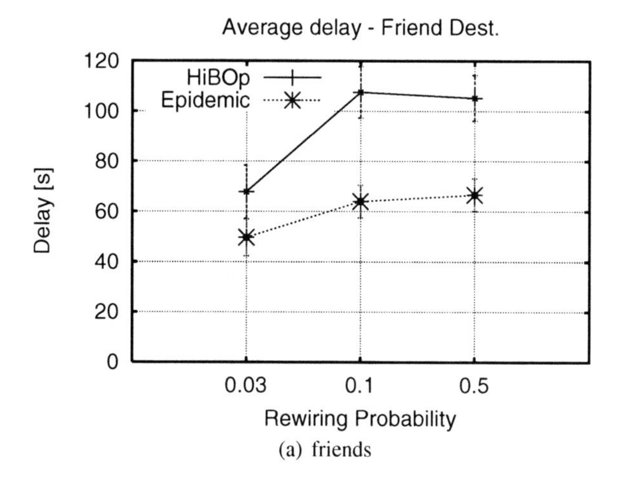

(a) friends

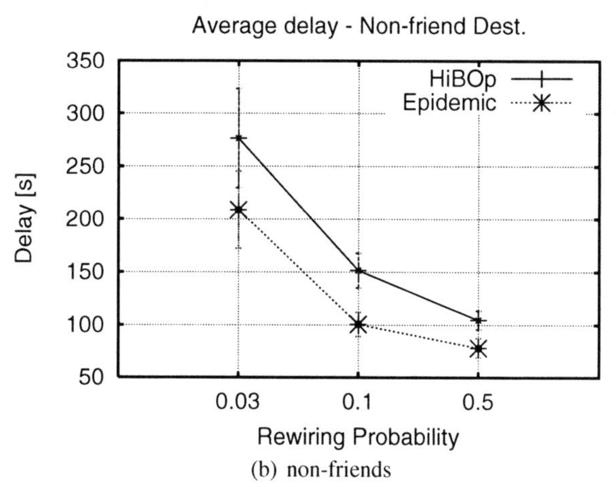

(b) non-friends

Fig. 16 Average delay (focus on the rewiring parameter)

Table 4 Users QoS (closed groups)

	range (m)	HiBOp	Epidemic
ploss (%)	62.5	61.41 ± 10.16	0 ± 0
	125	0 ± 0	0 ± 0
	250	0 ± 0	0 ± 0
delay (s)	62.5	14732.57 ± 1242.74	535.50 ± 14.05
	125	576.40 ± 177.56	102.83 ± 1.82
	250	1.77 ± 0.55	23.58 ± 0.80

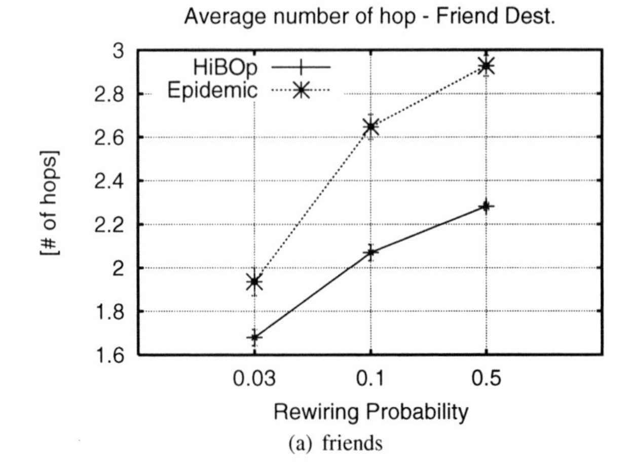

(a) friends

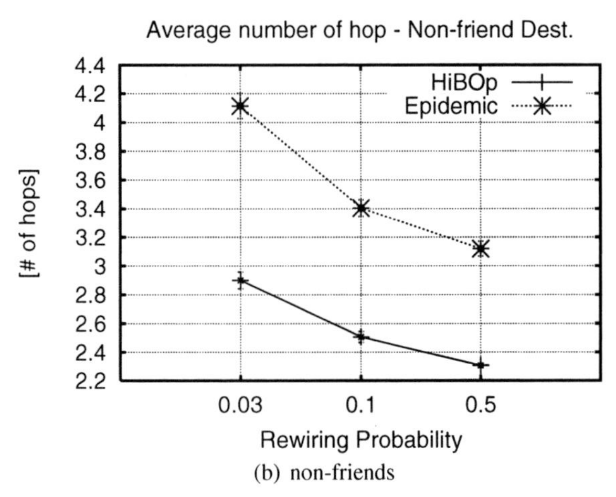

(b) non-friends

Fig. 17 Average number of hops (focus on the rewiring parameter)

6 Conclusions

In this chapter we have jointly presented results in two key research areas of the autonomic opportunistic networking research field. Specifically, we have considered mobility models based on users social relationships and behavior, and context-aware routing. Mobility models are a cornerstone to design and evaluate routing protocols, as users mobility is one of the key enabler of end-to-end communication in opportunistic networks.

We have discussed social-based mobility models in which users movements are based on social ties between people. We have highlighted that this information alone is not sufficient to model relevant scenarios, and that it should be complemented

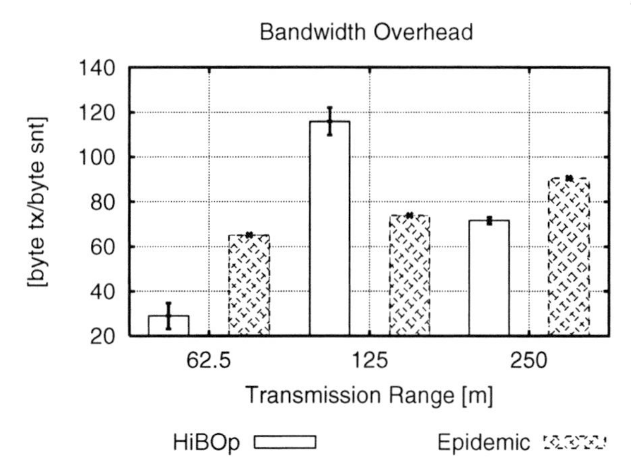

Fig. 18 Bandwidth overhead (closed groups)

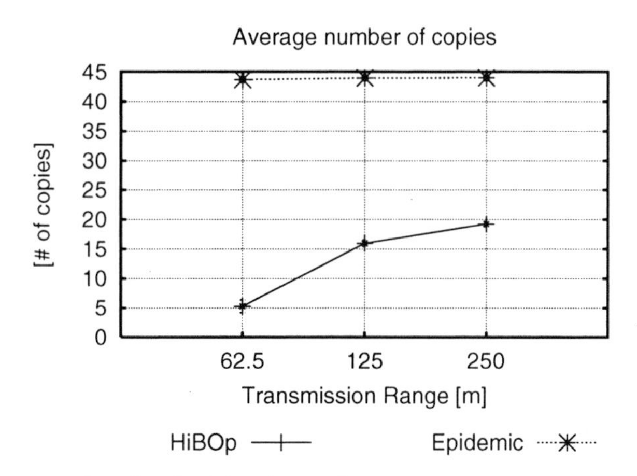

Fig. 19 Average number of copies (closed groups)

with information about the physical places where people spend their time due to their social behavior. We have presented a mobility model exploiting both types of information, and shown its advantages through an analytical model.

We have then considered routing issues, and how the social aspects of people behavior impact on context-aware routing protocols. Specifically, we have highlighted how different approaches to routing in opportunistic networks are able to autonomically adapt to the dynamic scenarios resulting from humans' mobility patterns. We have framed this work in the ongoing research on routing for opportunistic networks, and we have compared the performance figures of two protocols at the opposite ends of the spectrum as far as the use of context information is concerned, namely Epidemic and HiBOp.

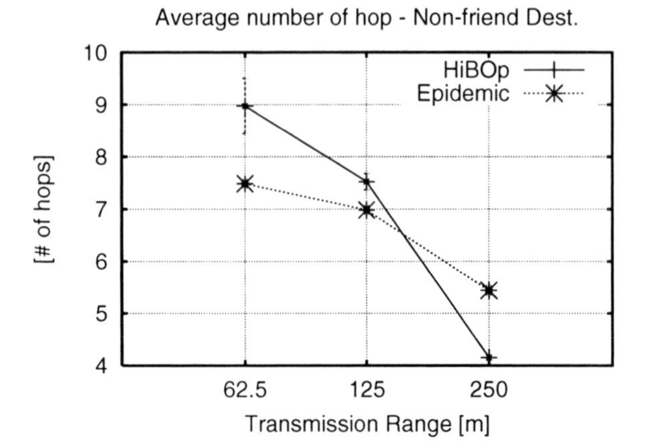

Fig. 20 Average number of hops (closed groups)

With respect to the routing aspects, the results we have presented can be summarized as follows. Context-based routing actually provides an effective congestion control mechanism, and, with respect to dissemination-based routing, provides acceptable QoS with drastically lower overhead, unless in very adverse scenarios. Indeed, HiBOp is able to automatically learn the connectivity opportunities determined by users movement patterns, and exploit them efficiently. This autonomic, self-learning feature is completely absent in dissemination-based routing schemes. The results also suggest a hybrid scheme for networks with varying levels of user sociability. When groups are very isolated, context data cannot circulate, and cannot be used for taking effective forwarding decisions. In such cases, dissemination-based schemes seem the only way to enable communication between groups. As soon as users become more social, context information spreads in the network, and context-based routing becomes a preferable solution. An interesting follow-up of this work is how to exploit context information to distinguish these different scenarios and select the appropriate routing scheme.

From a complementary standpoint, our results show that in opportunistic networks *user sociability helps routing*: users' relationships outside their "home" community allow context information to spread in the network, and make forwarding more and more efficient. These results open interesting research directions. Actually, since opportunistic networks build the network by exploiting mobile devices people carry with them, looking at social network theories to model users' social relationships and exploit these models for designing network protocols is a very interesting research direction. Indeed, the EC FET-PERADA SOCIALNETS project (started in February 2008) will be looking at these aspects. Other interesting research directions include providing privacy and security support through distributed and scalable systems in opportunistic networks. Also, another challenging research direction is how to integrate purely infrastructure-less opportunistic networks (like the ones we have considered in this chapter) with access points to the Internet infrastructure. Finally, designing data-management systems (built on top of opportunistic routing schemes) to improve data availability in opportunistic networks is another direction we find extremely important.

Acknowledgements This work was partially funded by the European Commission under the HAGGLE (027918) FET-SAC Project, and under the SOCIALNETS (217141) FET-PERADA Project.

References

1. Boldrini, C., Conti, M., Iacopini, I., Passarella, A.: Hibop: a history based routing protocol for opportunistic networks. Proc. of the IEEE International Symposium on a World of Wireless, Mobile and Multimedia Networks (WoWMoM 2007) (2007)
2. Boldrini, C., Conti, M., Passarella, A.: Impact of social mobility on routing protocols for opportunistic networks. In: Proc. of the First IEEE WoWMoM Workshop on Autonomic and Oppoortunistic Networking (AOC 2007) (2007)
3. Boldrini, C., Conti, M., Passarella, A.: Users mobility models for opportunistic networks: the role of physical locations. Proc. of the IEEE Wireless Rural and Emergency Communications Conference (WRECOM 2007) (2007)
4. Boldrini, C., Conti, M., Passarella, A.: Autonomic behaviour of opportunistic network routing. Int. J. Autonomous and Adaptive Communications Systems **1**(1) (2008)
5. Boldrini, C., Conti, M., Passarella, A.: Exploiting users social relations to forward data in opportunistic networks: The hibop solution. Pervasive and Mobile Computing (2008)
6. Borgia, E., Conti, M., Delmastro, F., Pelusi, L.: Lessons from an ad-hoc network test-bed: Middleware and routing issues. Ad Hoc & Sensor Wireless Networks, An International Journal **1**(1-2) (2005)
7. Burgess, J., Gallagher, B., Jensen, D., Levine, B.: Maxprop: Routing for vehicle-based disruption-tolerant networks. Proc. of the 25th IEEE Annual Joint Conference of the IEEE Computer and Communications Societies (INFOCOM 2006) (2006)
8. Burns, B., Brock, O., Levine, B.: Mv routing and capacity building in disruption tolerant networks. Proc. of the 24th IEEE Annual Joint Conference of the IEEE Computer and Communications Societies (INFOCOM 2005) (2005)
9. Camp, T., Boleng, J., Davies, V.: A survey of mobility models for ad hoc network research. Wireless Communication and Mobile Computing **2**(5) (2002)
10. Chaintreau, A., Hui, P., Crowcroft, J., Diot, C., Gass, R., Scott, J.: Impact of human mobility on opportunistic forwarding algorithms. IEEE Transactions on Mobile Computing **6**(6), 606–620 (2007)
11. Conti, M., Giordano, S.: Multihop ad hoc networking: The reality. IEEE Communications Magazine **45**(4), 88–95 (2007)
12. Conti, M., Giordano, S.: Multihop ad hoc networking: The theory. IEEE Communications Magazine **45**(4), 78–86 (2007)
13. D.J.Watts: Small Worlds The Dynamics of Networks between Order and Randomness. Princeton Studies on Complexity, Princeton University Press (1999)
14. Doria, A., Uden, M., Pandey, D.: Providing connectivity to the saami nomadic community. In: Proc. of the 2nd International Conference on Open Collaborative Design for Sustainable Innovation (DYD 2002) (2002)
15. Fall, K.: A delay-tolerant network architecture for challenged internets. Proc. of the 2003 ACM conference on Applications, technologies, architectures, and protocols for computer communications (SIGCOMM 2003) pp. 27–34 (2003)
16. Ghosh, J., Beal, M.J., Ngo, H.Q., Qiao, C.: On profiling mobility and predicting locations of wireless users. In: Proc. of ACM REALMAN (2006)

17. Grossglauser, M., Tse, D.N.C.: Mobility increases the capacity of ad hoc wireless networks. IEEE/ACM Trans. Netw. **10**(4), 477–486 (2002). DOI http://dx.doi.org/10.1109/TNET.2002.801403
18. Grossglauser, M., Vetterli, M.: Locating nodes with ease: last encounter routing in ad hoc networks through mobility diffusion. Proc. of the 22nd IEEE Annual Joint Conference of the IEEE Computer and Communications Societies (IEEE INFOCOM 2003) (2003)
19. Guo, S., Falaki, M.H., Oliver, E.A., Rahman, S.U., Seth, A., Zaharia, M.A., Keshav, S.: Very low-cost internet access using kiosknet. SIGCOMM Comput. Commun. Rev. **37**(5), 95–100 (2007). DOI http://doi.acm.org/10.1145/1290168.1290181
20. j. Hsu, W., Spyropoulos, T., Psounis, K., Helmy, A.: Modeling time-variant user mobility in wireless mobile networks. In: Proc. of IEEE Infocom (2007)
21. Hui, P., Crowcroft, J., Yoneki, E.: Bubble rap: social-based forwarding in delay tolerant networks. In: Proceedings of the 9th ACM international symposium on Mobile ad hoc networking and computing, pp. 241–250. ACM New York, NY, USA (2008)
22. Hui, P., Yoneki, E., Chan, S., Crowcroft, J.: Distributed community detection in delay tolerant networks. Proceedings of the 2nd ACM International Workshop on Mobility in the Evolving Internet Architecture (MobiArch 2007) (2007)
23. Jain, S., Fall, K., Patra, R.: Routing in a delay tolerant network. In: Proc. of the 2004 ACM Conference on Applications, technologies, architectures, and protocols for computer communications (SIGCOMM 2004), pp. 145–158. ACM, New York, NY, USA (2004). DOI http://doi.acm.org/10.1145/1015467.1015484
24. Jindal, A., Psounis, K.: Contention-aware analysis of routing schemes for mobile opportunistic networks. Proc. of the 1st international ACM MobiSys workshop on Mobile opportunistic networking (MobiOpp 2007) pp. 1–8 (2007)
25. Juang, P., Oki, H., Wang, Y., Martonosi, M., Peh, L., Rubenstein, D.: Energy-efficient computing for wildlife tracking: design tradeoffs and early experiences with zebranet. ACM SIGPLAN Notices **37**(10), 96–107 (2002)
26. Kim, M., Kotz, D., Kim, S.: Extracting a mobility model from real user traces. In: Proc. of Infocom (2006)
27. Leguay, J., Friedman, T., Conan, V.: Evaluating mobility pattern space routing for dtns. Proc. of the 25th IEEE Annual Joint Conference of the IEEE Computer and Communications Societies (INFOCOM 2006) pp. 1–10 (2006). DOI 10.1109/INFOCOM.2006.299
28. Leguay, J., Lindgren, A., Scott, J., Friedman, T., Crowcroft, J.: Opportunistic content distribution in an urban setting. Proc. of the 2006 SIGCOMM workshop on Challenged networks (CHANTS 2006) pp. 205–212 (2006)
29. Lelescu, D., Kozat, U.C., Jain, R., Balakrishnan, M.: Model t++: an empirical joint space-time registration model. In: Proc. of ACM MobiHoc (2006)
30. Lindgren, A., Doria, A., Schelen, O.: Probabilistic routing in intermittently connected networks. ACM Mobile Computing and Communications Review **7**(3), 19–20 (2003)
31. Musolesi, M., Hailes, S., Mascolo, C.: Adaptive routing for intermittently connected mobile ad hoc networks. Proc. of the IEEE International Symposium on a World of Wireless, Mobile and Multimedia Networks (WoWMoM 2005) pp. 183–189 (2005)
32. Pelusi, L., Passarella, A., Conti, M.: Beyond manets: Dissertation on opportunistic networking. Tech. rep., IIT-CNR, `http://bruno1.iit.cnr.it/~andrea/tr/commag06_tr.pdf` (2006)
33. Pelusi, L., Passarella, A., Conti, M.: Opportunistic networking: Data forwarding in disconnected mobile ad hoc networks. IEEE Communications Magazine **44**(11) (2006)
34. Pelusi, L., Passarella, A., Conti, M.: Handbook of Wireless Ad hoc and Sensor Networks, chap. Encoding for Efficient Data Distribution in Multi-hop Ad hoc Networks. Wiley and Sons Publisher (2007)
35. Pentland, A., Fletcher, R., Hasson, A.: Daknet: rethinking connectivity in developing nations. IEEE Computer **37**(1), 78–83 (2004)
36. Spyropoulos, T., Psounis, K., Raghavendra, C.: Efficient routing in intermittently connected mobile networks: The multiple-copy case. ACM/IEEE Transactions on Networking **16** (2007)

37. Tuduce, C., Gross, T.: A mobility model based on wlan traces and its validation. In: Proc. of IEEE Infocom (2005)
38. Vahdat, A., Becker, D.: Epidemic routing for partially connected ad hoc networks. Tech. Rep. CS-2000-06, CS. Dept. Duke Univ. (2000)
39. Widmer, J., Le Boudec, J.: Network coding for efficient communication in extreme networks. Proc. of the ACM SIGCOMM 2005 Workshop on Delay Tolerant Networking (WDTN 2005) pp. 284–291 (2005)
40. Yoneki, E., Hui, P., Chan, S., Crowcroft, J.: A socio-aware overlay for publish/subscribe communication in delay tolerant networks. Proceedings of the 10th ACM Symposium on Modeling, analysis, and simulation of wireless and mobile systems (MSWiM 2007) pp. 225–234 (2007)
41. Zhang, Z.: Routing in intermittently connected mobile ad hoc networks and delay tolerant networks: overview and challenges. IEEE Communications Surveys & Tutorials **8**(1), 24–37 (2006)

A Collaborative Knowledge Plane for Autonomic Networks

Maïssa Mbaye and Francine Krief

Abstract Autonomic networking aims to give network components self-managing capabilities. Several autonomic architectures have been proposed. Each of these architectures includes sort of a knowledge plane which is very important to mimic an autonomic behavior. Knowledge plane has a central role for self-functions by providing suitable knowledge to equipment and needs to learn new strategies for more accuracy. However, defining knowledge plane's architecture is still a challenge for researchers. Specially, defining the way cognitive supports interact each other in knowledge plane and implementing them. Decision making process depends on these interactions between reasoning and learning parts of knowledge plane. In this paper we propose a knowledge plane's architecture based on machine learning (inductive logic programming) paradigm and situated view to deal with distributed environment. This architecture is focused on two self-functions that include all other self-functions: self-adaptation and self-organization. Study cases are given and implemented.

1 Introduction

An autonomic element needs to know its internal state and the environment state. This ability of being aware is known as self-awareness and implies an ability of knowledge management. Building, using and managing network knowledge are cognitive processes that are not trivial. Hence intelligent capabilities are concentrated in functions that permit to acquire and maintain knowledge.

Maïssa Mbaye,
University of Bordeaux, LaBRI Laboratory, 351 Cours de la Libération 33400 Talence, Bordeaux, e-mail: maissa.mbaye@labri.fr

Francine Krief
University of Bordeaux, LaBRI Laboratory, 351 Cours de la Libération 33400 Talence, Bordeaux e-mail: francine.krief@labri.fr

A.V. Vasilakos et al. (eds.), *Autonomic Communication*, DOI: 10.1007/978-0-387-09753-4_3,

Knowledge plane has been proposed in [8] to provide an infrastructure for managing, sharing and reasoning about network's knowledge. It helps to close the "control loop" by giving autonomic element ability to automatically get experience and reliability during its activity. In most of current proposed autonomic architectures, knowledge plane is included because of its importance to achieve self-awareness and to manage all aspects of network's knowledge by using cognitive supports (learning and reasoning). However, interactions between elements of knowledge plane and the ones between two knowledge planes are somehow fuzzy or, sometimes, very different from one architecture to another. This difficulty has, as a consequence, appearance of new planes (Inference Plane [30], Information Plane [32]...) as part of (or out of) knowledge plane. A challenge in this topic is to design knowledge plane architecture that could be integrated in autonomic architecture instead of considering it as a new autonomic architecture itself ([1]).

This architecture should take into account important aspects:

- Distributed cognitive infrastructure: Machine learning and reasoning are traditionally centralized processes from AI [1] area and these processes are very important in knowledge plane's design. They need to be adapted to highly distributed environment of networking.
- Knowledge representation: Some efforts have been made for information models like CIM and MIBs. For knowledge plane context the problem is a little bit more complex. For example a piece of knowledge should be created by a cognitive process and evolved during activity to be more accurate. [12] and [27] identify a lot of problems on this topic.
- Knowledge Enrichment: It consists in discovering new knowledge from the existing ones to improve equipment efficiency.

In this paper we propose a knowledge plane architecture based on Machine learning tools and reasoning processes that can be integrated in most of existing autonomic architectures rather than creating a new autonomic architecture. In this architecture, knowledge plane element, learns strategies and enriches knowledge using inductive logic programming paradigm. By the way, we propose a high level knowledge representation considering it as state-strategy correspondences. Network knowledge is distributed between knowledge plane's elements and they share their knowledge with their neighborhood through situated views.

The remainder of this paper is organized in three parts: the first section presents autonomic networking's paradigm. The second section presents in detail our proposed knowledge plane architecture's elements and an application context (Diff-Serv) to show the self-adapting mechanisms (building blocks and interactions) and its benefits. And the last section shows collaborative processes (self-organization) of our architecture in details and our proposed knowledge sharing algorithm.

---

[1] Artificial Intelligence

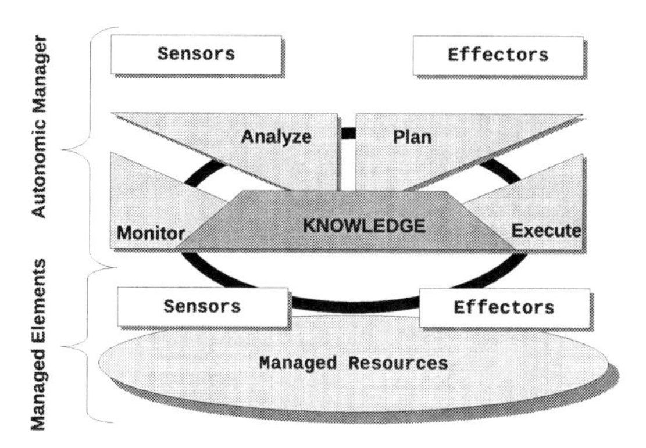

Fig. 1 Autonomic computing architecture

2 Autonomic Networking

2.1 Basic concepts

Autonomic networking is inspired from IBM's initiative ACI (Autonomic Computing Initiative) and aims to give network elements capacity to manage themselves (Fig. 1). This management consists of adapting themselves to their environment changing while achieving high-level goals given by administrators. This concept of autonomic networking will decrease network management costs and time tuning, by delegating management complexity to the equipment level. Basic concept of self-functions necessary to achieve autonomy has been defined. The core self-functions, as defined in IBM's initiative, are: self-configuring, self-healing, self-optimizing, and self-protecting. These core functions combined together enable an equipment to achieve autonomic behavior.

IBM has been the first to design an autonomic architectures in 2001 [17, 19] which is compound of a central element called autonomic element. Autonomic element is composed of an autonomic manager and one or more managed resources. Since this first architecture some other architectures have been proposed ([1, 2, 10]).

To manage knowledge, most of these architectures include cognitive supports that, brought together, have the same role as knowledge plane.

2.2 Related Work

Designing knowledge plane architecture has been central to a large number of articles. The first architecture is proposed by the knowledge plane inventor [8], but it doesn't define interactions between elements. It just defines what we need to solve the problem. FOCALE [29] (Fig. 2) is another architecture defining multiple control loops. But, model-based translation and combination of many control mechanisms make this architecture complex. Context-aware system is another approach for self-adapting architecture. But this architecture is highly oriented mobile services management and does not use Machine learning tools.

Autonomic Internet project [1, 3] also defines an architecture with adaptive loop [13]. Another work presents a knowledge plane combined with new planes such as action plane and execute plane [20].

Self-organization function involves mechanisms to realize distributed intelligence in autonomic networks [26]. Distributing intelligence has been studied since several decades by the artificial intelligence researchers with the multi-agent systems (MAS) [33]. However, the problematic of the collaboration with sharing knowledge still remains a challenge in spite of some propositions like [6, 22]. The network management researchers began to see in the MAS a means achieving the self-organization (example [23]) through a knowledge plane constituted by autonomous agents.

But in most of articles, the agents are dedicated to a specific task with a limited (or without in some cases) amount of exchanges among the agents [7, 22]. Then MAS's paradigm influences a lot the vision of the knowledge plane. Even besides the notion of situatedness, an important notion in the knowledge plan seems to be directly inspired from the definition of an autonomous agent. Situatedness in this context means simply that an agent has the knowledge of its environment and interacts with this last [14]. As we can notice the meaning of situatedness is a quite different from the one given above in networking context. The modification is due to the factor of scalability in networking. The ideal situation in a totally distributed intelligence is to use any calculus power available in the network. One node should construct its own knowledge and share it with its peers while propagating it to nodes where its knowledge is useful (the other nodes may not be in the same context for example).

Also some autonomic biologically-inspired architectures have been proposed in [4]. However, the mapping between biological systems and autonomic networks is still a challenge (because of complexity to determine best analogies between these domains). In [6] authors propose a peer-to-peer solution for knowledge management. The knowledge sharing also involves in some distributed knowledge management that is principally explored by the AI researchers. That guaranties the coherence of the knowledge. But there may be a difference (as in situatedness) of meaning of coherent knowledge between the two domains. In networking environment if the local knowledge is coherent the global knowledge could be considered as coherent for example. This is due to the possible heterogeneity through the network and that is the essence of the situatedness, which cares the context changing to limit the knowledge propagation.

Our architecture is focused on two aspects. On one side it defines how autonomic element infers and induces optimal strategies by learning from using information feedback. On the other side it defines how elements are organized to share their knowledge through situated views.

3 Collaborative knowledge plane architecture

3.1 Architecture overview

Since IBM autonomic initiative, a large number of self-functions have been listed among them: Self- and context-awareness, self-locating, self-negotiation and the list is far away from being exhausted. This growing number of self-functions rather than clarifying autonomic paradigm, makes its understanding more complex. We propose at first to reflect about the kernel self-functions that involves all other autonomic functions. Determining this kernel function simplifies number of self-functions to achieve before realizing autonomic networking. We think that self-adaptation and self-organization together embody all other self-functions. Self-adapting processes include all processes and mechanisms to adapt local resources to environment changes, and self-organizing includes all interactions between network element and its environment (neighbors). We base our knowledge plane architecture on these two self-functions and we represent them by two loops: Organization (or collaborative) loop and adaptation loop. Fig. 2 describes these two loops.

Self-organization is the horizontal loop. Organization loop involves global knowledge exchanges between elements composing knowledge plane. It is the knowledge sharing process in order each element takes into account its neighbors in its decision and having a situated view of their knowledge. Organization loop implements self-organization of network elements. It refers to distributed knowledge management. Self-adaptation is the process done by the vertical loop. Adaptation loop is the interaction between autonomic manager and managed resources. Autonomic manager gets information about managed resource state in order to analyze it and takes a strategy to adapt (if necessary) its configuration and behavior to current state. This vertical loop implements self-adaptation since locally each equipment manages it-

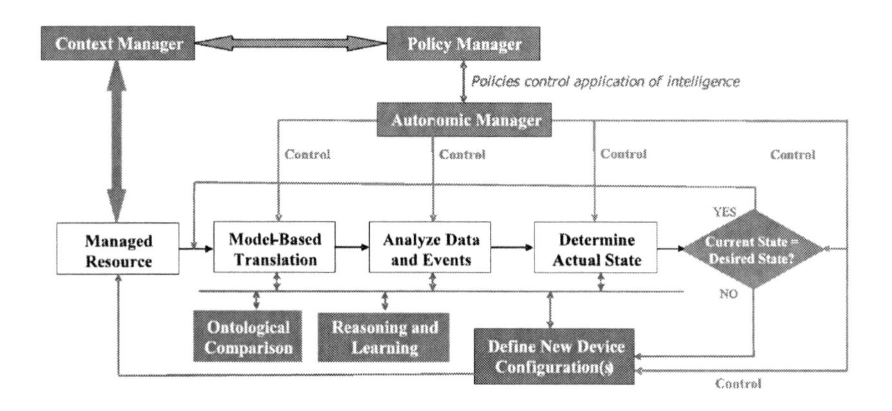

Fig. 2 FOCALE loops

self to be efficient according to the current state. Then knowledge plane induces best strategies for adapting equipment to the context. After this, feedback is used to evaluate the efficiency of strategy. An "example" that will be used by the learning algorithm is built from at the end of this loop (see section 4.1).

3.2 Basic Concepts

In this knowledge plane architecture (Fig. 3) many objects are manipulated by processes: Monitoring element manipulates information about equipment state, Action element manipulates strategies and other kinds of objects (facts, examples are manipulated by learning and reasoning processes). Defining these objects could clarify a better knowledge representation. Fig. 4 summarizes treatment of all objects from simple information to strategies (knowledge). During its activity, Machine learning process needs to evaluate its learning actions to be efficient. To do this we use observations set O (or facts) to have a feedback memory of actions performance. We propose a definition of observation for the context of the knowledge plane as a set of a triplet $<e_1, a, e_2>$ where e_1 and e_2 are equipment states and a an ordered set of actions. This triplet means that the equipment was, at time t_0, in the state e_1 and

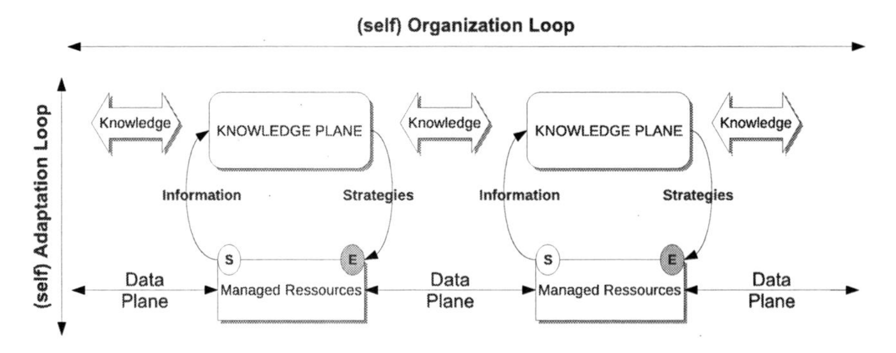

Fig. 3 Management loops

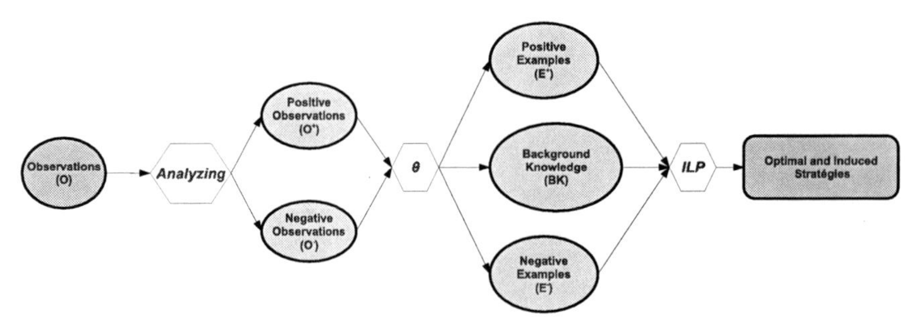

Fig. 4 Objects processing in knowledge plane

applied the set of actions a and its state moved to be e_2. Now we can define notions that will be useful for the remaining parts.

Definition 1. An **elementary action** of equipment is a predefined action that the equipment can execute without needing for a translation into another script format.

Definition 2. Given E the set of the possible states of an equipment, A its set of elementary actions. An observation is an element of the following set:
$O = A \times E \times A$. Where A^* contains all elements of all Cartesian powers of A.

$$A^* = A \times A^2 \times A^3 \times \ldots = a, \exists n \in N / a \in A^n \qquad (1)$$

The set E (of equipment sates) is ordered to help appreciation if an action brings to a better or a worst state. From this order we can define two kinds of observation: positive observations (O^+) and negative observations (O^-).

Definition 3. Given O the set of observations in equipment's knowledge plane. An observation $o = (e_1, a, e_2)$ is:

- a positive observation of fact if e_2 is better that e_1
- a negative observation otherwise

This partitioning is used to determine efficient (optimal) strategies for each internal state of equipment. We consider θ indicator which measures efficiency of an action in a state. An "atomic" strategy could be seen as a correspondence between a state and an elementary action. We give the following definition to fix these concepts.

Definition 4. A strategy s is a correspondence between an equipment state (E) and an ordered set of actions (A^*) to improve it. More formally, given S set of all strategies of an equipment we have:

- $s \in S \Rightarrow s \in E \times A^*$
- In a strategy an action cannot be executed more than a limited number of times.

Definition 5. A strategy $s = (e, a)$ is called elementary. That means action part (a) is compound of only one action.

Elementary strategies are evaluated by indicator θ:

$$\theta(e,a) = \frac{\left\|(e,a)^+\right\| + n_{(e,a)}}{\|(e,a)\| + m_e} \qquad (2)$$

- θ is probability law
- $\left\|(e,a)^+\right\|$ is the number of positive observations concerning e and a, and $\left\|(e,a)^-\right\|$ negative observations :
  $\|(e,a)\| = \left\|(e,a)^+\right\| + \left\|(e,a)^-\right\|$
- m_e is a Laplace regulator for initializing indicators, considering :
  $m_e = \sum_{(e,a) \in S} (n_{(e,a)})$

Machine learning process maintains θ indicators during activity. Then it builds incrementally correspondence between states and optimal actions. A second learning process induces new strategies from background knowledge and examples. We consider Inductive Logic Programming (ILP [25]) for doing this task. Examples are strategies, examples are said to be positive (in E^+) if $\theta < \frac{1}{2}$ and negative (in E^-) otherwise. For Inductive Logic Programming process positive examples are strategies to be used when inducing new strategies and negative strategies are considered to be inefficient (and not to be used at all).

At last it could be useful to remind the difference between data/information and examples and knowledge. In the human being case information can be transformed into knowledge by the simply keeping it in mind. However in the network domain and notably in a dynamic environment, information is to notify an event, which occurs internally or in the environment. Modifying a piece of information values changes its reliability radically in term of accuracy. Also the fact to deduce a piece of information from another is not always possible and in some cases has no sense. At the opposite, knowledge is a perception of the reality, it can be modified to better correspond to the reality. A fundamental difference also is genesis between information and knowledge. Information is a result of an event in an environment whereas knowledge is the result of a cognitive process such as Machine learning, generalization, deduction, etc. We did not treat the information sharing because the information plane [32] has been proposed intentionally to manage all the aspects related to the collection of information in the network. We only use information to determine equipment state.

3.3 Knowledge plane building blocks

Self-adaptation architecture needs knowledge plane in order to adapt efficiently. In order to realize self-adaptation we define building blocks compounding knowledge plane to better know how it works. In this section we present our vision of what should be building blocks compounding knowledge plane. Knowledge plane architecture we propose includes four elements (Fig. 5) that work together to realize self-adaptation and self-organization.

Reasoning engine

It enables to automatically enforce suitable actions and strategies. Decisions are taken according to the current state of knowledge base, equipment state and eventually information from neighbors in order to optimize working. Reasoning engine interacts out of knowledge plane with the Planning in IBM's architecture.

Knowledge base

It contains knowledge and information models, and represents experience gained by knowledge plane. It includes two kinds of knowledge: knowledge built locally from local experience and the knowledge from sharing process. All this knowledge needs to be validated incrementally with during activity, what is combination of knowledge exchange process and Machine learning engine. Defining knowledge form is one of the most important challenges for researchers. Ontologies have been widely studied and seem to be very complex to manage. We consider that all knowledge is not useful for autonomic tasks. To realize control loop an autonomic component only needs some strategies for acting and information about state of its state and environment.

Machine learning algorithm

This is the most important part because it products knowledge in knowledge base and makes the reasoning (which is related to learning process [11]) engine working efficiently. A unique definition of Machine learning does not exist but in cognitive sciences there is an accepted definition: "Learning is capacity to improve performance during an activity" [9]. It gives autonomy by validating local knowledge and discovering new knowledge by enrichment process (see section 4.1) and knowledge sharing process. From managed resource activity and alarms, Machine learning algorithm builds knowledge about optimized built-in knowledge and discovers another one. Machine learning updates knowledge base during its activity.

Knowledge sharing processes

They consist, in one side, of collaboration between knowledge plane elements and in other side, of making the Machine learning algorithm to be distributed. Knowledge sharing processes help equipments to cooperate in their knowledge validation and strategies sharing. Autonomic elements need taking into account one to others to have a coherent global behavior. This process is presented in section 5.2.

This Knowledge plane architecture interacts with autonomic entity's elements which are Monitoring, Analyzing, Planning and Executing to close the adaptation and organization loops. When Monitoring gets information about the managed resource state it transmits it to Analyzing. Analyzing interacts with the Machine learning module to detect if it should plan a set of actions and chooses them if necessary. The Machine learning uses the reasoning engine to extract the most pertinent strategy for current situation.

This strategy is passed to Planning and enforced by Executing through effectors. The feedback is treated by the Machine learning algorithm that evaluates the result of the enforced strategy. From the past and current strategy it stores an example and updates the knowledge base. In this architecture, Analyzing function is to map information from monitoring to learning and reasoning tools. Planifier has reverse

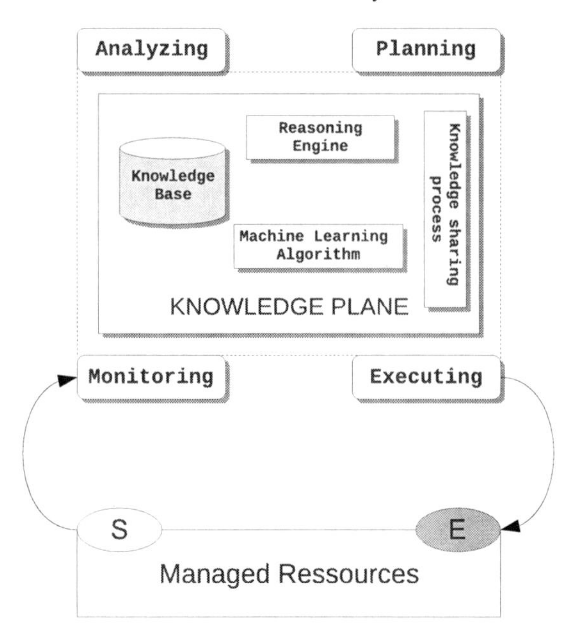

Fig. 5 Knowledge plane building blocks

function. It maps strategies from reasoning tool to actions that can be executed by managed resources.

4 Self-adaptation loop

4.1 Machine learning algorithm for self-adaptation

Machine learning in knowledge plane environment is a very important part because it is in charge of maintaining experience of equipment during activity. This maintenance as we said earlier is twofold. On the one hand we have the maintenance of knowledge by consolidating it or invalidating it. On the other hand, it consists of generalizing specific knowledge, enriching knowledge. The first aspect of this problem is to learn a classification that is a correspondence function from the set of states to the set of strategies. Each strategy is efficient in a limited set of states. And feedback of its past enforcement is used to update knowledge base. In this case incremental Machine learning is combined with reinforcement one. The second aspect is the knowledge enrichment. This knowledge enrichment is necessary to manage foreign situation, for example a state with which knowledge plane has not a suitable strategies. Inductive logic programming is very suitable for this task.

Algorithm 1: Generic Machine Learning Algorithm

```
   Data:
   K : Knowledge base;
   I :ILP engine ;
1  begin
2      while True do
3          E1 ← getState();
4          S ← K.getStrategy(E1);
           ;                        /* strategy depends on the state */
           ;                         /* uses planning to plan action */
5          planifier.plan(S);
           ;                                     /* getting feedback */
6          E2 ← getFeedback();
           ;  /* example building and use of ILP system to create
           new knowledge eventually */
7          example buildExample(E1,S,E2);
8          H=I.assert(example);
           ;                    /* update example and revise theory */
9          K.append_update(H);
10     end
11 end
```

Algorithm 1 pseudo-code describes the generic Machine learning algorithm in knowledge plane. General principle of this algorithm is to insert an experience gaining process in the "control loop". The process of experience building is to memorize the results of strategy enforcement in a knowledge base.

First, the Machine learning algorithm gets the state of the local managed equipment. From this information it can extract suitable strategy from the knowledge base and interacts with the Planning to enforce this strategy. After this step the efficiency of this strategy is done with feedback and by comparing the two states before and after the enforcement. An example is built and sent to the inductive logic programming module which incrementally updates its positive and negative examples. As a result it revises proposed theories (enriched strategies) before and returns them to the Machine learning process. This last algorithm finally appends this knowledge if it does not already exist in the knowledge base and updates it otherwise.

4.2 *Study Case: self-adaptation of a DiffServ router*

4.2.1 Context

This section presents a case of illustration of the Knowledge Plane in order to self-optimize the configuration of a DiffServ [5] router. The main idea is to use the

Machine Learning tools to automatically optimize the configuration of a DiffServ router. It consists of self-optimizing router configuration according to its state. Characteristics of the network selected to define the states of the network are: the loss rate and the bandwidth use ratio. The states of the network can be divided into two categories: state of crisis (packet loss for AF [15] traffics increases) and states of optimization (no packet loss for AF traffics). Alarms give an indication about the changes which occur and enable the learning module to react quickly and in the best way. Two important aspects of alarms must be defined: their nature and their frequency. Concerning their nature, alarms are $n-$tuples containing the following information: the DiffServ class concerned with alarm, the previous and current states, and configuration policy rule that generated it. The frequency of alarms release is also an important factor. An alarm is sent to the module each time the equipment state changes. In this context, four states of equipment are considered: `UnderUseBW`, `NeedBW`, `SlightCongestion`, `Congestion`, `SevereCongestion`. Each of them is defined according to loss rate and self-adapting Machine learning algorithm takes action considering DiffServ router state. Elementary actions are manipulations of the Peer-Hop-Behavior [5]:

- `ChangeQueuer` helps changes current Queuer to the most adapted accordingly to the current state
- `QueueLength` manipulation manipulates queue length to avoid sometimes loss and decreases this length to its minimal size when queue is underused.
- `ChangeScheduler` is equivalent to changeQueuer for scheduler.

Self-adapting mechanisms tries to build a correspondence of most fit action for each equipment state. This correspondence serves for reactive loop. In this context, a strategy is a subset of this correspondence. Self-organization mechanisms could share from time to time some part of their strategies. Fig. 6 shows interactions between elements of the Machine learning module. Autonomic Manager (AM) acquires information from managed resources and then checks its knowledge base to decide if a reconfiguration is necessary. Each reconfiguration provides a result and this reconfiguration feedback is sent as an example to Aleph which induces new adaptation strategies from these examples and the background knowledge. Aleph [28] is an ILP (Inductive Logic Programming tool [21, 24, 25]). This knowledge plane architecture has been tested for an autonomic DiffServ network and the next section describes this use case.

4.2.2 Testbed

The knowledge plane architecture has been implemented on J-Sim [16] using DiffServ [5] network context as an example. We extends J-Sim framework with a number of classes represented in Fig. 7.

The AM class represents the autonomic manager, which coordinates interactions between elements of the Machine learning . Knowledge and KB classes store experience gained during this activity. Finally, AlephInterface is an API used to interact with Aleph.

To do this, AM interacts with Aleph to enable the Diffserv network to have an autonomic behavior. Test bed is composed of two edge routers and a core router that has an autonomic manager (entity with Machine learning) in order to self-adapt according to its state. Test bed (Fig. 8) involves four classes of traffic: AF11, AF12, EF [18] and BE. EF has highest guaranties while BE has none. Principle of the algorithm is to guaranty service for EF and AF classes.

The traffic model used is Poisson and the packet size is variable in order to simulate an unstable traffic. Fig. s 9, 10 and 11 describe comparison of throughput for flows with and without autonomic manager. Fig. 9 and 10 show that adaptation, with autonomic manager can improve throughput of AFxy classes.

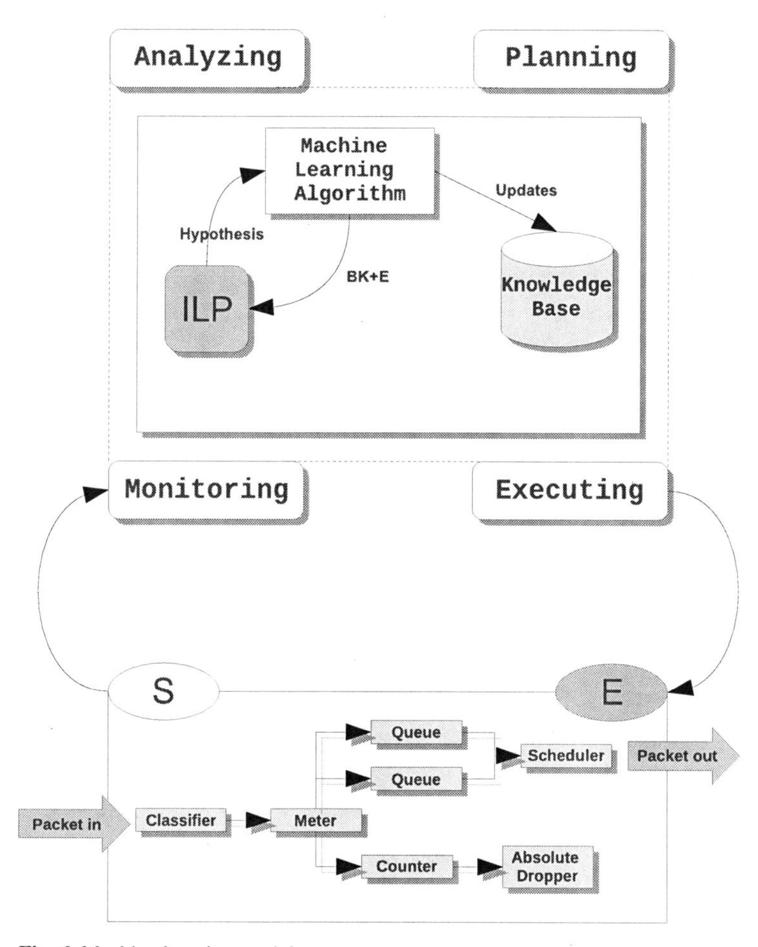

Fig. 6 Machine learning module

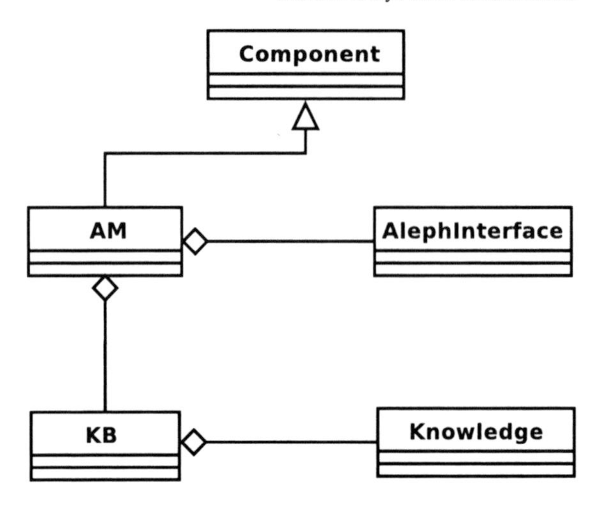

Fig. 7 Machine learning module

We notice that traffic improvement consists of two phases: an initial phase during which the autonomic manager learns action that improves the throughput. The two graphics, with or without autonomic manager are almost the same. In the second phase, the autonomic manager has found optimal action that allows an improvement. Autonomic manager maintains that action until flows reach an optimal value and then stabilizes the traffic flows by minimizing loss rate. Fig. 11 shows the degradation of BE flow, with and without autonomic manager. We can see that this degradation starts with enforcement of the optimal strategy. Autonomic manager improves throughput of protected traffics (AFxy) by degrading BE traffic. At the same time, it tries not to degrade too much the BE traffic by re-allocating to it a portion of bandwidth when it can do.

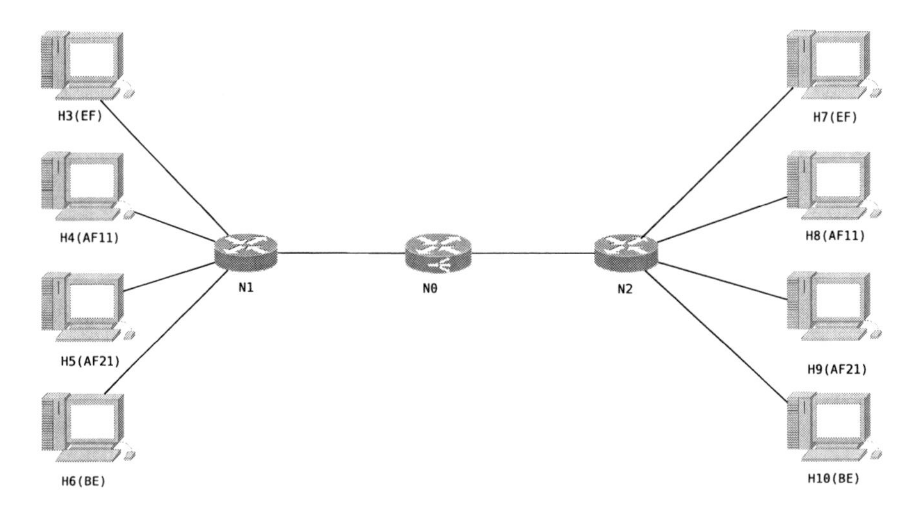

Fig. 8 Testbed scenario

5 Collaborative loop

Collaboration between elements in knowledge plane is an important aspect for scalability of communication mechanisms. Knowledge plane components should collaborate through situated views instead of using simple broadcasts. In this section we present self-organization mechanisms of knowledge plane architecture. First we present concept of situated view, then we describe algorithms and mechanisms for knowledge sharing in knowledge plane. We finish by giving some theoretical guaranties of these mechanisms.

5.1 Situated View and Basic concepts

As we said earlier, collaboration between elements in an autonomic environment is a very important aspect. This collaboration consists in information/data and knowledge sharing through situated views.

This collaboration presents several interests:

- The initialization of new elements in the network: New elements initialize their knowledge base by taking benefits in their neighbor's experience.
- Equilibration of the examples distribution between the different learning elements: The autonomic elements may not learn all to the same rhythm. The el-

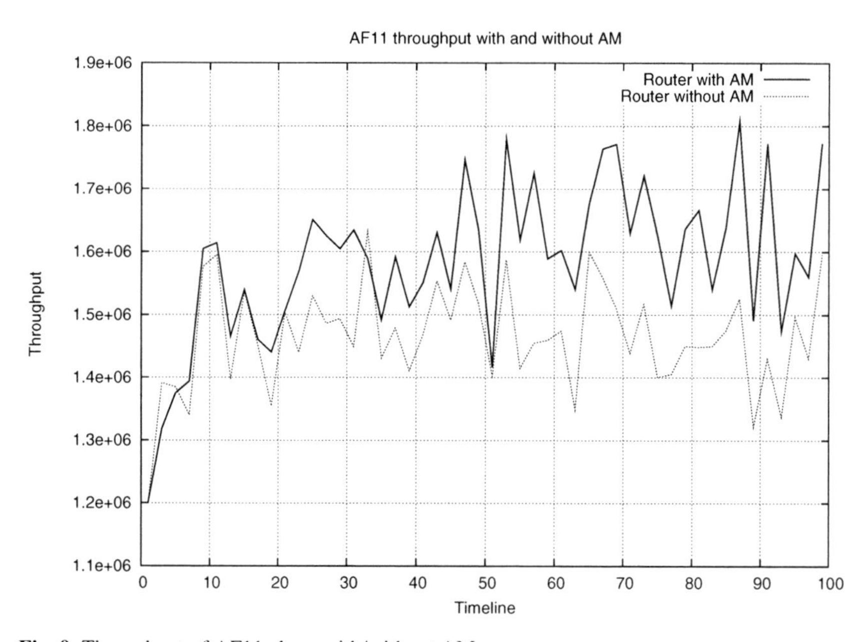

Fig. 9 Throughput of AF11 class, with/without AM

ements situated in a very dynamic area should have more examples to validate their experience than the remaining others.

- Distributed knowledge validation: The mutual critique of knowledge with the peers will permit to validate distributed knowledge. Sharing knowledge among elements will give all of them a comparison of their knowledge and neighbor's knowledge.
- Distributed knowledge management: In a network environment, the distribution of tasks is important allowing scalability. Managing distributed knowledge by keeping locally on the elements only the necessary knowledge and making some exchanges if necessary are two necessary approaches for knowledge plane.

Beyond the above cited interests of sharing knowledge, there is an objective of each autonomic system with knowledge plane, that is to be able to determine the domain limits where a piece of knowledge is useful and then should be propagated. This domain (area) is known as "situated view" or situatedness. Situated view is the limit of the domain in which a node can have interest to share its knowledge. This limit can be determined at a thin grain level of knowledge that consists of computing the view for each piece of knowledge. Situatedness also can be calculated by aggregation consisting for each element to determine progressively nodes with which it has interest to share its knowledge.

In this work we introduce the computation of view by the method of thin grain and without conservation of any information about the situated view. It is created on distributed way, implicitly and automatically. Expected result (with knowledge shar-

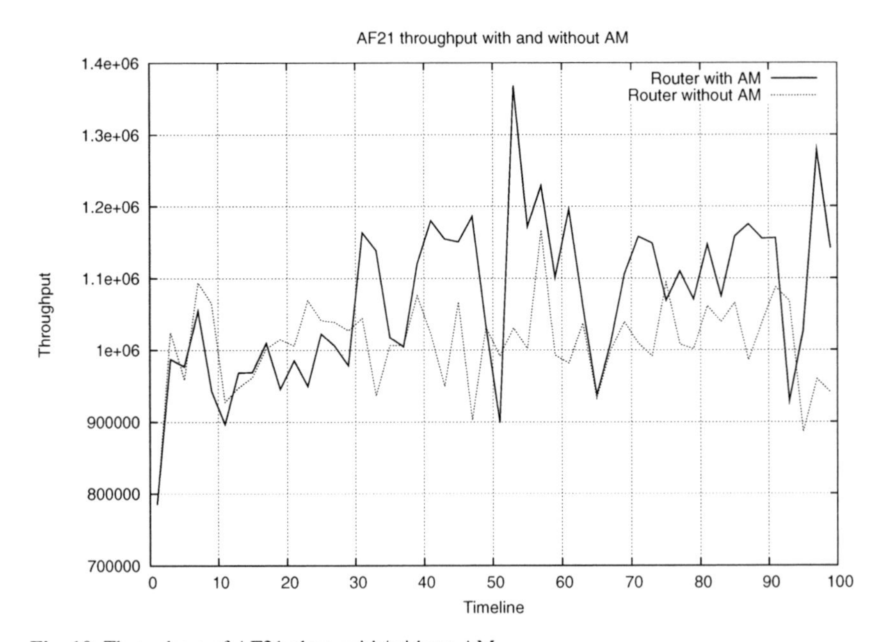

Fig. 10 Throughput of AF21 class, with/without AM

ing) is a decision-making to adapt equipment's behavior to environment's changing. This decision-making will help realizing dynamic adaptation according to the environment state changing by choosing best strategies and providing reactivity improvement by local reasoning. Then strategies are sharpened and validated with the local learning and the knowledge sharing.

An important concept related to situatedness is the notion of neighborhood. Indeed, the knowledge plane is a distributed structure whose elements essentially exchange knowledge with their neighbors. But it would be somewhat reducing to make this notion corresponding exclusively to the physical (accessible media) neighborhood notion. Several criteria can be used to characterize this notion such as role, power of calculation, technologies implemented on the node, etc. We propose the following definition:

Definition 6. Two nodes n_0 and n_1 are neighbors if only if there is a path between them that did not pass by any of their neighbors.

The definition is deliberately recursive because according to neighborhood's criteria we can have two types of neighborhood: physical neighborhood and logical neighborhood. In the first case, the criterion is the accessibility by the physical link (Fig. 12.a) whereas the second case is rather a logical criterion (Fig. 12.b) based on a consideration or a parameter (Example: the type of equipment, the role of the equipment, etc.). We also can notice that the physical neighborhood can be considered as a particular case of logical neighborhood. Fig. 12.a and 12.b give examples

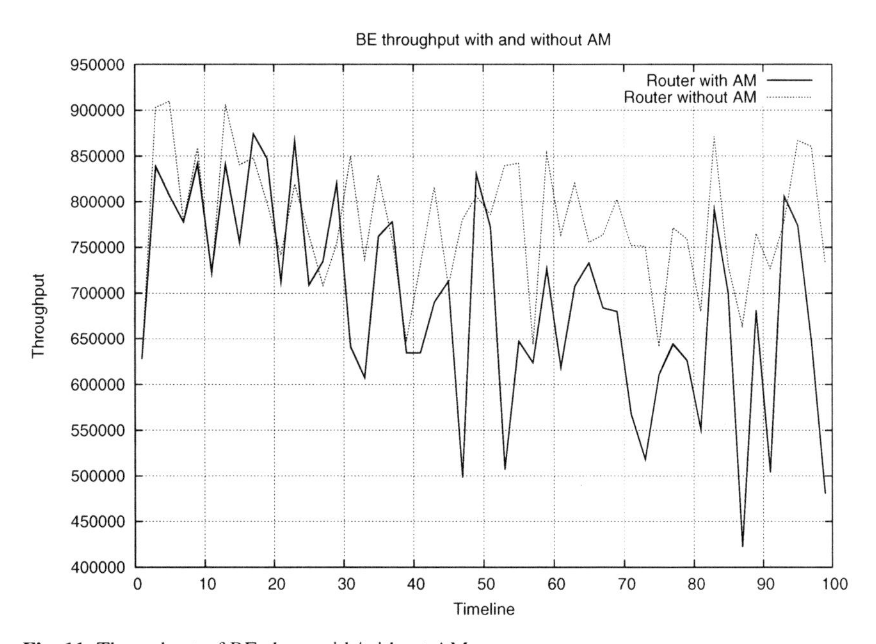

Fig. 11 Throughput of BE class, with/without AM

of neighborhoods of a node. We can notice that between two logical neighbors we can have one or several physical neighbors.

Elements in knowledge plane are supposed to share their knowledge (through a situated view) to the limit of knowledge's validity and relevance. Then elements have to identify their direct neighbors. A process of logical neighborhood's discovery is described in the algorithm 2. In a real world situation this process must be repeated periodically to update the neighborhood table in order to face the possible topology changes during the time. We assume that the context change of the nodes is not very fast. This hypothesis of stability is necessary because Machine learning and knowledge validation are processes that can take a more or less long time. We can notice that knowledge sharing between nodes helps to reduce this time by discovering the knowledge from neighbors.

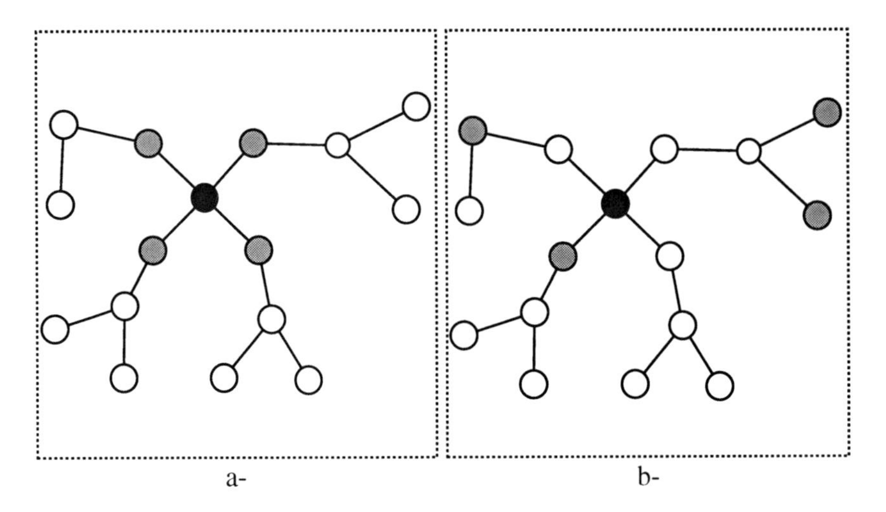

Fig. 12 The two kinds of neighborhoods : a- Physical neighborhood, b- logical neighborhood.

5.2 Situated Knowledge sharing algorithm

This section details principle of our knowledge sharing algorithm. The starting point of this algorithm is the node that initiates the knowledge sharing. Each node participating in this process has three knowledge bases (strategy bases):

- B_L (For local Base) is the knowledge base keeping strategies learned locally;
- B_S (For Share Base) contains the set of strategies resulting from knowledge sharing process;

- B_T (for Temporary Base) is a temporary knowledge base. Knowledge sharing algorithm is mainly founded on distributed algorithmic techniques. Algorithm 2 describes its global working principle.

The generic principle stresses on pro-activity of nodes that share their knowledge with their environment (neighborhood).

Algorithm 2: Knowledge sharing algorithm with situated view

1 PROCEDURE 1 /* Initiation of propagation */
2 **begin**
3 Send $< id_i, k >$ to all neighbors ; /* k is a piece of knowledge */
4 **end**
5 PROCEDURE 2 /* Receiving at each node */
6 **begin**
7 **When** receive $< id_i, k >$
8 **if** $id_{local} \neq id_i$ **then**
9 **if** $k \in B_T \cup B_S \cup B_L$ **then**
10 ignore k
11 **else if** $(k \cup B_T)$ *consistent* **then**
12 $B_T \leftarrow B_T \cup k$; /* Put it on a temporary knowledge base */
13 **if** *ACCEPT (k)* **then**
14 $B_S \leftarrow B_S \cup k$; /* Integrate it in the knowledge base */
15 **end**
16 Send $< (id_i, k >$ to all neighbors except to the destination id_i
17 **else**
18 Send $< \hat{e}, (id_i, k >$ to destination id_i ; /* $\hat{e}$ is a counter example */
19 **end**
20 **end**
; /* If knowledge returns back to sender it ignores it */
21 **end**
22 PROCEDURE 3 /* Initiation of propagation */
23 **begin**
24 **When** receive $< \hat{e}, (id_i, k >$
25 Store $\hat{e}$ in counterexample knowledge base. ; /* will be used by a learning process */
26 **end**

PROCEDURE 1 initiates knowledge sharing. It is executed by a node that wants to share a piece of knowledge it has learned by itself (from B_L). It sends this piece of knowledge to all its neighbors. Knowledge sharing does not concern knowledge bases B_S and B_T because knowledge in these bases are already known by neighbors.

PROCEDURE 2 is executed by all nodes which receive a piece of knowledge. This procedure allows these nodes to propagate or not the piece of knowledge received. At first each node verifies if it already has this knowledge in one of its bases. If it is the case, it ignores this knowledge because it has already treated it. If knowledge is not in its bases, it propagates it to all its neighbors after verifying that this knowledge is coherent with its local knowledge (in B_S). By this way the situated view is constructed progressively and no information about its borders is stored.

Algorithm 3: ACCEPT trigger

```
1  On event : localstateChaged to e_c
2  begin
3  |  s ← EtractOptimalStrategy(B_T) ;          /* which has biggest θ */
4  |  a ← getActions(s)
5  |  enforce(a)
6  |  e'_c =newlocalstate()
7  |  begin
8  |  |  if e'_c better then e_c then
9  |  |  |  return True
10 |  |  else
11 |  |  |  return False
12 |  |  end
13 |  end
14 end
```

PROCEDURE 3 treats counterexamples received by the node that initiates knowledge sharing. We remind that a counterexample is a negative observation (see section 3.2). The notion of acceptance that appears in PROCEDURE 2 (Algorithm 2) is necessary to determine if knowledge is pertinent or not. It is the test done by each node that receives knowledge to determine if it is relevant to keep in the share knowledge base or not. Algorithm 3 describes the acceptance function that returns true if the strategy succeeded in ameliorating the element state.

Fig. 13 is a snapshot of the execution of our algorithm with VISIDIA [31]. In this picture, element that shares knowledge is the node with the label A. Elements that are not concerned by the situated view are labelled N and X-labelled nodes are at the bounderies of the situated view.

5.3 Performance and guarantees

The propagation problematic in a networked structure is a classic problem of the distributed algorithmic. So we evaluate theoretical performances of our algorithm by using the same constraints as in distributed algorithmic. Algorithm described earlier is guaranteed to end after a certain time. This guarantee is given by the fact that a piece of knowledge is treated only one time by a given node. In PROCEDURE 2 a node that receives twice a piece of knowledge ignores it the second time. So it retransmits it to its neighbors only one time. The algorithm is finished when all nodes have treated the piece of knowledge.

Now we focus on the number of messages needed for executing this knowledge sharing algorithm. For a given node n_0, having a finished number ei of states, that propagates a piece of knowledge c_{i_0} crossing the set M_{i_0} of links in the network. The number of necessary messages to propagate c_{i_0} is given by the following equation:

$$\delta(c_o) = \sum_{l_k \in M_{i_0}} (f_0(l_k) + \lambda(e_0)) \tag{3}$$

Where $f_0(l_k)$ is the number of messages that crosses the link l_k during the propagation of the piece of knowledge c_{i_0} and $\lambda(e_0)$ is the number of counterexamples received by n_0.

This formula could be translated literally by: the sum of the total number of messages sent in the network is equal to the number of times when a piece of knowledge c_{i_0} is transmitted on each link l_k plus, the number of counterexamples received by the node n_0. Knowing that each node treats (propagates, gives a counterexample) a piece of knowledge only one time, the number of knowledge messages transmitted on a link is less or equal than two. Considering that only the node that initiates the knowledge sharing, receives the counterexamples we can bound by this inequation (4):

$$\delta(c_o) \leq (2m + deg(n_0)) \tag{4}$$

Where $deg(n_0)$ is the number of neighbors of n_0 and m_i is the number of links crossed by c_{i_0}. In summary, the algorithm of knowledge propagation ends at a relatively short time because it has a linear complexity in term of total number of messages depending on the number of links involved in the situated view.

The worst scenario we could have is when all the nodes are propagating their knowledge in a completely meshed network. Considering that the number of links

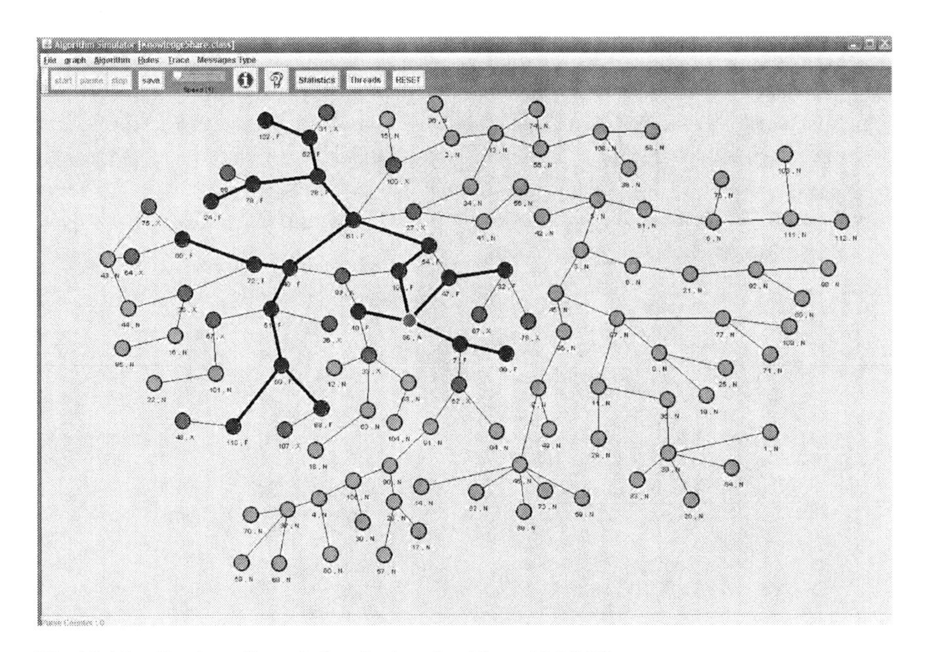

Fig. 13 Visualization of knowledge sharing algorithm with Visidia

in a network is bounded by a quadratic polynomial of the number of nodes and considering formula (4): The total number of messages is bounded by the following expression (formula 5)

$$\Delta(C) \leq n^2(n-1) \tag{5}$$

Where n is the number of nodes in the network and C the set of all propagated knowledge. The second term ($deg(n_0)$) disappeared because in this scenario all the nodes accept knowledge of their neighbors and no counterexample is sent. We can conclude that we have a good complexity which is polynomial (cubical more precisely) related to the number of nodes. That means that this knowledge sharing algorithm is scalable. If we consider consistency of knowledge bases, it is automatically held by test process of acceptance (algorithme 2 PROCEDURE 2). The knowledge bases considered separately are consistent (except for B_T which is not really a knowledge base). The local knowledge is privileged to the situated knowledge because we think that it corresponds more to the local realities.

6 Conclusion

We presented in this paper a generic architecture for knowledge plane based on Machine learning tools. This architecture is composed of two loops: self-organization (collaborative) loop and self-adaptation loop. Self-adaptation adapts managed resources behavior to environment changes. Self-organization includes knowledge sharing processes to construct situated views. By using this situated view, knowledge is shared only where it is relevant and suitable. Then, we presented a generic Machine learning algorithm and mechanisms for self-adaption and a knowledge sharing distributed algorithm for self-organization. Finally, we show some results in a specific study case for self-adaptation function of our architecture. In future we intend to show how self-organization mechanisms in our architecture could be applied to sensor network environment.

References

1. Abid, M., Berl, A., al.: Autonomic internet initial framework. Deliverable d6.1, Autonomic Internet (autoi) project (August 2008)
2. Agoulmine, N., Balasubramaniam, S., Al.: Challenges for autonomic network management. In: 1st IEEE International Workshop on Modeling Autonomic Communications Environments (MACE) (2006)
3. autoi: Atonomic internet project. Home page http://ist-autoi.eu/autoi/
4. Balasubramaniam, S., Botvich, D., Donnelly, W., Foghlú, M., Strassner, J.: Biologically inspired self-governance and self-organisation for autonomic networks. In: BIONETICS '06: Proceedings of the 1st international conference on Bio inspired models of network, information and computing systems, p. 30. ACM, New York, NY, USA (2006). DOI http://doi.acm.org/10.1145/1315843.1315880

5. Blake, S., Black, D., Carlson, M., Davies, E., Wang, Z., Weiss, W.: Rfc 2475 : An architecture for differentiated services. In: RFC (1998). Http://www.ietf.org/rfc/rfc2475.txt
6. Bonifacio, M., Bouquet, P., Mameli, G., Nori, M.: Kex: a peer-to-peer solution for distributed knowledge management. In: Fourth International Conference on Practical Aspects of Knowledge Management (PAKM-2002, pp. 490–500 (2002)
7. Brügge, B., Renner, P., Strassberger, M., Adamski, M.: Medusa - framework for the secure peer-to-peer sharing of topic map based knowledge. In: Conference on knowledge sharing and collaborative engineering (KSCE 2004) (2004)
8. Clark, D., Partridge, C., Ramming, C.J., Wroclawski, J.T.: A knowledge plane for the internet. In: SIGCOMM '03: Proceedings of the 2003 conference on Applications, technologies, architectures, and protocols for computer communications, pp. 3–10. ACM Press, New York, NY, USA (2003). DOI http://dx.doi.org/10.1145/863955.863957. URL http://dx.doi.org/10.1145/863955.863957
9. Cornuéjols A., M.L.: Apprentissage artificiel - Concepts et algorithmes. Eyrolles, Paris (2003)
10. Curran, K., Mulvenna, M., Nugent, C., Galis, A.: Challenges and research directions in autonomic communications. pp. 3–17. Inderscience Publishers, Inderscience Publishers, Geneva, SWITZERLAND (2007). DOI http://dx.doi.org/10.1504/IJIPT.2007.011593
11. Dietterich, T.G.: Learning and reasoning. Tech. rep., School of Electrical Engineering and Computer Science, Oregon State University (2004)
12. Dietterich, T.G., Langley, P.: Machine learning for cognitive networks: Technology assessment and research challenges. In: Cognitive Networks. ISBN: 9780470061961 (2007)
13. Fahy, C., Davy, S., Boudjemil, Z., al.: Towards an information model that supports service-aware, self-managing virtual resources. In: MACE'08: Proceedings of the 3rd IEEE international workshop on Modelling Autonomic Communications Environments, pp. 102–107. Springer-Verlag, Berlin, Heidelberg (2008)
14. Florian, R.V.: Autonomous artificial intelligent agents. Tech. rep., Technical Report Coneural-03-01 (2003)
15. Heinanen, J., Baker, F., Weiss, W., Wroclawski, J.: Rfc 2597 : Assured forwarding phb group (1999). Http://www.ietf.org/rfc/rfc2597.txt
16. Hung-Ying, T.: J-sim home page. Please see home Page http://www.j-sim.org/
17. Jacob, B., Lanyon-Hogg, R., Nadgir, D.K., Yassin, A.F.: A practical guide to the ibm autonomic computing toolkit. In: Proceedings of 18th International Conference on Logic Programming (ICLP 2002). IBM (2004)
18. Jacobson, V., Nichols, K., Poduri., K.: Rfc 2598: An expedited forwarding phb. In: RFC editor (1999). Http://www.ietf.org/rfc/rfc2598.txt
19. Kephart, J.O., Chess, D.M.: The vision of autonomic computing. pp. 41–50. IEEE Computer Society Press, Los Alamitos, CA, USA (2003)
20. Latré, S., Simoens, P., Meerssche, W., al.: Automated generation of knowledge plane components for multimedia access networks. In: MACE '08: Proceedings of the 3rd IEEE international workshop on Modelling Autonomic Communications Environments, pp. 50–61. Springer-Verlag, Berlin, Heidelberg (2008). DOI http://dx.doi.org/10.1007/978-3-540-87355-6_5
21. Lavrac, N., Dzeroski, S.: Inductive Logic Programming: Techniques and Applications. Ellis Horwood, New York (1994). URL http://www-ai.ijs.si/SasoDzeroski/ILPBook/
22. Li, J.: Agent organization and request propagation in the knowledge plane. Tech. rep., CSAIL Technical Reports. (2007)
23. Mola, F.D., l. Quitadamo, R.: Towards an agent model for future autonomic communications. In: Proceedings of the 7th WOA 2006 Workshop From Objects to Agents (2006)
24. Muggleton, S.: INDUCTIVE MACHINE LEARNING New generation computing. 8(4):295-318. (2002)
25. Muggleton, S., Raedt, L.D.: Inductive logic programming: theory and application. In: Journal of logic programming (1994)
26. Prehofer, C., Bettstetter, C.: Self-organization in communication networks: principles and design paradigms. pp. 78–85 (2005). DOI http://dx.doi.org/10.1109/MCOM.2005.1470824. URL http://dx.doi.org/10.1109/MCOM.2005.1470824

27. Quirolgico, S., Mills, K., Montgomery, D.: Deriving knowledge for the knowledge plane (2003)
28. Srinivasan, A.: Aleph (a learning engine for proposing hypotheses) manual page (2008). Http://www.comlab.ox.ac.uk/activities/machinelearning/Aleph/
29. Strassner, J., Agoulmine, N., Lehtihet, E.: Focale a novel autonomic networking architecture. In: International Transactions on Systems, Science, and Applications (ITSSA) Journal, *ISSN 1751-1461*, vol. Vol. 3, No 1, pp. 64–79 (2007)
30. Strassner, J., Foghlu, M.O., Donnelly, W., Agoulmine, N.: Beyond the knowledge plane: An inference plane to support the next generation internet. In: Global Information Infrastructure Symposium, 2007. GIIS 2007. First International, pp. 112–119 (2007). DOI http://dx.doi.org/10.1109/GIIS.2007.4404176. URL `http://dx.doi.org/10.1109/GIIS.2007.4404176`
31. VISIDIA: The visidia project: Visualization and simulation of distributed algorithms. In: Please see http://www.labri.fr/projet/visidia/ (2008)
32. Wawrzoniak, M., Peterson, L., Roscoe, T.: Sophia: An information plane for networked systems. In: In HotNets-II (2003)
33. Wei, G.: A multiagent perspective of parallel and distributed. In: Machine Learning", Proceedings of the 2nd International Conference on Autonomous Agents, pp. 226–230 (1998)

A Rate Feedback Predictive Control Scheme Based on Neural Network and Control Theory for Autonomic Communication

Naixue Xiong, Athanasios V. Vasilakos, Laurence T. Yang, Fei Long, Lei Shu, and Yingshu Li

Abstract The main difficulty arising in designing an efficient congestion control scheme lies in the large propagation delay in data transfer which usually leads to a mismatch between the network resources and the amount of admitted traffic. To attack this problem, this chapter describes a novel congestion control scheme that is based on a Back Propagation (BP) neural network technique. We consider a general computer communication model with multiple sources and one destination node. The dynamic buffer occupancy of the bottleneck node is predicted and controlled by using a BP neural network. The controlled best-effort traffic of the sources uses the bandwidth, which is left over by the guaranteed traffic. This control mechanism is shown to be able to avoid network congestion efficiently and to optimize the transfer performance both by the theoretic analyzing procedures and by the simulation studies.

Naixue Xiong, and Yingshu Li
Department of Computer Science, Georgia State University, Atlanta, USA, e-mail: \{nxiong, yli\}@cs.gsu.edu

Athanasios V. Vasilakos
Department of Department of Electrical and Computer Engineering, University of Western Macedonia, Greece, e-mail: vasilako@ath.forthnet.gr

Laurence T. Yang
Department of Computer Science, St. Francis Xavier University, NS, Canada, e-mail: lyang@stfx.ca

Long Fei
Department of Computer Science, Tsinghua University, Beijing, 100084, China, e-mail: longf05@mails.thu.edu.cn

Lei Shu
Digital Enterprise Research Institute, National University of Ireland, Galway, Galway, Ireland, e-mail: lei.shu@deri.org

A.V. Vasilakos et al. (eds.), *Autonomic Communication*, DOI: 10.1007/978-0-387-09753-4_4,

1 Introduction

With the rapid development of computer networks, more and more severe autonomic congestion problems have occurred. Designing efficient autonomic congestion control scheme is, therefore, a crucial issue to alleviate network congestion and to fulfill data transmission effectively. The main difficulty in designing such scheme lies in the large propagation delay in transmission that usually leads to a mismatch between the network resources and the amount of admitted traffic. The crucial issue of the network control is that we should adapt the controllable flows to the changing network environment, so as to achieve the goal of the data transfer and to alleviate network congestion. Congestion is the result of a mismatch between the network resources capacity and the amount of traffic for transmission.

Many research are provided on the autonomic communication [22-25, 27-28] and the network autonomic congestion problems. The paper [1] reviews all kinds of congestion control schemes having been proposed for computer networks. Among these schemes, the representative one, which is in common use, is the rate-based congestion control (see, e.g., [2-3]). The basic techniques include the Forward Explicit Congestion Notification (FECN) and the Backward Explicit Congestion Notification (BECN) [3-4].

The time delay in data transmission will result in slow transient behavior of buffer occupancy. The responsiveness of the congestion control scheme is crucial to the stability of the whole network system. The non-stability of dynamic network influences the network's performance. To deal with this difficulty, the authors in [5] suggest using the method of fuzzy control to realize the rate-based network congestion control, and the application of heredity algorithm in queue strategy is presented in [6-7]. Furthermore, the recent papers [8-9, 18-22, 26, 30] use a multi-step neural predictive technique to predict the congestion situation in computer networks, but the longer predictive steps has still existed and the effectiveness is greatly limited in existed papers. And yet the responsiveness of the congestion control scheme is crucial to the stability of the whole network system and the relevant performance, this issue is, however, not considered in these works. So this chapter aims to improve the predictive scheme. We implement the neural predictive controller at the sources rather than at the switch. This is due to the fact the less prediction horizon usually leads to better accuracy, whereas in the proposed scheme the predictive horizon is linked with the network structure. Under the same circumstance, we use less predictive steps than that in [8-9], this then usually brings forth better performance in terms of predictive accuracy and efficiency.

Our main contribution is the significant development of a multi-step neural network predictive technique for the congestion control. Through simulations of actual trace data from the real-time traffic, we demonstrate that the technique improves the control performance. Compared with the methods discussed in [8-9], this chapter introduces a BP neural network, analysis the neural network architectures and evaluates control performance.

The rest of this chapter is organized as follows: In section 2, we introduce a novel improved congestion control scheme based on neural networks. In section 3, we describe the predictive control scheme for resource management and in section 4, we use simulation to validate and evaluate the performance of our scheme. Finally, in section 5, we present the conclusions and the future work.

2 Congestion Control Model

The congestion control technique in this chapter provides such an approach for the dynamic evaluation of the low priority traffic in the network: the evaluation and distribution functions which compute the rate allocated to each individual source are based on a neural network control strategy, and the functions control the filling level of the low priority traffic buffer.

The chapter considers a general model as shown in Figures 1-2 with different connections and with various traffic requirements being mapped into different classes. The rate control algorithm computes the low priority bandwidth $\lambda_L(t)$ left by the sum of the highest priority traffic λ_{H1} and the higher priority traffic λ_{H2}. $x_L(t)$ is the number of λ_L packets waiting at time t in the queue, $x_0(t)$ is queue threshold at time t, usually $x_0(t)$ is a constant [8-9].

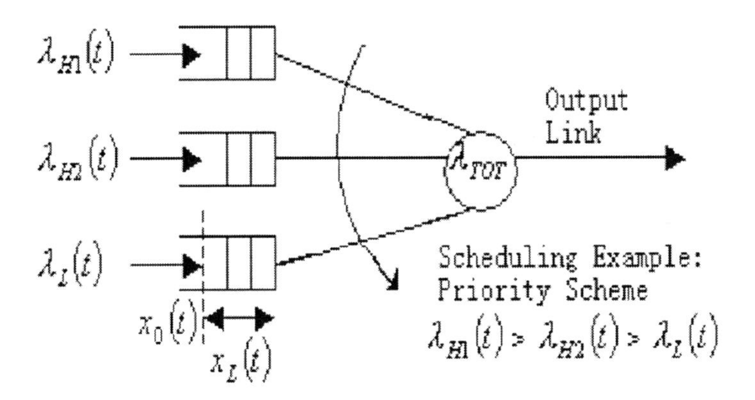

Fig. 1 A simple model of one source,λ_L is controlled

2.1 The Predictive Control Model of a Bottleneck Buffer

It describes the control procedures for multiple sources transmitting data to the buffer of a common bottleneck node. A control algorithm running at the source node evaluates the resource need of each source and distributes the estimated available resources accordingly [8].

In modeling the traffic through these nodes, one has to know the number of source/destination pairs and the rates at which these sources send control packets (CPs) to the network. It's assumed to be N though the number of active sources denoted by M may vary with time t. The switching node has a finite buffer space K to store the incoming CPs and has an output link to serve them at a constant data rate of v.

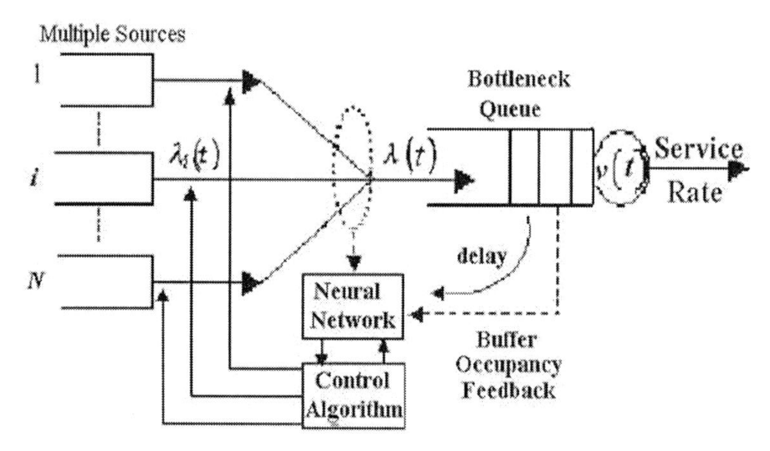

Fig. 2 A model of multiple sources and a bottleneck with controller

The control procedure works in the following manner: each source sends data to the bottleneck node at regular intervals. According to the current loading state, the bottleneck node feedbacks the information to the source along the original route [29-30]. According to this reception information, the sources can decide the most suitable amount of resources that each source should be available. Thus, the sources can adjust sent-out rates correspondingly. It is clear that the key point of this control architecture lies in the control algorithm that is employed at the source node.

Under the above notations and assumptions, the dynamic system of a switching node in a network can be described by the following non-linear time-variant and time-delayed equation [10-11].

$$\dot{x}(t) = Sat_K\{\sum_{i=1}^{N} e_i\lambda_i(t-\tau'_{1i}) - v\}, \tag{1}$$

where K is the buffer size, $\dot{x}(t)$ is the buffer occupancy at time t, and

$$e_i = \begin{cases} 1, active source; \\ 0, otherwise. \end{cases}$$

$$Sat_K\{x\} = \begin{cases} K, x > K; \\ x, 0 \le x \le K; \\ 0, x < 0. \end{cases}$$

If a feedback control is applied to the above system, we assume the signals get sampled every T seconds. It is reasonable because one can always add a small delay to the input delay so that it is a multiple of T when timing. So we can come to the virtual connection (VC) delay d_i = the input delay τ_{1i} from the ith source node to the switching node + the feedback delay τ_{2i} from the switching node to the ith source node. $\lambda_{iL}(n)$, $\lambda_{iH2}(n)$ and $\lambda_{iH1}(n)$ respectively denote the low priority traffic rate, the higher priority traffic rate and the highest priority traffic rate from the ith source, and $\lambda_i(n)$ denotes the sending rate of source i, i.e., $\lambda_i(n) = \lambda_{iL}(n) + \lambda_{iH1}(n) + \lambda_{iH2}(n)$.The low priority traffic can only be transmitted when no congestion appears in the network. Furthermore, we assume that the service is FCFS (first-come-first-served) and the packet length is constant. The buffer occupancy $x(n)$ is measured, the CPs are sent back to the controlled sources every T seconds. The rate control algorithm computes the low priority traffic rate $\lambda_L(n)$, i.e., the rate left by the high priority traffic $\lambda_{H1}(n)$ and $\lambda_{H2}(n)$.

When N sources transmit data towards a single bottleneck node, there is a control-loop delay between each source and the bottleneck node. The round trip delay (RTD), d, is set to be a single representative value $d = min(d_1, d_2, ..., d_N)$, and the input representative delay, τ_1, is set as $\tau_1 = min(\tau_{11}, \tau_{12}, ..., \tau_{1N})$. So $d = \tau_1 + \tau_2(\tau_2$ is the backward path delay). The best result in system performance is taken for granted the minimum delay [11]. Let $\lambda_i(n) = T \cdot \lambda_i(nT)$ denote the total numbers of data packets flowing into the destination node from the ith VC during the nth interval of T. The component $\mu = T\nu$ denotes the number of packets sent out from the switching destination node during the nth interval of T. The equation can be written into

$$x(n+1) = Sat_K\{x(n) + \Sigma_{i=1}^{N} e_i \lambda_i(n - \tau_{1i}) - \mu\}. \quad (2)$$

The control algorithm employs the following four steps [8-9]:

(i) Predict the buffer occupancy $\hat{x}(n+1)$ using the multi-step predictive technique.

(ii) Compute the total expected rate of the all sources $\lambda(n)$ at the time n and $\lambda(n) = \Sigma_{i=1}^{N} \lambda_i(n)$. This value varies dynamically with the buffer occupancy.

(iii) Compute the proportion of each source,$\delta_i(n)$, which is the most efficient share of the available resources to be attributed to source number i, ($1 \leq i \leq N$, $\sum_{i=1}^{N} \delta_i(n) = 1$), $\delta_i(n) = \lambda_i(n)/\lambda(n)$.

(iv) Compute the adjusted low priority traffic rate $\lambda_{iL}(n)$. In this section, every source equally shares the available network bottleneck bandwidth, $\lambda_i(n)$ can be expressed as: $\lambda_i(n) = \delta_i(n) \cdot \lambda(n)$. Based on the equation (4), the source i regulates the lowest priority traffic rate $\lambda_{iL}(n)$.

3 The Predictive Control Technique

3.1 The BP Neural Network Architecture

The BP neural network algorithm is introduced into this chapter as a predictive mechanism. We assume the number of input neuron is N, and the number of sample study group is M_0. The sample study groups are independent from each other. We further assume the output of the study sample group (teaching assigns) is $R_j^{(k)}$ ($j \in [0, N]$, $k \in [1, M_0]$), and the actual output for output element j in the network is $O_j^{(k)}$. So $E^(k)$ is set to be the kth group input goal function. Therefore, we have $E^{(k)} = \Sigma_j(R_j^{(k)} - O_j^{(k)})^2/2$. The total goal function is $J = \Sigma_k E^{(k)}$. If $J \leq \varepsilon_0$, ε_0 is a constant that is small enough and $\varepsilon_0 > 0$, then the algorithm is terminated; Otherwise adjust the weight W between the implicit layer and output layer until it satisfy the expected difference value [12-15].

3.2 Multi-step Neural Predictive Technique

We apply a neural network technique to determine how a BP-based algorithm satisfies its data transfer requirement by adjusting its data transfer rate in a network. As shown in Figure 2, the BPNN predictive controller is located at the sources. In order to predict the buffer occupancy efficiently, the neural model for the unknown system above can be expressed as:

$$\hat{x}(n+1) = \hat{f}[x(n), ..., x(n-l+1), \lambda(n-\tau_1-1), \\ ..., \lambda(n-\tau_1-m-L)], \quad (3)$$

where $x(n-i)$ $(1 \leq i \leq l-1)$is the history buffer occupancy and $\lambda(n-j)(\tau_1+1 \leq j \leq \tau_1+m+L)$ is the history sending rate of the source j. L is predictive step,$L = \tau_1+1$, and L, m are constant integers. $\hat{f}[\cdot]$ is the unknown function, which may be expressed by the neural network. The explicit mechanism of BP neural network L-step ahead prediction is shown in Figure 3, the value of buffer occupancy $x(n)$ and the history value (the past buffer occupancy: $x(n-1), ...x(n-l+1)$; the past source sending rates:$\lambda(n-\tau_1-1), ..., \lambda(n-\tau_1-m-L)$) are used as the known inputs of neural network. Every layer denotes one-step forward predictive, so $\hat{x}(n+L)$ in the output layer is the L-step prediction of $x(n)$.We can compute the expected total rate $\hat{\lambda}(n)$ of the N sources using the following equation:

$$\hat{x}(n+L) = Sat_K\{\hat{x}(n+L-1) + \hat{\lambda}(n) - \mu\}, \quad (4)$$

Based on the rate $\hat{\lambda}(n)$ the source i adjusts the sending rate $\lambda_{iL}(n) = \hat{\lambda}(n)\delta_i(n) - \lambda_{iH1}(n) - \lambda_{iH2}(n)$, and $\delta_i(n)$ is a factor of share the available resources to source i $(1 \leq i \leq N)$. The specific algorithm is given in the following (Figure 4), At the next

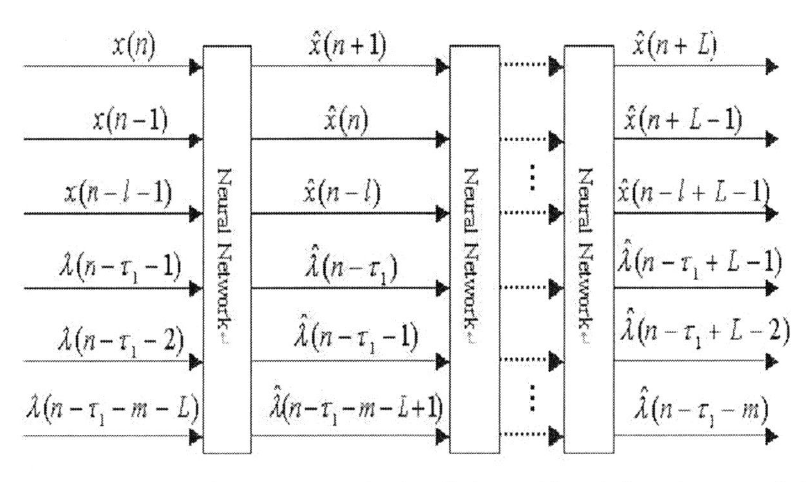

Fig. 3 The Back Propagation (BP) L-step ahead prediction, and $\hat{x}(n+L)$ is the L-step predictions of $x(n)$.

instant $n+1$, we can get new real measured value $x(n+1)$ and new history measure values: $x(n),...,x(n-l+2)$; $\lambda(n-\tau_1),...,\lambda(n-\tau_1-m-L+1)$ which can be used as the next instant inputs of neural network. Then the buffer occupancy $\hat{x}(n+L+1)$ can be predicted.

4 The Simulation Results

To evaluate the performance of the proposed congestion control method based on neural network, we focus upon the following simulation model with eleven sources and one switch bottleneck node (Figure 5), and assume that the sources always have data to transmit. The congestion controller is used to adjust sending rate over time in sources. The higher priority traffic, i.e., the sum of λ_{iH1} and λ_{iH2} traffic in source i with multiplexing of actual trace data, is acquired from the real time traffic.

As shown in Figure 6, the maximum sending rate of every source is $\lambda_0 = 15.5Mbps$. We use a simple resource sharing policy, i.e., the network bottleneck node equally shares the available bandwidth among every source. The sources start to transmit data at time $t = 1msec$ together. We assume the sending rate of the switch node is $v = 155Mbps$. The sampling time T is $1msec$ and the congestion threshold is set as $x_0 = 1000Kb$.

We propose to use a direct multi-step neural predictive architecture with 3 layer neural network, wherein the number of the input data, the input neurons, the hidden neurons and the output neurons are all $(L+m+l)$. There are $l(l=8)$terms of buffer occupancy x and $(L+m)$terms of the total input μ. The prediction horizon is $L = \tau_1 + 1$, and the control horizon is $N = L - \tau_1 + 1 = 2$.

Procedure Neural_Network_Prediction$(x, \lambda, \hat{x}, \hat{\lambda})$

{ *At time instant* n :

Step 1: Get value of buffer occupancy $x(n)$, *and the history measurement value by the controller:*

the past buffer occupancy: $x(n-1), \ldots, x(n-l+1)$;

the past total sending rate: $\lambda(n-\tau_1-1), \ldots \lambda(n-\tau_1-M-L)$.

All values above are used as available known input of neural network.

Step 2: Predict L *step* $\hat{x}(n+L)$.

(1) Predict $\hat{x}(n+i)$, $i \in [1, L]$;

$[\hat{x}(n+i-1), \hat{x}(N+i-2), \ldots, \hat{x}(N+i-l),\quad]$

as the neural network input. These predictions are to be used to predict the next set.

(2) Train neural network computes the goal function J *and adjust the weight;*

(3) Back propagation for the next L step ahead prediction;

(4) Go ahead until L step ahead predictions. Then, $\hat{x}(n+L)$ *is the L-step prediction.*

Step 3:Compute $\hat{\lambda}(n), \hat{\lambda}_i$ and $\hat{\lambda}_{id}(n)$.

Step4: At the next instant $n+1$, *update the history value,* $(n+1)$ *takes the place of* n, *go to step 1.*

Return $(\hat{x}(n+L) \in (\hat{x}, \hat{u}))$. }

Fig. 4 Algorithm for on-line control and neural network training at sources

To investigate the performance of this model, we set the distance from sources to switch node to be $300Km$ with the forward path delay and the feedback path delay being $\tau_{1i} = 3msec$, $\tau_{2i} = 3msec$ $(i = 1, 2, \ldots, 11)$ respectively. Therefore the RTD is

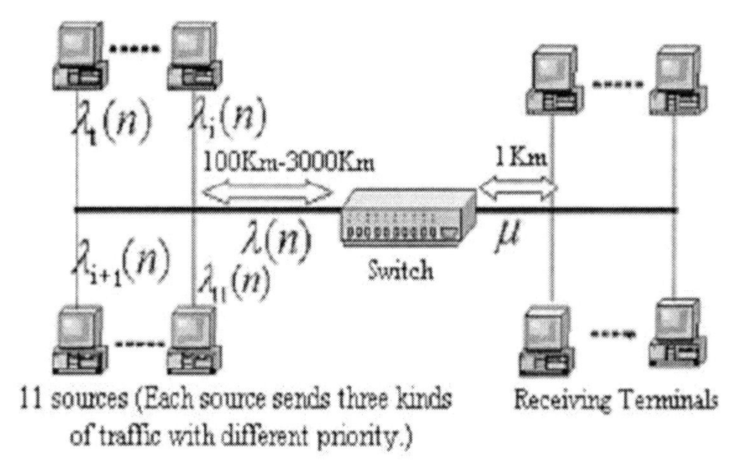

Fig. 5 A simulation model of multiple sources single buffer network

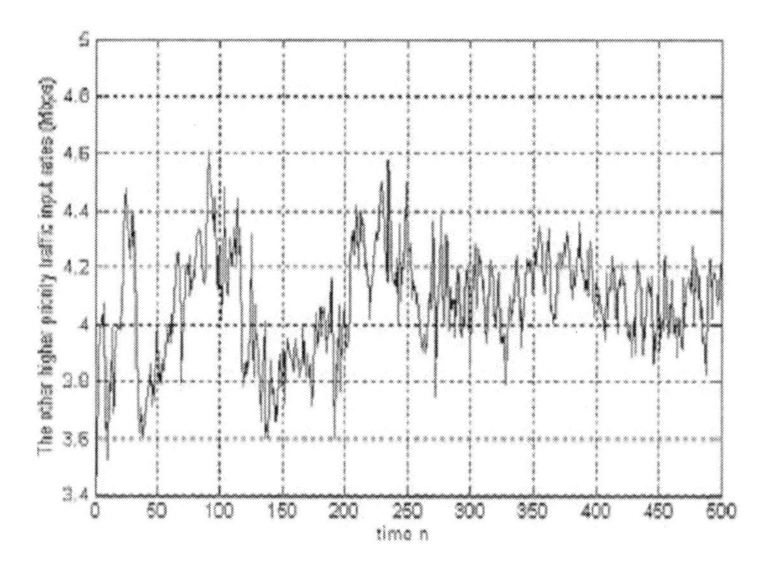

Fig. 6 High priority traffic rate sampled from the real time video traffic.

$d = 6msec$.We assume that the RTD is dominant compared to other delays such as processing delays and queuing delay, etc.

For this case, the prediction horizon is $L = 4$, and $m = 4$. Figure 6 shows the rate of higher priority $(\lambda_{H1} + \lambda_{H2})$ traffic. The dynamic of buffer occupancy is shown in Figure 7, where the predictive buffer occupancy and the actual buffer occupancy are described with broken line and real line respectively. The predictive value of the buffer occupancy is acquired beginning from the time $(\tau_1 + L + 9)$. Figure 8 shows the transmitting rate of the lowest priority traffic, which is yielded on the

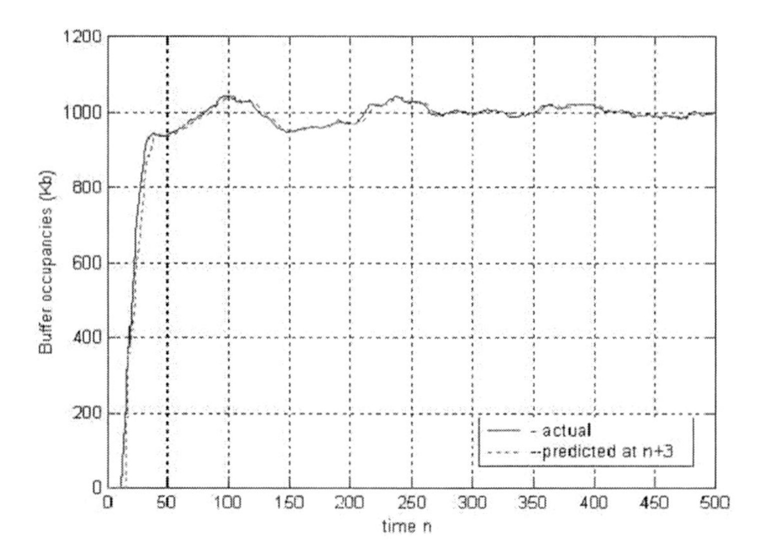

Fig. 7 The buffer occupancy for $L = 4$ step prediction.

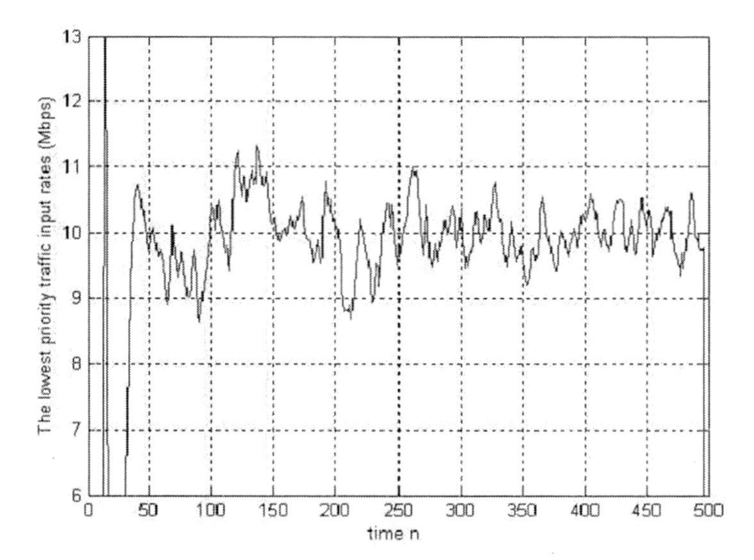

Fig. 8 The lowest priority traffic rate for $L = 4$ step prediction, based on the predictive value in Figure 7.

basis of the equation (1) and the predicted buffer occupancy from the time slot 12 to $(500 - \tau_1 - L) = 493$, and Figure 9 shows the total input rates.

From Figure 7, one observes that buffer occupancy is acquired beginning from the time slot $n = 16$ and that the queue size is maintained to be close to the threshold of $1000Kb$ by the proposed neural networks predictive technique. The average

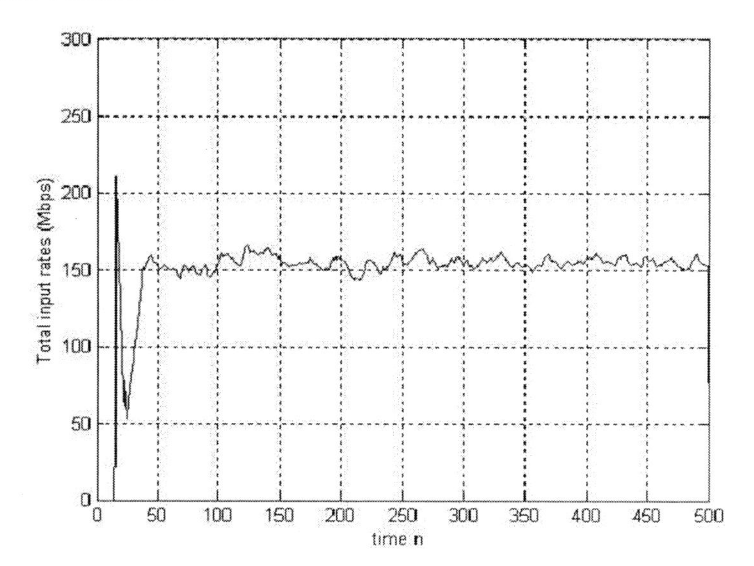

Fig. 9 The total input rates of $L = 4$.

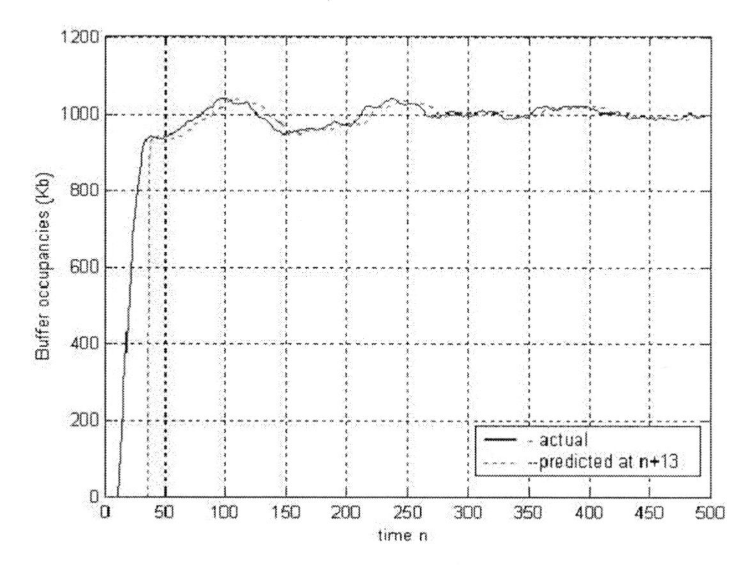

Fig. 10 The buffer occupancy for $L = 14$ step prediction.

relative error between the predictive buffer occupancy and actual buffer occupancy is 1.5099e-002, which is excellent in terms of accuracy.

Figure 10-12 show the performance that we set the sources $2600Km$ away from the switch node, and assume the forward delay and the feedback delay being $\tau_{1i} = 13msec$, $\tau_{2i} = 12msec$,$(i = 1,2,...,11)$ respectively. Therefore the RTD is $d = 25msec$. We take the prediction horizon $L = 14$ and $m = -6$. Figure 10 shows the

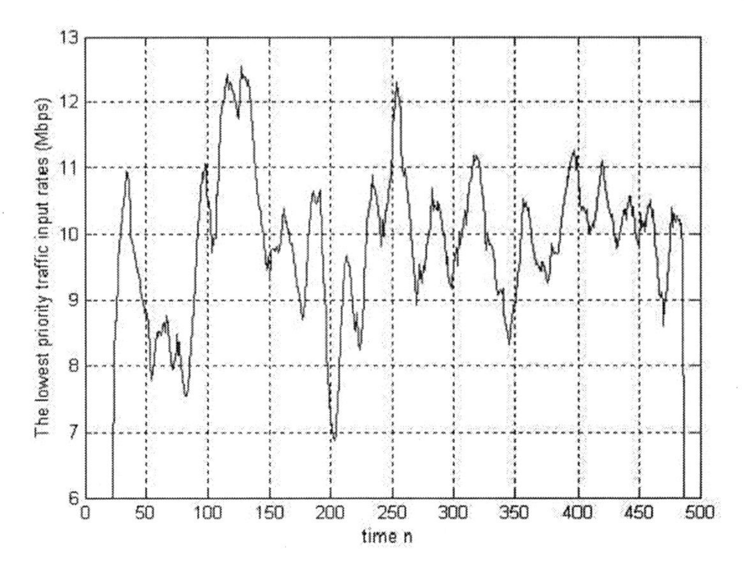

Fig. 11 The lowest priority traffic rate for $L = 14$ step prediction, based on the predictive value in Figure 10.

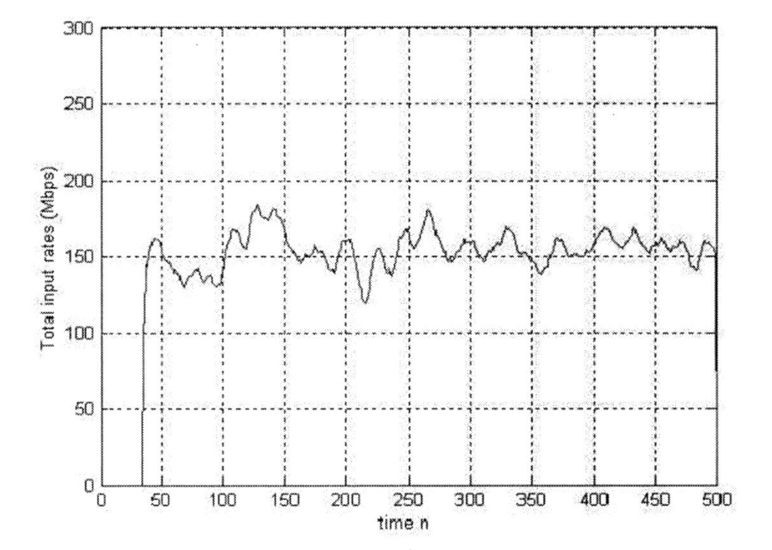

Fig. 12 The total input rates of $L = 14$.

buffer occupancy has the value that begins from the time slot at $n = 36$. The neural predictive congestion control technique is also able to maintain the queue size close to the threshold of $1000Kb$, and the average relative error between the predicted buffer occupancy and the actual buffer occupancy is 3.7026e-002. Figure 11 shows the lowest priority traffic rate for $L = 14$ step prediction, and it is yielded on the basis of the original flow equation (1) and the predicted buffer occupancy, and Figure 12 shows total input rate prediction.

The performance of the system is excellent for queue service rates. However, the performance is found to be better in the 4-step prediction than in 14-step prediction case. This is probably due to the fact that the less prediction horizon usually leads to better accuracy, whereas in our scheme the predictive horizon is linked with the forward path delay τ_1 .

To compare our algorithm with the conventional approaches like in [8-9], the following remarks can be given.

(i)This chapter introduces a new congestion control model based on neural network. The BP network model and algorithm develop the ideas and methods in [8-9].

(ii)The quicker transient response of the source rates is acquired in our mechanism. Under the same circumstance, we use less predictive steps than that in [8-9], because in this chapter the neural predictive controller is located at the sources rather than at the switch, this usually brings forth the better performance in terms of prediction accuracy.

(iii)The authors of [14 -17] suggest that only one implicit layer is enough, and it could be randomly mapped into R^m space. With the same number of the implicit layer node, the algorithm will be more efficient if there are less layers. So the implicit layer of BP algorithm in this chapter has just one layer and it could improve study efficiency with reasonable study accuracy.

(iv)We have explored the relevant theory on BPNN multi-step predictive architecture and training algorithm, and give relevant simulation analysis.

5 Conclusion

This chapter has described a dynamic resource management mechanism for computer communication networks on the basis of an adapting BP neural network control technique. Also we further explored the relevant theoretic foundations as well as the detailed implementation procedure for congestion control. The simulation results demonstrate that the proposed neural network architecture and training algorithm are excellent from the point of view of the system response, predictive accuracy and efficiency, and that it well adapts the data flows to the dynamic conditions in the data transfer process. We believe that the neural network predictive mechanism provides a sound scheme for congestion control in communication networks.

Areas for further research would cover, for example, the issue of congestion control for multicast communication systems by using the neural network predictive method to deal with the challenge of low responsiveness, which is due to the heterogeneous multicast tree structure.

Acknowledgment

This research has been supported by the US National Science Foundation CAREER Award under Grant No. CCF-0545667. We would like to thank many colleagues and anonymous reviewers for their constructive criticism and helpful suggestions for improving the overall quality of this chapter.

References

1. C. Q. Yang, A. A. S.Reddy (1995) A taxonomy for congestion control algorithms in packet switching networks. IEEE Network Magazine, Vol. 9, No.5, pp.34 - 45.
2. S. Keshav (1991) A control-theoretic approach to flow control, in: Proceedings of ACM SIGCOMM'91, Vol. 21, No. 4, pp.3-15.
3. D. Cavendish (1995) Proportional rate-based congestion control under long propagation delay, International Journal of Communication Systems, Vol. 8, pp. 79-89.
4. R. Jain, S. Kalyanaraman, S. Fahmy, R. Goyal (1996) Source behavior for ATM ABR traffic management: an explanation, IEEE Communication Magazine, Vol. 34, No. 11, pp. 50-57.
5. Rose Qingyang Hu and David W. Petr (2000) A Predictive Self-Tuning Fuzzy-Logic Feedback Rate Controller, IEEE/ACM Transactions on Networking, Vol. 8, No. 6, pp. 689 - 696.
6. Giuseppe Ascia, Vincenzo Catania, and Daniela Panno (2002) An efficient buffer management policy based on an integrated Fuzzy-GA approach, IEEE INFOCOM 2002, New York, No.107.
7. G. Ascia, V. Catania, G. Ficili and D. Panno (2001) A Fuzzy Buffer Management Scheme for ATM and IP Networks, IEEE INFOCOM 2001, Anchorage, Alaska, April 22-26, 2001, pp.1539-1547.
8. J. Aweya, D.Y. Montuno, Qi-jun Zhang and L. Orozco-Barbosa (2000) Multi-step Neural Predictive Techniques for Congestion Control -Part 2: Control Procedures, International Journal of Parallel and Distributed Systems and Networks, Vol. 3, No. 3, pp. 139-143.
9. J. Aweya, D.Y. Montuno, Qi-jun Zhang and L. Orozco-Barbosa (2000) Multi-step Neural Predictive Techniques for Congestion Control -Part 1: Prediction and Control Models, International Journal of Parallel and Distributed Systems and Networks, Vol. 3, No. 1, pp. 1-8.
10. L. Benmohamed and S. M. Meerkov (1993) Feedback Control of Congestion in Packet Switching Networks: The Case of Single Congested Node, IEEE/ACM Transaction on Networking, Vol. 1, No. 6, pp. 693-708.
11. J. Filipiak (1988) Modeling and Control of Dynamic Flows in Communication Networks, Springer Verlag Hardcover, New York.
12. S. Jagannathan, and G. Galan (2003) A one-layer neural network controller with preprocessed inputs for autonomous underwater vehicles, IEEE Trans. on Vehicular Technology, Vo. 52, no. 5.
13. D. H. Wang, N. K. Lee and T. S. Dillon (2003) Extraction and Optimization of Fuzzy Protein Sequence Classification Rules Using GRBF Neural Networks, Neural Information Processing - Letters and Reviews, Vol.1, No.1, pp. 53-59.
14. R. Yu and D. H. Wang (2003) Further study on structural properties of LTI singular systems under output feedback, Automatica, Vol.39, pp.685-692.
15. S. Jagannathan and J. Talluri (2002) Adaptive Predictive congestion control of High-Speed Networks, IEEE Transactions on Broadcasting, Vol.48, no.2, pp.129-139.
16. Simon Haykin (1998) Neural Networks: A Comprehensive Foundation ,(2nd Edition), Prentice Hall, New York, July 6, 1998.

17. F. Scarselli and A C Tsoi (1998) Universal Approximation Using FNN: A Survey of Some Existing Methods and Some New Results, Neural Networks, Vol. 11, pp. 15-37.
18. J. Alan Bivens, Boleslaw K. Szymanski, Mark J. Embrechts (2002) Network congestion arbitration and source problem prediction using neural networks, Smart Engineering System Design, vol. 4, NO. 243-252.
19. S. Jagannathan (2001) Control of a class of nonlinear systems using multilayered neural networks, IEEE Transactions on Neural Networks, Vol.12, No. 5.
20. P. Darbyshire and D.H. Wang (2003) Learning to Survive: Increased Learning Rates by Communication in a Multi-agent System, The 16th Australian Joint Conference on Artificial Intelligence (AI'03), Perth, Australia.
21. Lin, W. W. K., M. T. W. Ip, et al. (2001) A Neural Network Based Proactive Buffer Control Approach for Better Reliability and Performance for Object-based Internet Applications, International Conference on Parallel and Distributed Processing Techniques and Applications (PDPTA 2001), Las Vegas, Nevada, USA, CSREA Press.
22. S. Dobson, S. Denazis, A. Fernĺéndez, D. Gaiti, E. Gelenbe, F. Massacci, P. Nixon, F. Saffre, N. Schmidt, F. Zambonelli (2006) A survey of autonomic communications, ACM Transactions on Autonomous and Adaptive Systems (TAAS), Vol. 1 , No. 2, pp. 223 - 259.
23. Jeffrey O. Kephart , David M. Chess (2003) The Vision of Autonomic Computing, Computer, vol. 36, no. 1, pp. 41-50, January 2003.
24. N. Laoutaris, O. Telelis, V. Zissimopoulos, I. Stavrakakis (2006) Distributed Selfish Replication, IEEE Transactions on Parallel and Distributed Systems, vol. 17, no. 12, pp. 1401-1413.
25. G. Acampora, M. Gaeta, V. Loia, and Athanasios V.Vasilakos (2009) Ubiquitous Findability of Fuzzy Services for Ambient Intelligence Applications, ACM Transactions on Autonomous and Adaptive Systems (TAAS), to appear.
26. Athanasios V. Vasilakos, W. Pedrycz (2006) Ambient Intelligence, Wireless Networking and Ubiquitous Computing, ArtechHouse, MA, USA.
27. N. Xiong, A. V. Vasilakos, L. T. Yang, L. Song, P. Yi, R. Kannan, and Y. Li. (2009) Comparative Analysis of Quality of Service and Memory Usage for Adaptive Failure Detectors in Healthcare Systems, IEEE Journal on Selected Areas in Communications (IEEE JSAC), to appear.
28. R. Quitadamo, F. Zambonelli (2008) Autonomic communication services: a new challenge for software agents, Autonomous Agents and Multi-Agent Systems, IEEE Transactions on automatic control, vol. 17, no. 3, pp. 457–475.
29. C. Park, D.J. Scheeres, V. Guibout, A. Bloch (2008) Global Solution for the Optimal Feedback Control of the Underactuated Heisenberg System, IEEE Transactions on automatic control, vol. 53, no. 11, pp. 2638-2642.
30. S. S. Ge, C. Yang, T. H. Lee (2008) Feedback-Linearization-Based Neural Adaptive Control for Unknown Nonaffine Nonlinear Discrete-Time Systems, IEEE Transactions on neural networks, vo. 19, no. 9, pp. 1599-1614.

Part II
Autonomic Communication Services and Middleware

Hovering Information – Self-Organizing Information that Finds its Own Storage

Alfredo A. Villalba Castro, Giovanna Di Marzo Serugendo, and Dimitri Konstantas

Abstract Hovering information is a mobile computing paradigm where pieces of self-organizing information are responsible to find their own storage on top of a dynamic set of mobile devices. Once deployed, the hovering information service acts as a location-based service for disseminating Geo-localized information generated by and aimed at mobile users. It supports a wide range of pervasive applications, from urban security to stigmergy-based systems. A piece of hovering information is attached to a geographical point, called the anchor location, and to its vicinity area, called the anchor area. A piece of hovering information is responsible for keeping itself alive, available and accessible to other devices within its anchor area. It does not rely on any central server. This chapter presents the hovering information model and results of simulations performed using replication and caching algorithms involving up to 200 distinct pieces of hovering information in a small geographic area.

1 Introduction

User generated content is taking a large part of the Internet with social networking web sites such as YouTube or MySpace. The equivalent of these sites for mobile users, such as the GyPSii[1] social networking web site, now combine both user-

---

Alfredo A. Villalba Castro
Centre Universitaire d'Informatique, University of Geneva, Battelle bâtiment A, 7 route de Drize, CH-1227 Carouge, Switzerland, e-mail: alfredo.villalba@unige.ch

Giovanna Di Marzo Serugendo
School of Computer Science and Information Systems, Birkbeck, University of London, Malet Street, London WC1E 7HX, UK, e-mail: dimarzo@dcs.bbk.ac.uk

Dimitri Konstantas
Centre Universitaire d'Informatique, University of Geneva, Battelle bâtiment A, 7 route de Drize, CH-1227 Carouge, Switzerland, e-mail: dimitri.konstantas@unige.ch

[1] www.gypsii.com

A.V. Vasilakos et al. (eds.), *Autonomic Communication*, DOI: 10.1007/978-0-387-09753-4_5,

generated content and location-based services. Among other, they allow groups of mobile users to share dynamic real-time content or retrieve themselves on a map.

Location-based services usually rely on base stations and from there possibly to the whole Internet to provide some requested information to a mobile user (e.g. what is the nearby Chinese restaurant or where is my friend's car for a user-generated content). This solution has clear advantages such as providing access to large computing capabilities and broadband network access that go beyond those of mobile phones or PDAs.

However, it is not always possible or desired to rely on a central server in particular for user-generated content: extra-terrestrial systems need local communication infrastructures, they cannot communicate with an Earth-based server, or if in a hostile environment cannot rely entirely on a single server; after a natural disaster, when no more infrastructure is available, local communications among available devices help coordination among emergency services; finally, for reliability reasons, it is not always possible to rely on a centralized server representing a single point of failure.

Hovering Information [15] is a concept characterizing self-organizing information responsible to find its own storage on top of a highly dynamic set of mobile devices. This is a location-aware service for mobile users (people, cars, robots, etc.) that supports dissemination of user-generated Geo-localized data among a highly mobile set of devices. This service exploits the mobile devices themselves as a physical support and do not make use of a server. The main requirement of a single piece of hovering information is to keep itself stored in the vicinity of some specified location, which we call the anchor location, despite the unreliability of the device on which it is stored. Whenever the mobile device, on which the hovering information is currently stored, leaves the area around the specified anchor location, the information has to hop - "hover" - to another device.

Current services supporting Geo-localized data, are deployed using one of the following approaches: centralized servers, virtual structured overlay network offering a stable virtual infrastructure, or direct communication among the mobile nodes themselves. In all these approaches, the mobile nodes decide when and to whom the information is to be sent. Here we take the opposite view; it is the information that decides upon its own storage and dissemination. This opens up other possibilities, not available for traditional MANET services, such as different pieces of hovering information all moving towards the same location and (re-)constructing there a coherent larger information for a user, e.g. TV or video streaming on mobile phones.

A piece of hovering information is a *self-organizing* user-defined piece of data which does not need a central server to exist. Individual pieces of hovering information each use local information, such as direction, position, power and storage capabilities of nearby mobile devices, in order to select the next appropriate location. Hovering information benefits from the storage space and communication capacities of the underlying mobile devices.

Main dependability requirements of hovering information are *survivability*, *availability* and *accessibility*. Survivability means that the information is alive somewhere in the environment (i.e. it is stored in some device) but not necessarily close to its anchor location. Availability means that the information has found storage in the vicinity of the anchor location. Accessibility combines both availability and communication range of wireless mobile devices, and represents the possibility for a user located in the anchor location to access hovering information stored on nearby devices.

The hovering information service requires the following: mobile nodes with computing capacity; direct wireless communications among mobile node such as Bluetooth or Wi-Fi; and a location tracking capability such as a GPS (mobile PDAs), relative distances calculations or light-color tracking (robots).

This chapter presents the hovering information model as well as replication and caching algorithms allowing multiple pieces of hovering information to get attracted to their respective anchor locations.

Section 2 discusses potential applications of this concept. Section 3 presents the hovering information concept and model. Section 4 discusses replication and caching algorithms, in particular the Attractor Point Algorithm that we have designed where the information is "attracted" by the anchor location and keeps coming back to this location, and the Location Based Caching algorithm aiming at reducing the number of hovering information stored in the different nodes, when memory is limited. Section 5 reports on simulation results involving up to 200 distinct pieces of hovering information. Finally Section 6 compares our approach to related works, and Section 7 discusses some future works.

2 Applications

This section highlights some future applications in very different areas that could all be developed from the concept of hovering information.

Urban Security The environment considered for this application is a dense urban area where each person carries a GPS enabled device. A hovering information service is available on the device, which allows users to enter comments or warnings related to dangers in the urban environment. Different types of information can be disseminated by the user: warnings about holes in the road or about the existence of thieves (pick-pockets); or comments like "this corridor is dark and I feel a danger". Each person entering the area where the information is kept will potentially receive it. Each user has a "profile" and chooses what types of dangers are relevant to her. For example a blind person will be interested in holes on the road; a weight lifter is not really concerned to be attacked by a thief, while an elderly will find these two pieces of information particularly relevant for her. Similarly, policeman and security guards operating in the same local urban area with high-rate crime could exchange information to each other. Each user attracts different information depending on his profile as soon as it enters the region where the information is located. For such an application the trust/security aspect becomes then crucial, since any GPS owner (including a criminal) may enter fraudulent information. Although we are also working on trust and security issues related to hovering information, this discussion is beyond the scope of this chapter.

Archaeological Sites The environment in this application is a real archaeological site, most likely an outdoor site, rather large, where visitors just move freely inside it. Users are the visitors of the site. Fixed sensors placed at several locations on the archaeological site provide either current information on actual weather and temperature or historical information about the different locations in the archaeological site. Users visiting the archaeological site wear mobile devices carrying information specific to the user itself (e.g. adult, child, man, woman, teacher, etc) or more likely about the virtual character they want to learn more about. The user receives on his mobile device a virtual reconstructed version of the site as it would be on the day they are visiting: with the same weather conditions, targeted in content to the character they wanted to learn more about, and populated with the other characters that are currently visiting the same location. For instance, consider a group of 3 people (characters) visiting a house in an archaeological site: a cook, a child, and a house's owner. All of them would have a virtual view of how the house looked like at that time. If it is sunny then the view shows a sunny area, if it is cold it could show heating aspects. Each visitor would have a specific tailored explanation (cooking, playing, and ownerŠs information) and could visualize the avatars of the others on his mobile device while moving around. There are different types of hovering information going around: visitor-dependent information, weather information, and archaeological/historical information of a specific location in the archaeological site. Personalized information is attracted by the corresponding user's device, aggregates there and shows some virtual view of the site (audio only or both audio/video).

Self-Generative Art Self-generative art [8] refers to art practice where inputs from the creator of the piece of art are assembled together according to some rules (algorithm), such that the resulting piece of art, generated by a computer, is a real-time unfolding work which may display randomness, evolutionary aspects, or self-organizing (swarm) behavior. The piece of art may be music, painting, 3D construction, writing, etc. In this case, the users/creators would then be the multiple visitors of a "learning art experience center". A piece of art could be a large scale 3D virtual shape produced by inputs provided by each visitor: location in the experience area, weight/height and behavior (jumping/walking), preferred color or shape. The virtual shape would have holes were people are currently placed, and bumps where they have left; or heart beating bumps if they are jumping. Rules for combining the different inputs could vary: assembling the virtual surface according to actual or relative distances in the real world, summing up the colors and weights according to different algorithms, keeping visitors input for a random amount of time after they have left the experience area, etc. Sensors are required to determine the weight/size of the person. This value will then have an impact on the final virtual surface. Heavier people or groups of people will provide heavier holes, etc. People moving across the surface create temporary paths across it (that would dissolve completely after a certain time). People carry mobile device through which they provide additional personalized information:

preferred color and shape. Each input provided by the different visitors is a piece of hovering information whose goal is to travel to the center of the experience area and aggregate there with the others in order to provide a visual 3D shape.

Intravehicular Networks Virtual tags are inserted at specific locations on roads or motorways either by cars' drivers or traffic management staff. The purpose here is to provide information to cars' drivers about road conditions, accidents, etc. In such a scenario, using the notion of hovering information, such tags will not be stored on a specific server and made available to users when they reach the zone of interest of the information. Instead the tags are locally stored in the cars and made available through wireless channels to nearby cars. Since data have a meaning for the specific location they have been attributed, data will have to "change" car as soon as the car they are currently stored in leaves the area of the anchor location. The data will then hop from one car to the next one.

Emergency Scenarios In an emergency scenario, virtual data present before a disaster may want to "survive" by using emergency crew or survivors devices. This data can also present useful information for emergency services. Additionally, disaster's survivors may want to indicate their position by placing the appropriate hovering information attaching it to their own location. Emergency crew member can place hovering information to areas where survivors have been found or where there is a chance to find some survivors. In this case, the information will hop from one emergency/survivor device to another one.

Stigmergy Stigmergy is an indirect communication mechanism among individual components of a self-organizing system. Communication occurs through modification brought to local environment. The use of ant pheromone is a well known example of stigmergy. Users that communicate by placing hovering information at a Geo-referenced position, which is later on retrieved by other users is also an example of stigmergy. The hovering information concept, using an infrastructure free storage media, naturally supports stigmergy-based applications that need to be deployed on an ad hoc manner (e.g. unmanned vehicles or robots). As pointed out by [13] most stigmergy-based deployed systems use a server to support diffusion of artificial pheromone, they do not actually attach the pheromone to physical supports. Hovering information provides a way to store digital pheromone among a group of robots or unmanned vehicles using the robots themselves as physical support for the pheromone. Robots, producing pheromone that needs to be deposited at a certain geographical area, will deposit it under the form of a piece of hovering information. It will then stay located where it has been produced by hovering among the robots present in this area making it available for those robots wishing to retrieve it.

3 Hovering Information Concept

This section formally defines the notion of hovering information system, as well as the three main dependability requirements: survivability, availability and accessibility.

3.1 Coordinates, Distances and Areas

We denote by $\mathbb{E}$ the set of all pairs of geographic coordinates,

$$\mathbb{E} = [-90, 90] \times [-180, 180].$$

A *geographic coordinate* a is a pair:

$$a = (lat, long), \text{ and } a \in \mathbb{E}.$$

North latitude and East longitude are positive coordinates, while South latitude and West longitudes are negative coordinates. We do not consider here depths and heights.

An *area* $A(a, r)$ is defined as the disk whose center is the geographic coordinate a and has a positive *radius* $r \in \mathbb{R}^+$:

$$A(a, r) = \{b \in \mathbb{E} \mid dist(a, b) < r\}.$$

We consider $dist(a, b)$ to be the distance in meters between two locations on a sphere, provided by any reliable method. See for instance[2].

3.2 Mobile Nodes

Mobile nodes represent the storage and motion media exploited by pieces of hovering information. They are defined as follows. A *mobile node* n is a tuple:

$$n = (id, loc, speed, dir, r_{comm}),$$

where:

[2] http://www.fcc.gov/mb/audio/bickel/distance.html

$$\begin{aligned} id &\in \mathcal{I} \text{ is a mobile node identifier,} \\ loc &\in \mathbb{E} \text{ is a geographic coordinate location,} \\ speed &\in \mathbb{R}^{+} \text{ is a speed in } m/s \\ dir &\in \mathbb{E} \text{ is a relative geographic coordinate location,} \\ r_{comm} &\in \mathbb{R}^{+} \text{ is the communication radius in meters.} \end{aligned}$$

We denote by $\mathcal{I}$ the set of all mobile nodes identifiers. When referring to the *id*, *loc* or other field of a mobile node n, we will use the following notation $id(n)$, $loc(n)$, etc. Field $loc(n)$ represents the current location of node n, while $dir(n)$ is a vector representing the direction of its most recent movement. The range of communication r_{comm} is the maximum distance in meters within which the mobile node may communicate wirelessly with another mobile node.

Let's consider $\mathcal{N}$ a set of mobile nodes, we consider that identifiers of mobile nodes are unique, and we will say that $\mathcal{N}$ is *well defined* if:

$$\forall n_1, n_2 \in \mathcal{N}, (id(n_1) = id(n_2)) \Rightarrow (n_1 = n_2).$$

Given a mobile node n with location $loc(n)$ and communication radius $r_{comm}(n)$, the *communication area* of n, $A_C(n)$, is the subset of $\mathbb{E}$ given by:

$$A_C(n) = A(loc(n), r_{comm}(n)).$$

Figure 1 shows three mobiles nodes, m, n, and p. While the communication range of m is enough to let it be in range of both n and p, the communication range of n and p being much smaller prevents them to be directly in range.

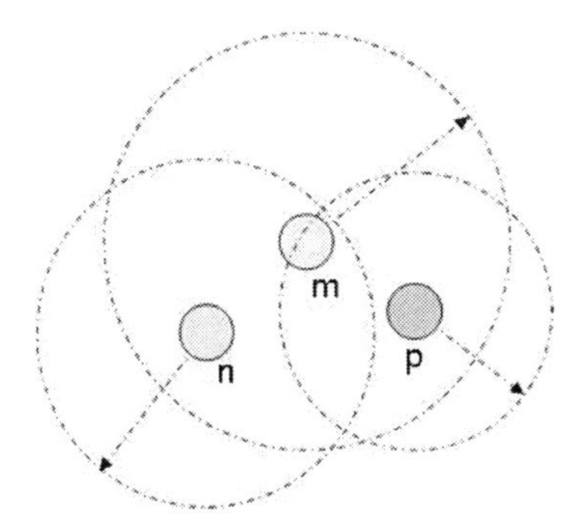

Fig. 1 Mobile Nodes and Communication Range

3.3 Hovering Information

Let $\mathcal{N}$ be a well defined set of mobile nodes. A *piece of hovering information h* is a tuple:

$$h = (id, a, r, n, data, policies, size),$$

where:

$$\begin{aligned} id &\in \mathcal{J} \text{ is a hovering information identifier,} \\ a &\in \mathbb{E} \text{ is the anchor location,} \\ r &\in \mathbb{R}^{+} \text{ is the anchor radius,} \\ n &\in \mathcal{N} \text{ is the mobile node where } h \text{ is currently located,} \\ data &\text{ is the data carried by } h, \\ policies &\text{ are the hovering policies of } h, \\ size &\in \mathbb{N}^{+} \text{ is the size of } h \text{ in bytes.} \end{aligned}$$

We denote by $\mathcal{J}$ the set of all hovering information identifiers. When referring to the *id*, *a* or other field of a hovering information *h*, we will use the following notation *id*(*h*), *a*(*h*), etc. Policies stand for hovering policies stating how and when a piece of hovering information has to hover. The size is an important element of a single piece of hovering information; however the simulation algorithms presented in this chapter are not yet using this notion.

A piece of hovering information *h* is a piece of data whose main goal is to remain stored in an area centered at a specific location called the *anchor location* $a(h)$, and having a radius $r(h)$, called the *anchor radius*.

The *anchor area* of *h*, $A_H(h)$, is the disk whose center is the anchor location $a(h)$ and whose radius is $r(h)$:

$$A_H(h) = A(a(h), r(h)).$$

Let $\mathcal{H}$ be a set of pieces of hovering information. We consider that identifiers of pieces of hovering information are unique, but replicas (carrying same data and anchor information) are allowed on different mobile nodes, and we will say that $\mathcal{H}$ is *well defined* if:

$$\begin{aligned} &\forall h_1, h_2 \in \mathcal{H}, (h_1 \neq h_2) \Rightarrow \\ &\quad (id(h_1) \neq id(h_2)) \vee \\ &\quad ((id(h_1) = id(h_2)) \wedge (a(h_1) = a(h_2)) \wedge \\ &\quad (r(h_1) = r(h_2)) \wedge (data(h_1) = data(h_2)) \wedge \\ &\quad (n(h_1) \neq n(h_2))). \end{aligned}$$

Let $\mathcal{H}$ be a well defined set of pieces of hovering information. Let $h \in \mathcal{H}$ be a hovering information, a *replica* h_r of *h* is a piece of hovering information $h_r \in \mathcal{H}$ such that:

$$id(h) = id(h_r) \wedge n(h) \neq n(h_r).$$

From now on, we will consider only well defined sets $\mathcal{H}$ of pieces of hovering information, where pieces of hovering information with the same *id* are either the same or a replica of each other. We also consider that there is only one instance of a hovering information in a given node n, any other replica resides in another node.

Figure 2 shows a piece of hovering information (blue hexagon) and two mobile nodes (yellow circles). One of them hosts the hovering information whose anchor location, radius and area are also represented (blue circle). The communication range of the second mobile node is also showed.

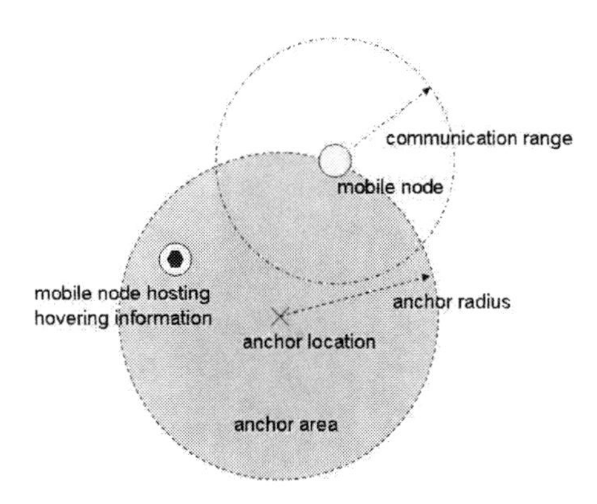

Fig. 2 Mobile Nodes and Hovering Information

Definition 1 (Hovering Information System at time t**).** A hovering information system at time t, $HoverInfo_t$, is a tuple:

$$HoverInfo_t = (\mathcal{N}_t, \mathcal{H}_t),$$

where $\mathcal{N}_t$ is a well defined set of mobile nodes, $\mathcal{H}_t$ is a well defined set of hovering information over $\mathcal{N}_t$:

$$\forall h \in \mathcal{H}_t \Rightarrow n(h) \in \mathcal{N}_t.$$

A hovering information system at time t is a snapshot (at time t) of the status of the system. Mobile nodes can change location, new mobile nodes can join the system, others can leave. New pieces of hovering information can appear (with new identifiers), replicas may appear or disappear (same identifiers but located on other nodes), hovering information may disappear or change node.

Figure 3 shows two different pieces of hovering information h_1 (blue) and h_2 (green), having each a different anchor location and area. Three replicas of h_1 are currently located in the anchor area (nodes n_2, n_3 and n_4), while two replicas of h_2 are present in the anchor area of h_2 (nodes n_2 and n_5). It may happen that a mobile device hosts replicas of different pieces of hovering information, as it is the case in the figure for the mobile node n_2 that is at the intersection of the two anchor areas.

The arrows here also represent the communication range possibilities among the nodes.

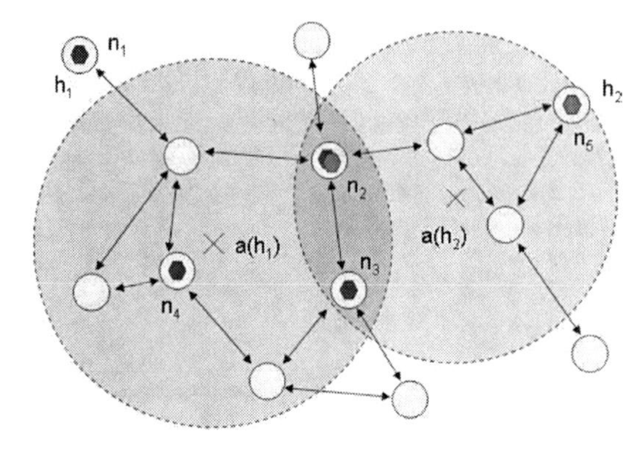

Fig. 3 Hovering Information System at time t

3.4 Notations

Before defining the notions of survivability, availability and accessibility, we will define the following additional notations.

Let's consider $HoverInfo_t = (\mathcal{N}_t, \mathcal{H}_t)$, a Hovering Information System at time t, let h be a piece of hovering information, and $n \in \mathcal{N}_t$ be a mobile node at time t, we denote:

$$R_H(h,t) = \{k \in \mathcal{H}_t \mid id(h) = id(k)\},$$

the set of replicas of a piece of hovering information h at time t;

$$R_N(n,t) = \{h \in \mathcal{H}_t \mid n(h) = n\},$$

the set of pieces of hovering information in node n at time t;

$$P_N(n,t) = loc(n), \text{the position of node } n \text{ at time } t;$$

$$N_N(n,t) = \{m \in \mathcal{N}_t \mid (dist(loc(m), loc(n)) < r_{comm}(n)) \vee (dist(loc(m), loc(n)) < r_{comm}(m)),\}$$

the set of neighboring nodes of node n at time t;

$$S(X) = \text{the surface area of a surface } X.$$

Since $\mathcal{H}_t$ is well defined, there are no two different pieces of hovering information (with different data) referred by the same identifier. It is also important to notice that h does not necessarily belong to $\mathcal{H}_t$, it may have disappeared from the system, but some of its replicas are still located in some mobile nodes. We consider that

$R_N(n,t)$ is actually a set (not a multi-set), i.e. there are no copies of the same hovering information stored at the same location. The neighboring nodes in $N_N(n,t)$ are those in range of communication to n.

3.5 Properties - Requirements

3.5.1 Survivability

A hovering information is alive at some time t if there is at least one node hosting a replica of this information.

Definition 2 (Survivability of Hovering Information h at time t). Let $HoverInfo_t = (\mathcal{N}_t, \mathcal{H}_t)$ be a Hovering Information System at time t. Let h be a piece of hovering information, the survivability of h at time t is given by the boolean value:

$$sv_H(h,t) = \begin{cases} 1 & \text{if } \exists n \in \mathcal{N}_t, R_H(h,t) \cap R_N(n,t) \neq \emptyset \\ 0 & \text{otherwise.} \end{cases}$$

The survivability along a period of time is defined as the ratio between the amount of time during which the hovering information has been alive and the overall duration of the observation.

Definition 3 (Rate of Survivability of Hovering Information h at time t). Let h be a piece of hovering information, the survivability of h between time t_c (creation time of h) and time t is given by:

$$SV_H(h,t) = \frac{1}{t - t_c} \sum_{\tau = t_c}^{t} sv_H(h,\tau).$$

3.5.2 Availability

A hovering information is available at some time t if there is at least a node in its anchor area hosting a replica of this information.

Definition 4 (Availability of Hovering Information h at time t). Let $HoverInfo_t = (\mathcal{N}_t, \mathcal{H}_t)$ be a Hovering Information System at time t. Let h be a piece of hovering information, the availability of h at time t is given by:

$$av_H(h,t) = \begin{cases} 1 & \text{if } \exists n \in \mathcal{N}_t, (P_N(n,t) \in A_H(h)) \wedge (R_H(h,t) \cap R_N(n,t) \neq \emptyset) \\ 0 & \text{otherwise.} \end{cases}$$

The availability of a piece of hovering information along a period of time is defined as the rate between the amount of time along which this information has been available during this period and the overall time.

Definition 5 (Rate of Availability of Hovering Information *h* at time *t*). Let h be a piece of hovering information, the availability of h between time t_c (creation time of h) and time t is given by:

$$AV_H(h,t) = \frac{1}{t - t_c} \sum_{\tau=t_c}^{t} av_H(h,\tau).$$

3.5.3 Accessibility

We distinguish availability from accessibility in the following way: a piece of hovering information (or one of its replica) present on some node located in the anchor area is said to be available. However, such a piece of hovering information may not be accessible to a mobile which is far apart from the mobile node where the hovering information (or its replica) is actually stored.

A hovering information is accessible by a node n at some time t if the node is able to get this information. In other words, if it exists a node m being in the communication range of the interested node n and which contains a replica of the piece of hovering information.

Definition 6 (Accessibility of Hovering Information *h* for node *n* at time *t*). Let $HoverInfo_t = (\mathcal{N}_t, \mathcal{H}_t)$ be a Hovering Information System at time t. Let h be a piece of hovering information, let $n \in \mathcal{N}_t$ be a mobile node, the accessibility of h for n at time t is given by:

$$ac_H(h,n,t) = \begin{cases} 1 & \text{if } \exists m \in \mathcal{N}_t, (m \in N_N(n,t)) \wedge (R_H(h,t) \cap R_N(m,t) \neq \emptyset) \\ 0 & \text{otherwise.} \end{cases}$$

We also define the accessibility of a piece of hovering information as the rate between the covered area by the hovering information's replicas and its anchor area.

Definition 7 (Accessibility of Hovering Information *h* at time *t*). Let $HoverInfo_t = (\mathcal{N}_t, \mathcal{H}_t)$ be a Hovering Information System at time t. Let h be a piece of hovering information, the accessibility of h at time t is given by:

$$ac_H(h,t) = \frac{S(\bigcup_{r \in R_H(h,t)} A_C(n(r)) \cap A_H(h))}{S(A_H(h))},$$

where $S(X)$ denotes the surface of X.

The accessibility along a period of time is defined as the average of the accessibility through that period of time.

Definition 8 (Rate of Accessibility of Hovering Information *h* at time *t*). Let h be a piece of hovering information, the accessibility of h between time t_c (creation time of h) and time t is given by:

$$AC_H(h,t) = \frac{1}{t-t_c} \sum_{\tau=t_c}^{t} ac_H(h,\tau).$$

Let us notice that an available piece of hovering information is not necessarily accessible and vice-versa, an accessible piece of hovering information is not necessary available. Figure 4 shows different cases of survivability, availability and accessibility. In Figure 4(a), hovering information h (blue) is not available, since it is not physically present in the anchor area, however it is survival as there is a node hosting it. In Figure 4(b), hovering information h is now available as it is within its anchor area, however it is not accessible from node n_1 because of the scope of the communication range. Finally, in Figure 4(c), hovering information h is survival, available and accessible from node n_1.

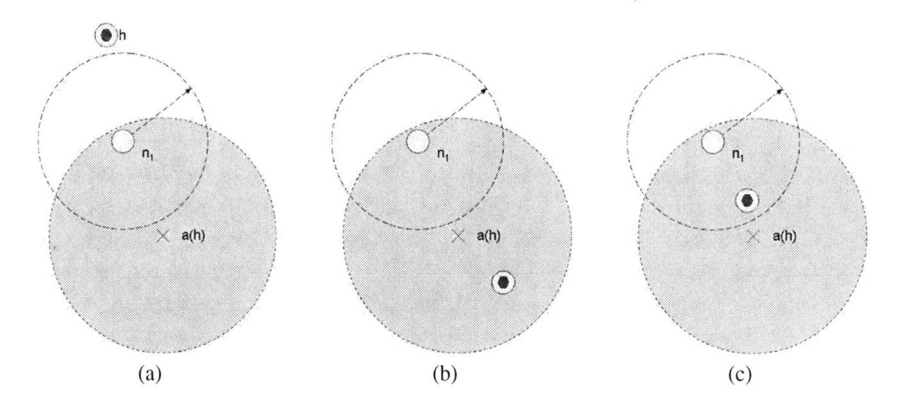

Fig. 4 Survivability, Availability and Accessibility

It is thus important to distinguish availability from accessibility: a piece of hovering information may be available (i.e. present) in the anchor area, but due to actual communication ranges among the nodes, it is not necessarily accessible for all nodes into the anchor area.

Similarly, it is interesting to note that in some situations a piece of hovering information even though not available at its anchor location could be accessible for some nodes provided there is a node in communications range hosting h.

4 Algorithms for Hovering Information

Survivability, availability, and accessibility are among the most fundamental issues of hovering information as we discussed in [15] and [3]. Security and trust issues are important issues when considering hovering information, however they go beyond the scope of this chapter, and will not be discussed here. Survivability addresses the problem of keeping a piece of hovering information alive as long as defined by the information itself. Availability deals with the problem of keeping the information present in its anchor area while accessibility relates to the possibility for a user to access a piece of hovering information stored on a device which is in communication range.

As mentioned in the previous section, these notions are closely related to each other, but none of them necessarily implies the others.

This chapter focuses on the study of the survivability and availability of pieces of hovering information. We propose an *Attractor Point* algorithm, whose aim is to keep the hovering information alive and available in its anchor area as long as possible. An anchor location a acts as an attractor point: all pieces of hovering information that have a as anchor location tend to converge towards a.

Besides the Attractor Point algorithm, we describe a *Broadcast* algorithm which is expected to have better survivability and availability performances than the Attractor Point, but at the cost of being more memory and network greedy. We use the Broadcast algorithm as a comparison threshold. Pieces of hovering information periodically broadcast (replicate) themselves to all the nodes in the communication range.

Mobile nodes have a limited memory and so cannot store an infinite number of hovering information replicas. We study two different caching policies: *Location-Based Caching* and *Generation-Based Caching*. The Location-Based Caching policy decides whether to remove or keep a replica on the basis of the current position of the node (or the replica), its proximity to the anchor location, and the portion of the anchor area covered by the communication area of the node. The Generation-Based Caching policy takes the decision of removing a replica based on the generation of the replica, removing those replicas that have been replicated most.

4.1 Assumptions

We make the following assumptions in order to keep the problem simple while focusing on measuring availability and resource consumption.

Limited memory   All mobile nodes have a limited amount of memory able to store hovering information replicas. The proposed algorithms take into account the remaining memory space.

Uniform size   All pieces of hovering information have the same size and the caching algorithms do not take in consideration the size as a criteria when removing a replica.

Unlimited energy   All mobile nodes have an unlimited amount of energy. The proposed algorithms do not consider failure of nodes or impossibility of sending messages because of low level of energy.

Instantaneous processing   Processing time of the algorithms in a mobile node is zero. We do not consider performance problems related to overloaded processors or execution time.

In-built Geo-localization service Mobile nodes have an in-built geo-localization service such as GPS which provides the current position. We assume that this information is available to pieces of hovering information.

Velocity vector service Mobile nodes have an in-built velocity vector service providing the instantaneous speed and direction of the node. We assume that this information is available to pieces of hovering information.

Neighbors discovering service Mobile nodes are able to get a list of their current neighboring nodes at any time. This list contains the position, speed, and direction of the nodes. As for the other two services, this information is available to pieces of hovering information.

4.2 Safe, Risk and Relevant Areas

Hovering policies are attached to pieces of hovering information. We consider here that all pieces of hovering information have the same hovering policies: active replication and hovering in order to stay in the anchor area (for availability and accessibility reasons), hovering and caching when too far from the anchor area (survivability), and cleaning when too far from the anchor area to be meaningful (i.e. disappearance). The decision on whether to replicate itself or to hover depends on the current position of the mobile device in which the hovering information is currently stored.

Given an anchor area $A(a,r)$, the *safe area* $A(a,r_{safe})$ is the disk whose center is the anchor location a and whose radius is the *safe radius* r_{safe}, a positive radius smaller than the anchor radius r, i.e. $r_{safe} < r$ and $r_{safe} \in \mathbb{R}^+$:

$$A(a,r_{safe}) = \{b \in \mathbb{E} \,|\, dist(a,b) < r_{safe}\}.$$

A piece of hovering information located in the safe area can safely stay in the current mobile node, provided the conditions on the node permit this: power, memory, etc.

Given an anchor area $A(a,r)$, a risk area is a ring centered at the anchor location, which overlaps with the anchor area and is limited by the safe area.

The *risk area* $R(a,r_{safe},r_{risk})$ is the ring given by;

$$R(a,r_{safe},r_{risk}) = A(a,r_{risk}) \backslash A(a,r_{safe}),$$

where $r_{safe} < r < r_{risk}$ and $r_{risk}, r_{safe} \in \mathbb{R}^+$.

A piece of hovering information located in the risk area should actively seek a new location on a mobile node going into the direction of the safe area. It is in this area that the hovering information actively replicates itself in order to stay available and in the vicinity of the anchor location.

The *relevant area* limits the scope of survivability of a piece of hovering information. The relevant area $A(a,r_{rel})$ is the disk whose center is the anchor location a and whose radius is the relevant radius r_{rel} bigger than the risk radius:

$$A(a, r_{rel}) = \{b \in \mathbb{E} \mid dist(a,b) < r_{rel}\},$$

where $r_{risk} < r_{rel}$, and $r_{rel} \in \mathbb{R}^+$.

The ring area $A(a, r_{rel}) \backslash A(a, r_{risk})$ represents the area where the hovering information seeks to survive but does not actively replicate itself (in order to avoid flooding). It may come back to the anchor area through mobile devices going in the direction of the anchor area.

The *irrelevant area* is all the area $U(a, r)$, outside the relevant area, it is given by:

$$U(a, r_{rel}) = \mathbb{E} \backslash A(a, r_{rel}).$$

A piece of hovering information located in the irrelevant area can disappear; it is relieved from survivability goals.

Figure 5 depicts the different types of radii and areas discussed above centered at a specific anchor location a. The smallest disk represents the safe area, the blue area is the anchor area, the ring limited by the risk radius and the safe radius is the risk area, and finally the larger disk is the relevant area.

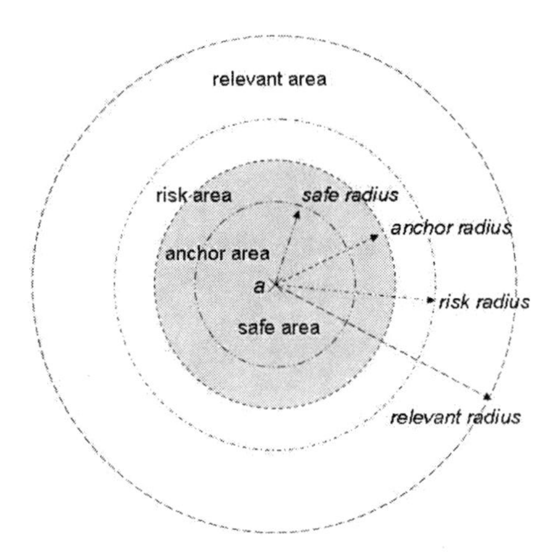

Fig. 5 Radii and Areas

The values of these different radii are different for each piece of hovering information and are typically stored in the Policies field of the hovering information. In the following algorithms we consider that all pieces of hovering information have the same relevant, risk and safe radius.

4.3 Replication

A piece of hovering information h has to replicate itself onto other nodes in order to stay alive, available and accessible. We describe two such replication algorithms for simulating two variants of these policies: the Attractor Point algorithm (AP) and the Broadcast-Based algorithm (BB). Both algorithms are triggered periodically each T_R (replication time) seconds and only replicas of h being in the risk area are replicated onto some neighboring nodes (nodes in communication range) which are selected according to the replication algorithm.

4.3.1 Attractor Point Algorithm

The anchor location of a piece of hovering information acts constantly as an attractor point to that piece of hovering information and to all its replicas. Replicas tend to stay as close as possible to their anchor area by jumping from one mobile node to the other.

Algorithm 4: Attractor Point Replication Algorithm

```
1  begin
2      pos ← NodePosition();
3      N ← NodeNeighbours();
4      P ← NeighboursPosition(N);
5      foreach repl ∈ REPLICAS do
6          a ← AnchorLocation(repl);
7          dist ← Distance(pos, a);
8          if (r_safe ≤ dist ≤ r_risk) then
9              D ← Distance(P, a);
10             M ← SelectKrClosests(N, D, k_R);
11             Multicast(repl, M);
12 end
```

Periodically and for each mobile node (see Algorithm 4), the position of the mobile node (line 2) is retrieved together with the list and position of all mobile nodes in communication range (lines 3 and 4). Hovering information replicas verify whether they are in the risk area and need to be replicated (line 8). The number of target nodes composing the multicast group is defined by the constant k_R (replication factor). The distance between each mobile node in range and the anchor location is computed (line 9). The k_R mobile nodes with the shortest distance are chosen as the target nodes for the multicast (lines 10). A piece of hovering information in the risk area multicasts itself to the k_R mobile nodes that are in communication range and closest to its anchor location (line 11).

Figure 6 illustrates the behavior of the Attractor Point algorithm. Consider a piece of hovering information h in the risk area. It replicates itself onto the nodes in communication range that are the closest to its anchor location. For a replication factor $k_R = 2$, nodes n_2 and n_3 receive a replica, while all the other nodes in range do not receive any replica.

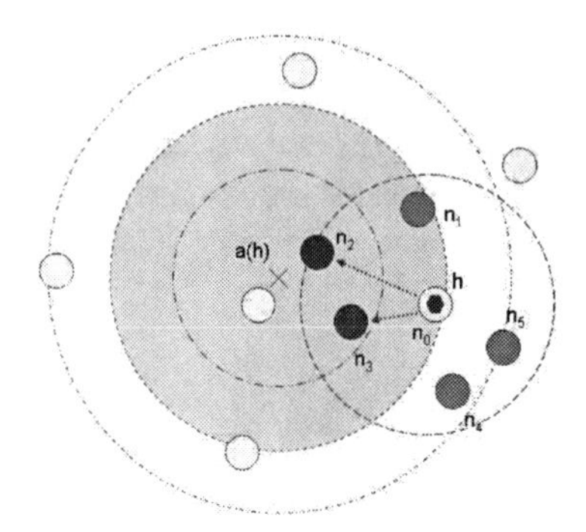

Fig. 6 Attractor Point Algorithm

4.3.2 Broadcast-Based Algorithm

The Broadcast-based algorithm (see Algorithm 5) is triggered periodically (each T_R) for each mobile node. After checking the position of the mobile node (line 2); pieces of hovering information located in the risk area (line 6) are replicated and broadcasted onto all the nodes in communication range (line 7). We expect this algorithm to have the best performance in terms of availability but the worst in terms of network and memory resource consumption.

Algorithm 5: Broadcast-Based Replication Algorithm

```
1 begin
2     pos ← NodePosition();
3     foreach repl ∈ REPLICAS do
4         a ← AnchorLocation(repl);
5         dist ← Distance(pos, a);
6         if (r_safe ≤ dist ≤ r_risk) then
7             Broadcast(repl);
8 end
```

Figure 7 illustrates the behavior of the Broadcast algorithm. Consider the piece of hovering information h in the risk area, it replicates itself onto all the nodes in communication range, nodes n_1 to n_5 (blue nodes).

4.4 Caching

In this chapter we assume that nodes have a limited amount of memory to store the pieces of hovering information (replicas). As the number of distinct hovering information increases, so will be the total number of replicas. The buffer of nodes will get full at some point and some replicas should have to be removed in order to store new ones.

We present two different caching policies. The first one, called the Location-Based Caching (LBC), decides whether to remove or keep a replica based on the current position of the node (or the replica), its proximity to the anchor location, and the portion of the anchor area covered by the communication area of the node. The second one, called the Generation-Based Caching (GBC), is based on the generation of replicas, the more a replica is old, the more it will have a tendency to disappear as the priority is given to younger replicas.

We compare these caching techniques with a simpler one which only ignores the incoming replicas as soon as there is no free space in the mobile device buffer.

Besides these caching algorithms, it is important to notice that we only consider the position and the generation of replicas. We do not take into consideration caching policies such as the priority, the time-to-live or the replicas size (since all replicas considered in this paper have the same size).

4.4.1 Location-Based Caching

At each node, this caching policy decides to remove a previously stored replica from the node's full buffer, and to replace it by the new incoming replica based on their respective location relevance value. We define the location relevance value of a replica, being this replica already stored in the node's buffer or being a new incoming replica, to its anchor location and area as it follows:

$$relevance = \alpha * area + \beta * proximity,$$

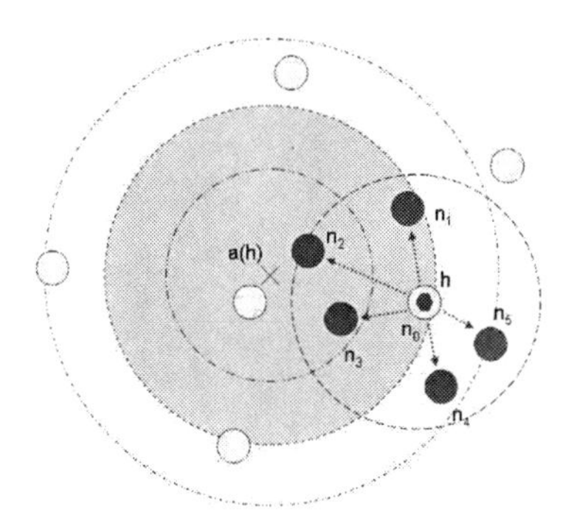

Fig. 7 Broadcast-Based Algorithm

where *area* is the normalized estimation of the overlapping area of the nodes' communication range area and the replica's anchor area, *proximity* is the normalized proximity value between the current position of the node and the anchor location of the replica, α and β are real coefficients having values between 0 and 1 and $\alpha + \beta = 1$.

Each time a new incoming replica arrives (see Algorithm 6), the least location relevant replica is chosen from all the replicas stored in buffer of the node (lines 2 to 10). The location relevance of the incoming replica is computed and compared to that of the least location relevant replica, whatever the original hovering information they refer to (lines 11 and 12). The least location relevant replicas is removed from the buffer and replaced by the incoming replica if the latter has a greater location relevance value (lines 12 to 14). Otherwise, the incoming replica is just discarded (line 16). In this way, the location-based caching algorithm will tend to remove replicas being too far from their anchor location or being hosted in a node covering only a small part of their anchor area.

The location relevance function (see Algorithm 7) computes the location relevance of a replica hosted in a node sitting in the anchor area of the replica. The distance between the location of the node and the anchor location of the replica are computed (lines 2 to 5). The overlapping area of the node's communication range area and the replica's anchor area is estimated (lines 6 to 13). Based on these two values, distance and estimated overlapping area, a normalized overlapping area and a normalized proximity values are computed (lines 14 and 15). Finally, the location relevance of the replica hosted in the node is computed using the previous formula with $\alpha = 0.8$ and $\beta = 0.2$ (lines 16 and 17). Figure 8 illustrates the notion of location relevance.

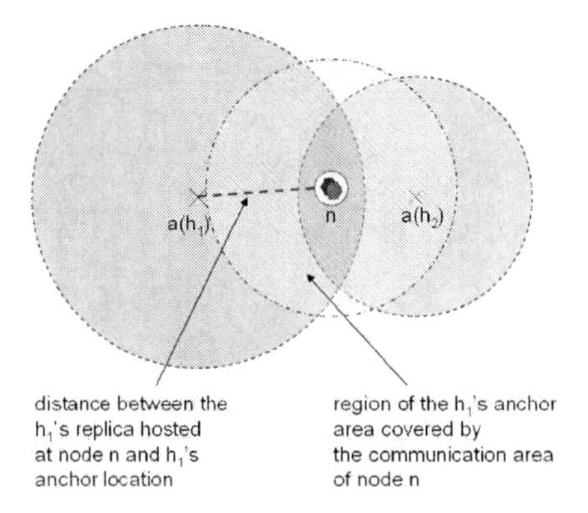

Fig. 8 Location-Based Caching Policy

4.4.2 Generation-Based Caching

We define the generation of a replica in the following way: the first replica created (normally by the user or user application) of a piece of hovering information has a generation 0, when this replica replicates itself then it creates new replicas having generation 1, and so on. The generation of a replica gives us an idea of the number of replicas existing as the process of replication follows an exponential growth. The generation-based caching algorithm tends to remove replicas having a high generation as there are likely more replicas leaving around than a replica having a lower generation.

Each time a new incoming replica arrives (see Algorithm 8), the oldest replica (the one having the highest generation value) is chosen from all the replicas stored in the buffer of the node (lines 2 to 10). The generation of the incoming replica is retrieved and compared to that of the oldest replica, whatever the original hovering information they refer to (lines 11 and 12). The oldest replica is removed from the buffer and replaced by the incoming replica if the latter has a smaller generation value (lines 12 to 14). Otherwise, the incoming replica is just discarded (line 16).

4.5 Cleaning

The cleaning algorithm periodically - each T_C (cleaning time) seconds - and for each node, removes the replicas that are too far from their anchor location, i.e. those replicas that are in the irrelevant area. This represents the cases where the replica

Algorithm 6: Location-Based Caching (LBC)

```
   input: replica
1  begin
2      repl_min ← null;
3      rele_min ← maxRelevanceValue;
4      foreach repl ∈ REPLICAS do
5          rele ← Relevance(repl);
6          if (rele ≤ rele_min) then
7              repl_min ← repl;
8              rele_min ← rele;
9      rele ← Relevance(replica);
10     if (rele > rele_min) then
11         Remove(repl_min, REPLICAS);
12         Insert(replica, REPLICAS);
13     else
14         Discard(replica)
15 end
```

Algorithm 7: Location Relevance Function

```
    input: replica
 1  begin
 2      a ← AnchorLocation(replica);
 3      r ← AnchorRadius(replica);
 4      pos ← NodePosition();
 5      dist ← Distance(pos, a);
 6      area ← 0;
 7      if (dist < (r_comm + r)) then
 8          if (dist > (r_comm − r)) then
 9              area ← ((r_comm + r) − dist)^2;
10          else
11              area ← (2 ∗ r)^2;
12      area ← area/(4 ∗ r^2);
13      proximity ← (e^(dist/100))^−1;
14      relevance ← 0.8 ∗ area + 0.2 ∗ proximity;
15      return relevance
16  end
```

considers itself too far from the anchor area and not able to come back anymore. This avoids as well the situation were all nodes have a replica.

Algorithm 8: Generation-Based Caching (GBC)

```
    input: replica
 1  begin
 2      repl_max ← null;
 3      gen_max ← minGenerationValue;
 4      foreach repl ∈ REPLICAS do
 5          gen ← Generation(repl);
 6          if (gen ≥ gen_max) then
 7              repl_max ← repl;
 8              gen_max ← gen;
 9      gen ← Generation(replica);
10      if (gen ≤ gen_max) then
11          Remove(repl_max, REPLICAS);
12          Insert(replica, REPLICAS);
13      else
14          Discard(replica);
15  end
```

5 Evaluation

We evaluated the behavior of the above described replication algorithms and caching policies under different scenarios by varying the number of nodes and the number of hovering informations.

For a single piece of hovering information, results reported in [16] show that the Attractor Point algorithm reaches availability levels of 80% when the number of nodes in the environment reaches 100 and 93% with 200 nodes. It is thus competitive to the Broadcast-Based algorithm while using much less memory and network traffic.

This Section reports results for multiple distinct pieces of hovering information co-existing at the same time. We measured the average survivability (Definition 3) and availability (Definition 5) as well as another performance metrics (cf. 5.2) such as the messages complexity, replication complexity, overflows, erased replicas and concentration .

We performed simulations using the OMNet++ network simulator (distribution 3.3) and its Mobility Framework 2.0p2 (mobility module) to simulate nodes having a simplified WiFi-enabled communication interfaces (not dealing with channel interferences) with a communication range of 121m.

5.1 Simulation Settings and Scenarios

The generic scenario consists of a surface of 500m x 500m with mobile nodes moving around following a Random Way Point mobility model with a speed varying from 1m/s to 10m/s without pause time. In this kind of mobility model, a node moves along a straight line with speed and direction changing randomly at some random time intervals.

In the generic scenario, pieces of hovering information have an anchor radius (r) of 50m, a safe radius (r_{safe}) of 30m, a risk radius (r_{riks}) of 70m, a relevance radius (r_{rel}) of 200m, and a replication factor of 4 (k_R).

Each node triggers the replication algorithm every 10 seconds (T_R) and the cleaning algorithm every 60 seconds (T_C). Each node has a buffer having a capacity to 20 different replicas. The caching algorithm is constantly listening for the arrival of new replicas. Table 1 summarizes these values.

Based on this generic scenario, we defined specific scenarios with varying number of nodes: from 20 to 200 nodes, increasing the number of nodes by 20; and varying number of different pieces of hovering information existing in the system: from 20 to 200 hoverinfos, increasing the number of pieces by 20. Each of this scenarios has been investigated with different replication algorithms and caching policies.

We have performed 20 runs for each of the above scenarios. One run lasts 3'600 simulated seconds. All the results presented here are the average of the 20 runs for each scenario, and the errors bars represent a 95% confidence interval. All the simulations ran on a Linux cluster of 32 computation nodes (Sun V60x dual Intel Xeon 2.8GHz, 2Gb RAM).

Blackboard	500mx500m
Mobility Model	Random Way Point
Nodes speed	1m/s to 10 m/s
Communication range (r_{comm})	121m
Buffer size	20 replicas
Replication time (T_R)	10s
Cleaning time (T_C)	60s
Replication factor (k_R)	4
Anchor radius (r)	50m
Safe radius (r_{safe})	30m
Risk radius (r_{risk})	70m
Relevant radius (r_{rele})	200m

Table 1 Simulation Settings

5.2 Metrics

In addition to the survivability and availability properties of a hovering information system, we also measured the performance metrics described below.

Messages Complexity. The messages complexity of a hovering information system at a given time t is defined as the total number of messages exchanged by nodes since the initial time. This metric provides as well a feasibility criterion and will serve as a basis to extrapolate actual implementation results, such as latency and overhead of a real system.

Replication Complexity. The replication complexity measures, for a given piece of hovering information h, the maximum number of replicas having existed in the whole system at the same time. In a real implementation this parameter will play an important role since we will prefer algorithms minimizing the replication complexity and maximizing the availability and survivability of hovering information.

Overflows. The overflows of a caching policy stands for the number of times that new incoming replicas have not found enough storage space in a node to be hosted. After an overflow happens, it is up to the caching policies to replace or not an existing stored replica by the new incoming one. We expect that the Location-Based Caching policy will generate less overflows than the other caching polcies.

Erased Replicas. The erased replicas of a caching policy represents the number of times that a node has had to remove a replica from its buffer to store a new incoming replica. The node takes this decision based on the caching policy. We also expect that the Location-Based Caching policy will erase less replicas than the other caching policies.

Concentration. The concentration of a given piece of hovering information h is defined as the rate between the number of replicas of h present in the anchor area and the total number of replicas of this hovering information in the whole environment. This metric shows how replicas are distributed around a geographical area.

5.3 Results

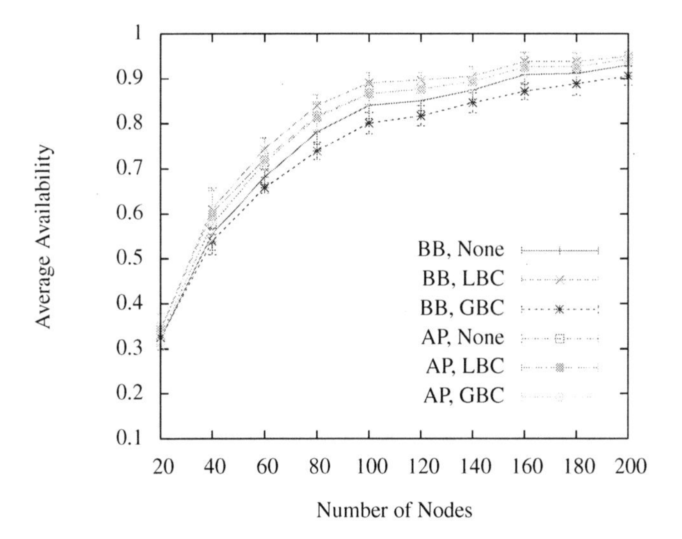

Fig. 9 Availability - 40 Hoverinfos

Figures 9, 10 and 11 show the average availability for the two replication algorithms: Broadcast-Based (BB) and Attractor Point (AP). Each algorithm uses three different caching policies: no caching (None), Location-Based Caching (LBC) and Generation-Based Caching (GBC). Each figure corresponds to a system containing 40, 120 and 200 pieces of hovering information. We observe that BB gets worse results than AP as the number of pieces of hovering information increases. Indeed, the BB tends to overload the system with an exponential growing number of replicas. As each node has a limited buffer size, the latter tend to get full and not all replicas can be accommodated within the buffer size and a large portion of them is discarded. On the other hand, AP manages to better administrate the buffer size producing less replicas which are stored in nodes closer to the anchor location of the replicas. The combination of the AP and LBC keep the information in its anchoring area in a much more optimal way than the other cases.

We also observe that algorithms using the LBC caching policy keep good levels of availability despite the number of nodes or the number of pieces of hovering

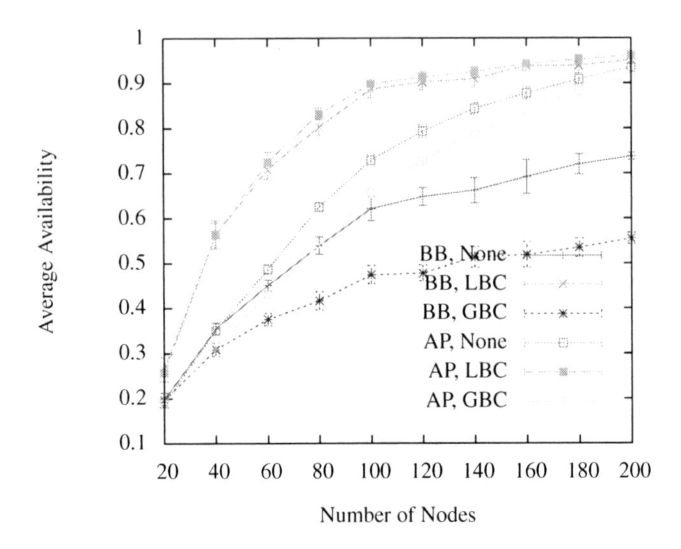

Fig. 10 Availability - 120 Hoverinfos

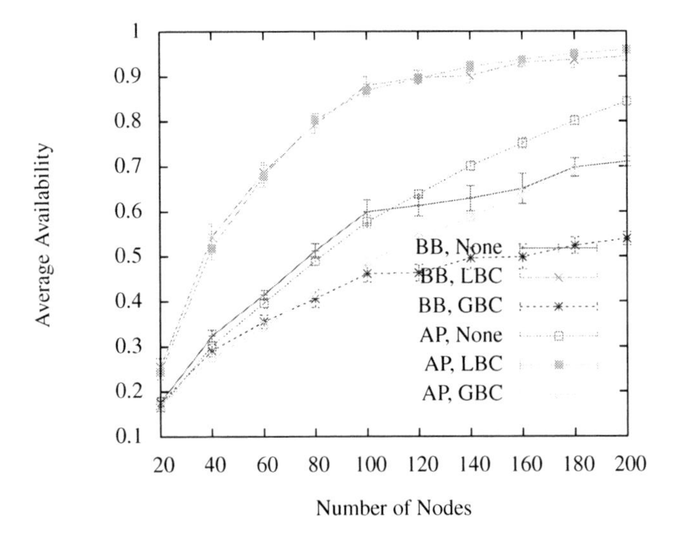

Fig. 11 Availability - 200 Hoverinfos

information, whereas it gets gradually worse for algorithms using the GBC or no caching policy.

Figure 12 depicts the average availability of the AP replication algorithm using the LBC caching policy for different numbers of nodes. For a number of nodes above 120, we notice that the availability is high enough (above 85%) and it keeps quite stable as the number of pieces of hovering information increases. We confirm from

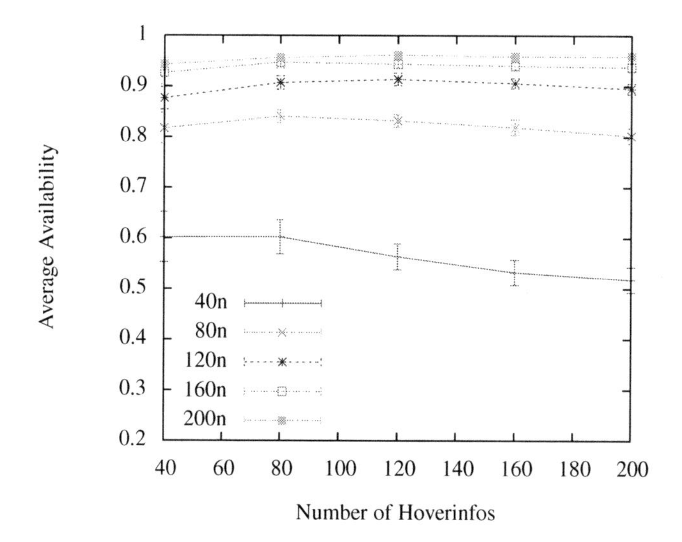

Fig. 12 Availability - Attractor Point with Location-Based Caching

this that the AP with LBC algorithms are scalable in terms of absorption of hovering information (number of distinct pieces of hovering information), since during the experiments with 120 nodes and more, up to 200 distinct hovering information pieces have been accommodated into the system with an availability above 85%. In the case when the number of nodes is 40, we can observe that the absorption limit, for this configuration, has been reached as the availability starts decreasing after 80 pieces of hovering information.

Figure 13 compares the survivability and availability for the AP algorithm in a system composed of 200 pieces of hovering information. As expected, the survivability is higher than the availability in all the cases. This proves the fact that an available piece of hovering information is survival but a survival one is not necessary available. We can also notice that these two metrics have the same shape (for the same algorithm) meaning that they are strongly related and consequently a piece of hovering information with a lot of chances to survive will have a lot of chances to keep itself available as well.

Figure 14 depicts the average number of overflows for the BB and AP using the three different caching policies: None, LBC and GC. We can observe that the BB produces around 10 times more overflows than the AP because of its exponential replication nature. We also see that the number of overflows for the AP tends to stabilize as the number of nodes grows. This is due to the controlled behavior of the AP algorithm that prevents exponential growth.

Figure 15 shows the average number of replicas erased from the nodesŠ buffer of nodes in order to store new incoming replicas. We observe that the AP algorithm erases much less replicas and in a more stabilized way than the BB algorithm. We also notice that the BB with GBC tends to erase replicas in an exponential way

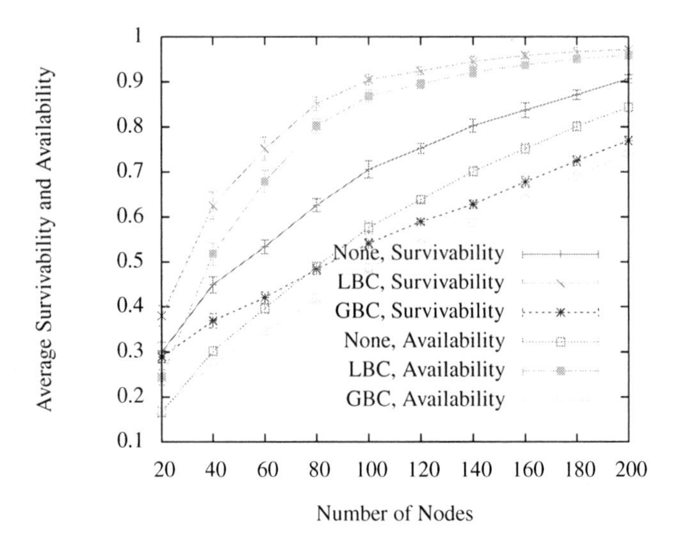

Fig. 13 Survavibility vs Availability - Attractor Point - 200 Hoverinfos

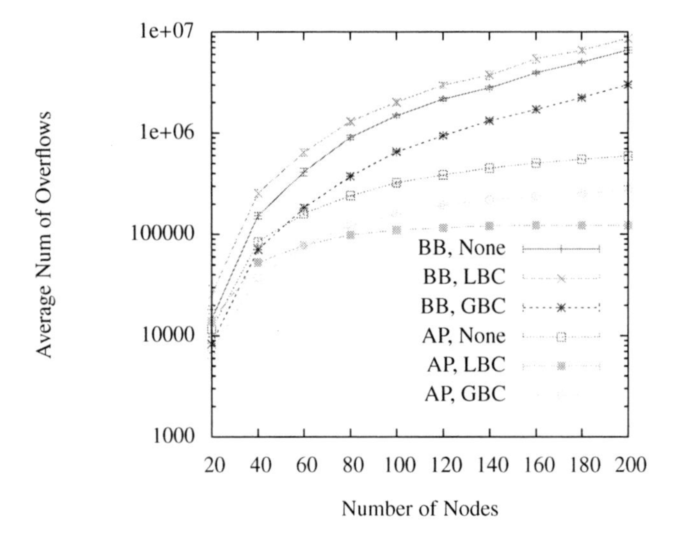

Fig. 14 Overflows - 200 Hoverinfos

which means that the generation-based caching policy combined with the exponential replication behavior of BB is not a good differentiation factor for caching replicas since this combination of algorithsm tends to insert and erase replicas permanently.

Figure 16 depicts the average number of messages sent by the different algorithms using the different caching policies. We notice that the algorithms using the

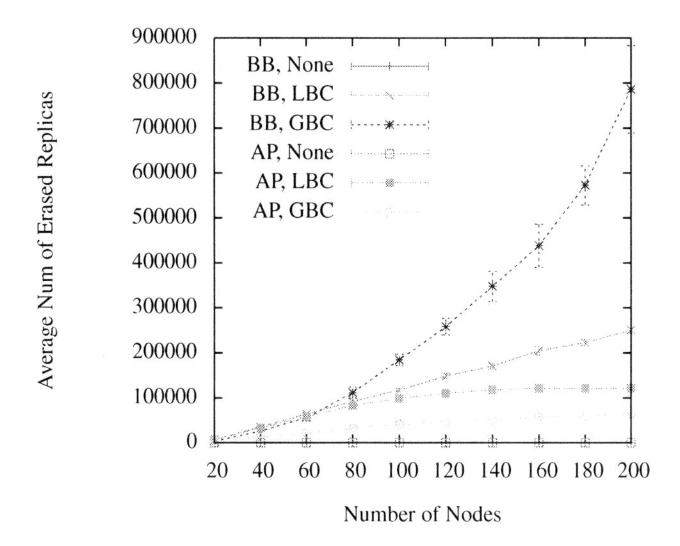

Fig. 15 Erased Replicas - 200 Hoverinfos

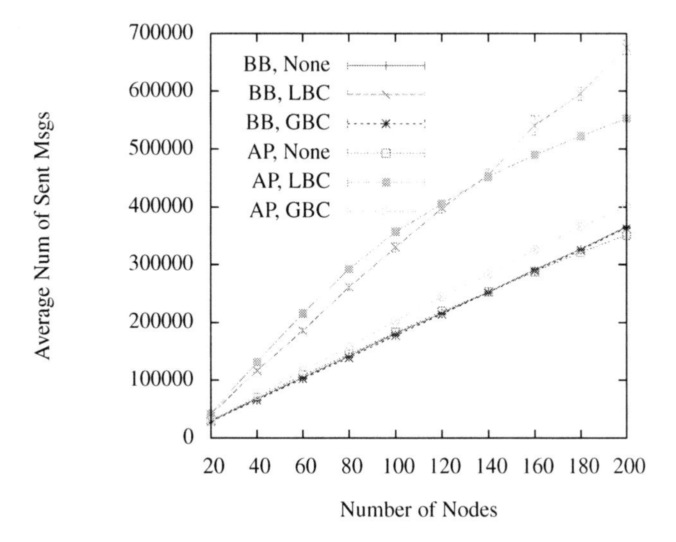

Fig. 16 Messages Complexity - 200 Hoverinfos

LBC caching policy generate more messages than the other cases. The reason of this behavior is the low availability performances for the algorithms not using the LBC caching in the presence of many pieces of hovering information. For the LBC algorithm, we can also observe that it tends to have slower growing gradient for the AP algorithm compared to that of the BB, which let us suppose that the messages complexity will get smaller as the number of nodes increases. This shows the scalability of AP with LBC algorithms in terms of network usage.

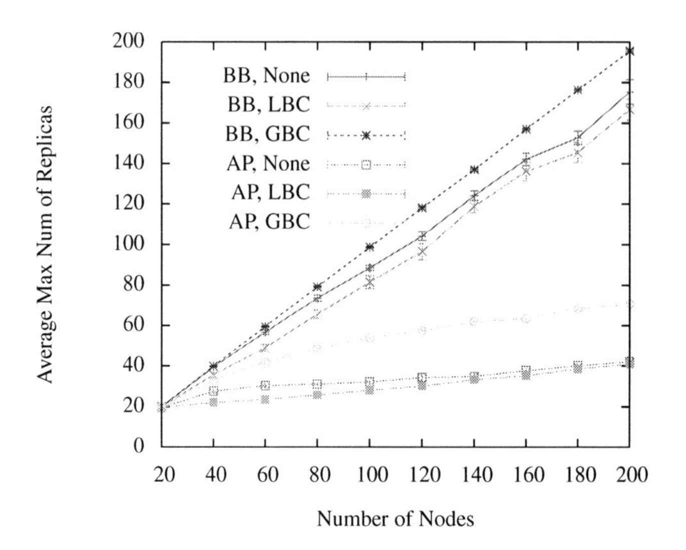

Fig. 17 Replication Complexity - 200 Hoverinfos

Figure 17 shows the replication complexity for all pieces of hovering information in the case of the AP with LBC. This confirms that the AP algorithm limits the number of replicas by concentrating them around the anchor area and not spreading them around all the system as the BB does. We also observe that the LBC caching policy tends to improve the convergence of replicas towards their anchor area.

Figure 18 shows the average of the maximal number of replicas having existed in the system for the AP using the LBC. It is interesting to see that it decreases as the number of pieces of hovering information increases. It means that the buffer resources are evenly shared among the different pieces of hovering information, while the availability still remains at high levels (see Figure 12). We conclude from this, that the AP with LBC succeeds to distribute the network resource in a fair way among all the pieces of hovering information, and that we probably observe an emergent load-balancing of the memory allocated to the different pieces of hovering information.

Finally, Figure 19 shows the average concentration for both algorithms using the different caching policies. We can observe that the LBC policy improves the concentration factor of both algorithms compared to the other caching policies. Particularly, the combination AP and LBC reaches a concentration factor of around 25%.

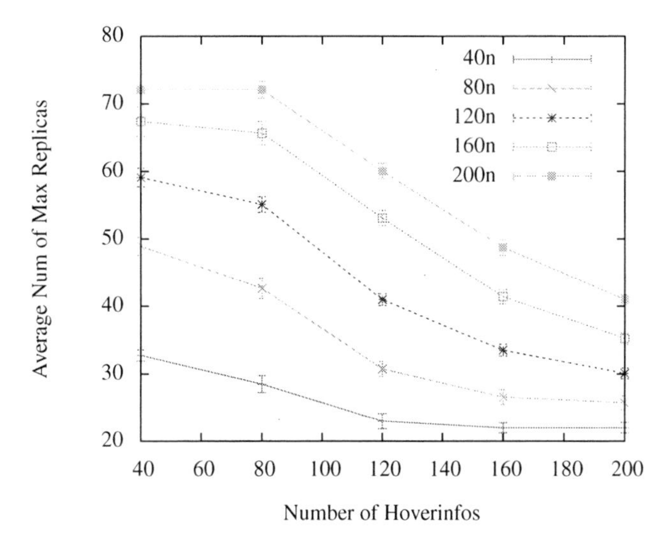

Fig. 18 Replication Complexity - Attractor Point with Location-Based Caching

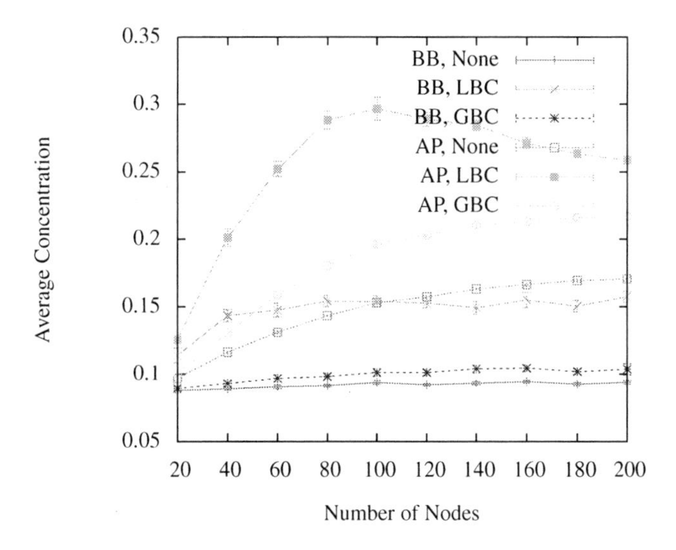

Fig. 19 Concentration - 200 Hoverinfos

6 Related Works

The Virtual Infrastructure project [4–7] defines virtual (fixed) nodes implemented on top of a MANET. This project proposes first the notion of an *atomic memory*, implemented on top of a MANET, using the notion of *quorums* or focal points where a reasonable amount of mobiles nodes intersect. Quorums work as atomic memory cells and ensure their persistency by replicating their state in neighboring mobile devices. This notion has been extended to the idea of *virtual mobile nodes* which are state machines having a fixed location or a well-defined trajectory and whose content is also replicated among the nearby mobile devices. Finally, this project provided the notion of a *timed I/O automaton mobile node* where virtual mobile nodes access a clock in order to perform real-time operations. The motivation behind this project is the development of a virtual infrastructure on top of which it will be easier to define or adapt distributed algorithms such as routing, leader election, atomic memory, motion coordination, etc. Hovering information shares similar characteristics, it tries to benefit from the mobility of the underlying nodes, but the goal is different. We intend to provide a hovering information service on top of which applications using self-organizing user-defined pieces of information can be built.

GeOpps [10] proposes a geographical opportunistic routing algorithm over VANETs (Vehicular Ad Hoc Networks). The algorithm selects appropriate cars for routing some information from a point A to a point B. The choice of the next hop (i.e. the next car) is based on the distance between that carŠs trajectory and the final destination of the information to route. The planned trajectory generated by a navigation system of the car is used when estimating the relevance of a car to route some information. This work focuses on routing information to some geographical location, it does not consider the issue of keeping this information alive at the destination, while this is the main characteristics of hovering information.

The work proposed by [11] aims to disseminate traffic information in a network composed by infostations and cars. The system follows the publish/subscribe paradigm. Once a publisher creates some information, a replica is created and propagated all around where the information is relevant. Cars having a replica periodically poll their neighboring cars, using a broadcast message, to know whether they are interested or not in the replica's information. If some cars reply in an affirmative way the information is sent to them. Based on these periodic polling, clusters are composed and replicas are removed or propagated to clusters where more subscribers and interested cars are situated. Replicas are also propagated to a randomly chosen car part of the cluster driving in the opposite direction to that of the current host in order to try to keep the information in its relevant area. Cars reply polling with their interests and also their direction. While the idea is quite similar to that of hovering information, keeping information alive in its relevant area, this study does not consider the problem of having a limited amount of memory to be shared by many pieces of information or the problem of fragmentation of information. It also takes the view of the cars as the main active entities, and not the opposite view, where it is the information that decides where to go.

The Ad-Loc project [1] proposes an annotation location-aware infrastructure-free system. Notes stick to an area of relevance which can grow depending on the location of interested nodes. Notes are kept in their relevance area is by periodically broadcasting location-aware information to neighboring nodes. This work also proposes to use this annotation system as a cache for Internet files in order to spare bandwidth. In this case, URLs are used as note identifiers. Similarly to the previous work, nodes are the active entities. In addition, in this case the size of the area of relevance grows as necessary in order to accommodate the needs of users potentially far from the central location. The information then becomes eventually available everywhere.

The ColBack system [2, 9] is part of the MoSAIC project and intends to set up a collaborative backup system for mobile devices. The two main issues the authors investigate are: fault- and intrusion-tolerant collaborative backup; and the self-carried reputation and rewards for collaboration. The environment consists of sporadically interconnected and mutually suspicious peer devices having no fixed infrastructure and access to trusted third parties. This system does not focus on Geo-localized information but replication strategies and replica scheduling and dissemination techniques could be used as inspiration for hovering information replication algorithms.

PeopleNet [14] describes a mobile wireless virtual social network which mimics the way people seek information via social networking. It uses the infrastructure to propagate queries of a given type to users in specific geographical locations called bazaars. Within each bazaar the query is further propagated between neighboring nodes via peer-to-peer connectivity until it finds a matching query. The proposed queries propagation inside bazaar techniques could be a source of inspiration when we will develop query to retrieve specific hovering information.

7 Conclusion

In this chapter we have defined the notion of hovering information in a formal way and we have defined and simulated the Attractor Point algorithm which intends to keep the information alive and available in its anchor area. This algorithm multicasts hovering information replicas to the nodes that are closer to the anchor location of the information. The performances of this algorithm have been compared to those of a Broadcast-Based algorithm which broadcasts replicas regardless of the proximity or not to the anchor location.

We have also defined and simulated two different caching polices, the Location-Based Caching and the Generation-Based Caching. Their performances have been compared under a scenario containing multiple pieces of hovering information and nodes having a limited amount of memory.

Results show that the Attractor Point algorithm with the Location-Based Caching policy is scalable in terms of the number of pieces of hovering information that the system can support (absorption limits). They also show the emergence of a load-balancing property of the buffer usage which stores replicas in an equilibrated and optimal way as the number of pieces of hovering information increases.

7.1 Future Works

Real mobility patterns We have tested the algorithms under a Random Way Point mobility model and under ideal wireless conditions. This is not characteristic of real world behavior. We are currently applying the different algorithms to scenarios following real mobility patterns (e.g. crowd mobility patterns in a shopping mall or traffic mobility patterns in a city) with real wireless conditions (e.g. channel interferences or physical obstacles).

Real wireless conditions The simulations performed have been done in an environment where there were neither wireless channel interferences nor physical obstacles. In real world scenarios, these two factors are inherent to wireless communications. It is thus very important to apply the Attractor Point algorithm in a more realistic environment taking in consideration these factors in order to measure the negative drawbacks on the availability performances.

Spatial Memory Service We are currently defining and implementing a distributed memory service, storing and retrieving pieces of hovering information, exploiting available (stationary and mobile) devices as the main storage medium.

Fragmentation and recombination (swarm) In this chapter we have considered atomic information only, but depending on the size of the hovering information (e.g. an image or even a video) it could be fragmented into smaller pieces to fit in multiple nodes' memory. A query of this information will require a recombination mechanism that recovers the different pieces and gets them reassembled to form the original information. We are currently considering this recombination process as a swarm of self-assembling information particles.

Movement speed and direction The current attractor point algorithm takes in consideration the position of the neighboring nodes only. A significant improvement will be achieved by taking into consideration the speed and direction of the nodes when choosing the nodes that will host replicas.

Coordinates precision The Attractor Point algorithm and the simulations performed do not consider issues related to the precision of Geo-localization service. Precision is an important factor in real world scenario, since it deeply affects the ideal communication range. In order to cope with reality, it is thus important to evaluate and adapt the algorithm in order to take into account problems related to precision

Other simulation environments Results vary when simulation environments change. In order to further validate and compare the results obtained using OMNet++, we will use other simulation environments such as Swarm [3] or attraction fields [12].

[3] http://www.swarm.org

References

1. Corbet, D.J., Cutting, D.: Ad loc: Location-based infrastructure-free annotation. In: ICMU 2006. London, England (2006). URL `citeseer.ist.psu.edu/759227.html`
2. Courtès, L., Killijian, M.O., Powell, D., Roy, M.: Sauvegarde coopérative entre pairs pour dispositifs mobiles. In: UbiMob '05: Proceedings of the 2nd French-speaking conference on Mobility and uibquity computing, pp. 97–104. ACM Press, New York, NY, USA (2005). DOI http://doi.acm.org/10.1145/1102613.1102635
3. Di Marzo Serugendo, G., Villalba, A., Konstantas, D.: Dependable requirements for hovering information. In: Supplemental Volume - The 37th Annual IEEE/IFIP International Conference on Dependable Systems and Networks (DSN'07), pp. 36–39 (2007)
4. Dolev, S., Gilbert, S., Lahiani, L., Lynch, N.A., Nolte, T.: Timed virtual stationary automata for mobile networks. In: OPODIS, pp. 130–145 (2005)
5. Dolev, S., Gilbert, S., Lynch, N.A., Schiller, E., Shvartsman, A.A., Welch, J.L.: Virtual mobile nodes for mobile ad hoc networks. In: DISC (2004)
6. Dolev, S., Gilbert, S., Lynch, N.A., Shvartsman, A.A., Welch, J.: Geoquorums: Implementing atomic memory in mobile ad hoc networks. In: DISC (2003). URL `citeseer.ist.psu.edu/dolev04geoquorums.html`
7. Dolev, S., Gilbert, S., Schiller, E., Shvartsman, A.A., Welch, J.: Autonomous virtual mobile nodes. In: DIALM-POMC '05: Proceedings of the 2005 joint workshop on Foundations of mobile computing, pp. 62–69. ACM Press, New York, NY, USA (2005). DOI http://doi.acm.org/10.1145/1080810.1080821
8. Galanter, P.: What is generative art? complexity theory as a context for art theory. In: Generative Art Conference (2003)
9. Killijian, M.O., Powell, D., Banâtre, M., Couderc, P., Roudier, Y.: Collaborative backup for dependable mobile applications. In: MPAC '04: Proceedings of the 2nd workshop on Middleware for pervasive and ad-hoc computing, pp. 146–149. ACM Press, New York, NY, USA (2004). DOI http://doi.acm.org/10.1145/1028509.1028517
10. Leontiadis, I., Mascolo, C.: Geopps: Opportunistic geographical routing for vehicular networks. In: Proceedings of the IEEE Workshop on Autonomic and Opportunistic Communications. (Colocated with WOWMOM07). IEEE Press, Helsinki, Finland (2007)
11. Leontiadis, I., Mascolo, C.: Opportunistic spatio-temporal dissemination system for vehicular networks. In: MobiOpp '07: Proceedings of the 1st international MobiSys workshop on Mobile opportunistic networking, pp. 39–46. ACM Press, New York, NY, USA (2007). DOI http://doi.acm.org/10.1145/1247694.1247702
12. Mamei, M., Zambonelli, F.: Field-Based Coordination for Pervasive Multiagent Systems. Springer Series on Agent Technology. Springer Verlag, Berlin (2005)
13. Mamei, M., Zambonelli, F.: Pervasive pheromone-based interaction with rfid tags. TAAS **2**(2) (2007)
14. Motani, M., Srinivasan, V., Nuggehalli, P.S.: Peoplenet: engineering a wireless virtual social network. In: MobiCom '05: Proceedings of the 11th annual international conference on Mobile computing and networking, pp. 243–257. ACM Press, New York, NY, USA (2005). DOI http://doi.acm.org/10.1145/1080829.1080855
15. Villalba, A., Konstantas, D.: Towards hovering information. In: Proceedings of the First European Conference on Smart Sensing and Context (EuroSSC 2006), pp. 161–166 (2006)
16. Villalba Castro, A., Di Marzo Serugendo, G., Konstantas, D.: Hovering information - self-organising information that finds its own storage. In: IEEE International Conference on Sensors, Ubiquitous and Trust Computing (SUTC'08) (2008)

The CASCADAS Framework for Autonomic Communications

Luciano Baresi, Antonio Di Ferdinando, Antonio Manzalini, and Franco Zambonelli

Abstract An interesting approach to the design and development of the future Internet foresees a networked service *eco-system* capable of seamlessly offering services for human-to-human, human-to-machine and machine-to-machine interactions.
This chapter builds in this direction by describing a distributed component-ware framework for autonomic and situation-aware communication developed within the CASCADAS project. The core of this framework is the *Autonomic Communication Element* (ACE), an innovative software abstraction capable of providing dynamically adaptable services that can be built, composed, and let evolve according to autonomic principles. Services are capable of adapting their logic to the dynamically changing context they operate in without human intervention. As a result, whenever the need arises, ACEs can be federated autonomously and produce new services on a situation-aware basis. Systems and, in particular, eco-systems can thus be conceived as collections of ACEs.
The chapter introduces the concept of ACE and its different facets. It also presents the architecture of a prototype ACE-based platform and exemplifies the different concepts through a future *Pervasive Behavioral Advertisement* scenario.

Luciano Baresi
Dipartimento di Elettronica e Informazione, Politecnico di Milano, e-mail: Baresi@elet.polimi.it

Antonio Di Ferdinando
Electrical and Electronic Engineering Department, Imperial College London, e-mail: A.Di-Ferdinando@imperial.ac.uk

Antonio Manzalini
Telecom Italia Lab, e-mail: Antonio.Manzalini@telecomitalia.it

Franco Zambonelli
Dipartimento di Scienze e Metodi dell'Ingegneria, Università di Modena e Reggio Emilia, e-mail: Franco.Zambonelli@unimore.it

A.V. Vasilakos et al. (eds.), *Autonomic Communication*, DOI: 10.1007/978-0-387-09753-4_6,

1 Introduction

Today's Internet is rapidly evolving towards a collection of highly distributed, pervasive, communication-intensive services [26]. In the next future, such services will be expected to (i) autonomously detect and organize the knowledge necessary to understand the context in which they operate, and (ii) self-adapt and self-configure, to exploit at their best any situation, to meet the needs of diverse users, in diverse situations, without explicit human intervention. These features will enable a wide range of new activities that are simply not possible or impractical now. However, the achievement of such capabilities requires a deep re-thinking of the current way of developing and deploying distributed systems and applications.

In this direction, a promising approach consists in conceiving services as part of an "ecology" within which they can prosper and thrive at the service of users (i.e., an *eco-system*). This vision is attractive because would provide better services to end-users while, at the same time, meeting the emerging economic urge for service provision and system management deriving by the higher level of dynamism and variability of communication systems. In addition, systems built in respect of this view are characterized by a flatter architecture, where services at Network and Transport level of the classical ISO OSI architecture (levels 3 and 4 respectively) are provided at the same level of application-oriented services, that is levels 5 to 7, and cross-layer interactions are a natural part of the ecology itself.

In this context, this chapter presents the CASCADAS Autonomic Service Framework (or the *Framework*, for short), capable of enabling the conception outlined above through the development of *autonomic applications*, that is, applications capable of dynamically adapting their plans to cope well with situations where the environment changes in uncertain ways [22]. The Framework represents the major outcome of the CASCADAS EU-IST project [1], and advances the field of autonomic communications with at least the two contributions of (i) providing a novel component model that facilitates the development of services as autonomic applications, and (ii) providing an environment that supports the evolution of such services in an autonomous fashion.

The Framework is conceived around a set of complementary key enabling features, namely *situation-awareness*, *semantic self-organization* and *self-similarity*, around which we believe any future communication services infrastructure should be conceived. The identification of these features, described in the following, starts from key state-of-the-art concepts in the area of modern distributed computing and communication systems, and tries to advance and generalize them to properly account for the specific characteristics of autonomic and situation-aware communication services. The features above are blended in a sound component model, which provides a robust and dynamic modular conceptual framework for building autonomic semantic services. This *Autonomic Component-ware* provides a high-level reference model for the production of a new generation of programmable communication elements that can be reused at different levels of the stack.

The Framework is centered around the fundamental abstraction of *Autonomic Communication Elements* (ACEs), and supports the vision of advanced autonomic ser-

vices as developed and deployed in the form of ACEs and networks of them. To this extent, the Framework provides an environment where basic services can be created and executed. This environment allows services to evolve autonomously, through the enhancement of the enabling features mentioned above, and according the local needs arising on a timely basis. Particular emphasis is put on the support for autonomously organizing services to compose more sophisticated ones: through this feature, in fact, it is possible to build systems (i.e. collections of ACEs) that "specialize" in a topic. Then, self-similarity allows these to be easily integrated into other systems recursively. This enhances modular flexibility in the Framework, which can be easily extended with components offering specific composite services.
The Framework is offered as a Java-oriented open source software development toolkit for situated autonomic communications (hereafter, the *Toolkit*), under *GNU General Public License* (GPL), and can be freely downloaded [34] through the Sourceforge website [15]. The rest of the chapter is structured as follows. Section 2 frames the problem of a framework for autonomic communications to set the motivations of our research. Section 3 introduces the CASCADAS Framework, its basic self-organization, self-awareness and self-similarity features, and describes the ACE component model. Sections 4 and 5 provide details on such basic features, while Sections 6 and 7 describe the way they have been used to equip the Framework with more sophisticated features. These sections exploit a *Pervasive Behavioral Advertisement* as running case study to better illustrate and exemplify the features introduced. Section 8 sketches the dynamics of a fully working ACE-based prototype eco-system for the case study. Finally, Section 9 concludes the chapter and outlines some future research directions.

2 Autonomic Communication Frameworks

In recent years, the body of research work in the area of autonomic systems has been growing [22], and a number of frameworks have been proposed. Each of these has different characteristics, with, for instance, some narrowing the target scenario to clustered services [5] or grid environments [23, 25, 31], while others designing containers through which non-functional properties can be injected and managed in legacy systems [4, 17, 28]. All of these, however, originate from the view of autonomic behaviour as a mean for reducing the management costs of complex IT systems achieved, in turn, by enhancing autonomous reconfiguration, optimization and management of network elements in the system [16].
This view, perfectly in line with IBMs *autonomic computing* initiative [24], implies autonomic behavior to be supported only at the level of computing resources. As a consequence, all of the frameworks above are (more or less) effective in enhancing network resources with self-configuration and self-management capabilities. However, none of them targets, nor foresees, the provision of "smart" services to the users, for which they all result ineffective. Quitadamo and Zambonelli [30] affirm that the main reason for this is the lack of a broader support for autonomic behavior

in the stack. This, in turn, motivates a broader approach known under the name of *autonomic communication*.
Autonomic communication takes the key motivations of autonomic computing and extends them to conceive the creation and provision of a new generation of autonomic services that can be made available, and usable, by end users. Dobson [16] define autonomic communication as "a more general thrust aimed at a deep foundational rethinking of communication, networking and distributed computing paradigms to face the increasing complexities and dynamics of modern network scenarios". Despite the evident similarities, the autonomic computing approach is significantly different from the autonomic communication one. In fact, while the former is more oriented towards the direct management of network resources, the latter is concerned with the provision of services and the management of resources at both infrastructure and user levels.
Let us take an example, in order to clarify. Let us consider a crowded venue, such as a museum, an airport, or a rail/metro station, with a number of public screens used to advertise events as well as third party commercial contents. As of today, the advertisement policy typically consists in a cyclical, sequential, display of content. This latter is typically preloaded, anticipately to the start of advertisement process, in a static fashion so that any modification aimed at either modifying the content itself or the sequence with which it is displayed is subject to explicit human intervention. More importantly, the advertisement process is carried out independently of the context it operates in, and therefore the content displayed on a screen remains the same regardless of the differences among people passing by. Exceptions can be eventually found in marginal cases, where the advertisement is based on timed priorities (i.e., higher priority to food advertisements when the time for a meal approaches). We call this category of advertisement techniques *audience-insensitive*, as they are insensitive to actual audience. Indeed, this highly static scenario results ideal to show the advantages that the introduction of autonomic technology might bring. However, as described below, these advantages change significantly according to the approach used.
According to its *manifesto* [24], an autonomic computing approach "would aim at facilitating configuration and management of the IT infrastructure to the extent of limiting, or even avoiding, explicit human intervention while also reducing the costs for its maintenance". As a result, enhancement of autonomic technology through the autonomic computing approach would produce an IT platform with interacting software components capable of reconfiguring and managing themselves in an autonomous fashion at runtime. However, no improvement would be seen at user level, where the service offered would be exactly the same. In particular, no enhancement would be observed in terms of autonomy of the service, which would still need explicit human intervention to, for instance, re-sequence and update contents to be shown.
On the other hand, a smart service might consider the presence of infrastructures such as wireless networks, RFID receptors, etc. These might enable access to pervasive services, such as for instance downloadable maps or events program for the venue, which can be accessed by personal mobile devices carried by the audience in

the venue. This might constitute an incentive for device holders to provide publicly accessible information (e.g. age slot, interests, gender, etc), which might be used to the extent of adapting the displayed contents on the basis of the peculiar interests of people detected. This would enhance effectiveness of the advertisement process, as the exposition of a product would be optimized towards people known to be interested (and thus more receptive), while enhancing the validity of the advertisers' business investment, since the system would be in the position to provide guarantees on the impact of the content. We call the category of strategies allowing for such enhancement *audience-sensitive*, as they are sensitive to the audience they are shown to. As a side remark, it is worth noting that audience-sensitive strategies would also provide an effective way of avoiding the display of inappropriate content to particularly sensitive audience, such as children, without repercussions on the actual cost of the infrastructure.
An autonomic communication approach would aim at providing all the necessary support for the construction of such a smart audience-sensitive service, and therefore its application has the capabilities to impact the scenario in such a way to provide a service whose benefits can be immediately made available to all actors involved. Unfortunately, to the best of our knowledge no framework fulfilling the characteristics of the autonomic communication area can currently be found in literature.
The considerations above constitute the rationale and leading motivation behind our work. The CASCADAS ACE Framework aims at filling the lack highlighted above by providing an environment, and all necessary tools, to build complex and highly dynamic applications in respect of the paradigms of situated autonomic communication. Central to the Framework is the concept of *Autonomic Communication Element* (ACE), an innovate software engineering abstraction that allows the construction, and supports, the provision of highly dynamic services through the development of autonomic applications. In the remainder of this chapter, we will use the terms 'application' and 'service' interchangeably.
With respect to our context, the scenario introduced above can be realized as a collection of ACEs, each of which providing basic services, autonomously evolving in the system according to changes in context they operate in, and the needs that arise locally at each ACE, in a way similar to living organisms in natural eco-systems do. A working prototype eco-system for such scenario, that we named *Pervasive Behavioural Advertisement* (PBA), has been fully developed. We will use its structure to better explain the characteristics of the Framework throughout this document, and finally detail the prototype.

3 CASCADAS Framework

The CASCADAS autonomic service framework foresees services to interact with service users according to motivations driven by the context the former ones operate in. Services are developed and deployed as according to the component-ware paradigm offered by the specifications of ACEs. Therefore, ACEs are effectively

system components, and in the remainder of this document we will use the terms 'component' and 'ACE' interchangeably.
ACEs are capable of autonomously interacting to the extent of providing the desired functionality in a situation-aware way with no, or very limited, configuration efforts. Interactions may be aimed at forming more sophisticated services than the ones currently available and, in this case, may involve the spontaneous formation of new components in the system through aggregation of the original ones. Indeed, enabling of these highly dynamic interactions requires the components to own a set of features, which we retain of paramount importance to enable the vision of ecosystems of services.
Self-organization is one of the key design principles underpinning the autonomic management of large populations, and is therefore an essential property for the Framework. **Semantic self-organization** aims at exploiting the potential of "classical" self-organization and self-aggregation as key enablers for service composition and aggregation, and allows the identification of self-organization models of semantic (i.e. cognitive) nature. In the Framework, semantic self-organization allows ACEs to identify and document local rule-sets through which they can aggregate and form more sophisticated services. This is achieved according to *clustering*, *differentiation*, and *synchronization* techniques aimed at autonomously reinforcing organizational links, differentiating according to resource management strategies and select ideal aggregation partners respectively. By doing so, the desired collective behavior can be promoted in groups of ACEs.
Situation-awareness takes from *context-awareness*, that is, the capability of sensing the environment and reacting accordingly [32, 33], and advances it with techniques to organize the amount of distributed information in proper, strongly distributed "knowledge networks" to support situated and adaptive service provisioning. This enables services to autonomously adapt their logic to the context from which they are requested and in which they execute.
In the Framework, this capability is offered through a dedicated service, accessible through aggregation, which allows components to obtain situation-aware information. Internally, this information is made available by a *Knowledge Network* (KN) [8], which is in charge of correlating small bits of local context-aware information, typically received through interaction with pervasive technology receptors, and organizing it into global knowledge. Other ACEs can, then, become situation-aware by aggregating to the KN *façade* ACE and observe the evolution of the system by receiving and interpreting the information made available through the aggregation.
Internally, the KN ACE is itself an aggregate of a number of ACEs, each dealing with isolated local views, which semantically self-organize in order to provide a global snapshot on the present situation. Aggregation of the ACEs involved in the KN service is made in full respect of self-similar constraints, and thus the ACE resulting from aggregation exports standard communication interfaces that hide its aggregate nature to the eyes of other ACEs, also facilitating further aggregations to other ACEs in the eco-system.
Self-similarity refers to the capability of individual components to self-organize and self-aggregate to reproduce nearly identical structures over multiple scales

[2, 14]. Besides scalability, self-similarity is indeed a key enabler for the composition of complex communication-intensive services and for the structuring of the possibly enormous and multi-faceted items they will have to exploit in the knowledge networks. The Framework supports self-similarity by imposing that ACEs formed as aggregation of other ACEs provide the same set of interfaces as the original ones, so that service invocations to aggregate ACEs do not need to be treated in a specific way by the invoker. Properties above represent fundamental building blocks for the Framework. These, in turn, have been used, combined and exploited in the process of equipping the Framework with more sophisticated self-* features.

- **Self-healing** refers to special-purpose management techniques aimed at detecting ACEs entering faulty states and finding corresponding reasons to restore the correct operation of the element. This principle takes from semantic self-organization and situation-awareness to build, configure, and drive a control infrastructure over interoperating ACEs. Again, self-healing materializes in the Framework as (logically) dedicated ACEs responsible for monitoring one or more ACEs to detect faulty states and undertake recovery actions, when needed.
- **Self-Preservation** is the natural instinct of preserving oneself from harm and destruction. In our context, this is translated into the capability of self-recognize the emergence of anomalies, and self-repair them, in a scenario populated by heterogeneous entities and characterized by the lack of a centralized organization. The approach towards these issues focuses on the provision of a framework for basic common security mechanisms, which can be aggregated in order to provide more complex mechanisms. In doing so, the **security** elements effectively work in synergy with the supervision structure above, which provides an ideal ground for enhancing *confidentiality*, *integrity*, *authentication*, and *non-repudiation* features in a distributed fashion. Then, more sophisticated services can be provided through structured aggregation, semantic self-organization, of these simple components.

3.1 ACE Component Model

The ACE (Autonomic Communication Element) forms the core of the Framework, and its component model enables the design of applications in a self-similar manner. Central to the component model is the concept of *organ*. An organ is an ACE internal operative component that, as the name suggests, behaves as an organ in the human body. Specifically, organs are able to harmonize their own behavior to the execution of other interacting organs and, more in general, to the context in which the ACE is operating. Each organ is responsible for a specific type of tasks, and the interaction among them allows the constitution of an ACE as a standalone component. As example, Figure 1 depicts the assembly of an ACE from a set of organs with clearly defined tasks that leads to a well-structured modularized component model.

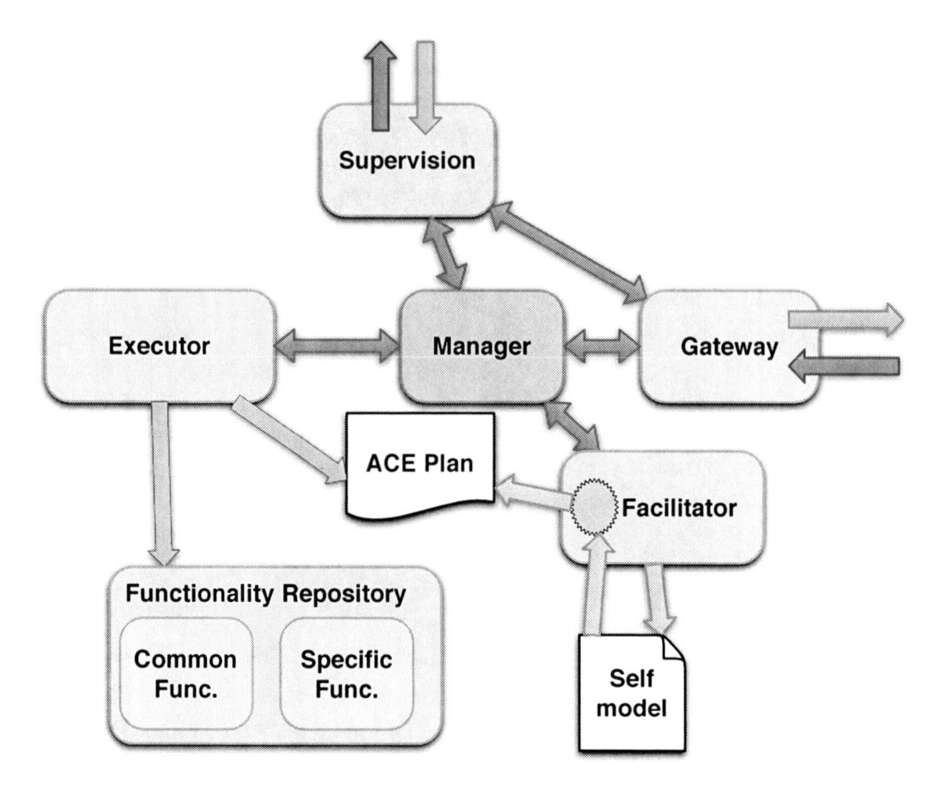

Fig. 1 ACE Component Model.

The *Gateway* organ is in charge of handling interactions with the external world. To this end, two different communication protocols are used: a connectionless protocol is used for initial service discovery through a publish-subscribe paradigm [18], supported by the *REconfigurable Dispatching System* [9] (REDS), under the name of *GN-GA* protocol [21]. On the other hand, a connection-oriented communication protocol is used for all other communications, where (one-to-one or one-to-many) communication channels are established through a contracting technique. The DIET [27] agent framework supports this communication.
The *Manager* organ is in charge of handling the internal communication among the organs and is responsible for the ACEs lifecycle management. With respect to this activity, any ACE must be in one of the four states of Figure 2: *inactive*, *running*, *prepared to move*, and *destroyed*. Different lifecycle actions are possible from these individual states.
The *Facilitator* is the organ that provides an ACE with capabilities for autonomously adapting its behavior to changes detected in the context. Adaptation is achieved through modification of the existing capabilities and/or addition of new ones. In more detail, ACEs are provided an original behaviour through a *self-model*, which defines the ACE's behavior with a number of *plans* each containing a set of states

along with rules to switch from one state to another. Plans can also include a set of *modification rules*, which allow to change the original behavior by modifying states and transitions in, or disabling, existing plans. In addition, modification rules might enable plans originally disabled or even create new plans. It is worth noting that modification rules are a major achievement in terms of situation-awareness, as in fact modifications are typically based on events occurring in the context the ACE is operative in. Semantic interpretation of plans, performed during initialization, leads to the creation of the original *ACE plan* that, in turn, allows creation of the ACE's initial behavior. During its operation, the Facilitator continuously evaluates the environmental conditions and the operations the ACE performs. Based on the outcome of the reasoning process, which are in turn influenced by the occurrence of contextual events, it subsequently performs, if needed, the specified modification actions on the basis of the peculiar events occurred. Changes made according to these, then, are submitted to the executor, described below, where they are made effective.
The *Executor* organ governs the evolution of the ACE according to the actions taken as per self-model. Its main role is to ensure that any reasoning and decision taken by the Facilitator is put in place in an effective and efficient way by ensuring that conditions are verified, actions are executed, and appropriate messages are exchanged.
Execution of plans may involve the use of specific capabilities provided by the

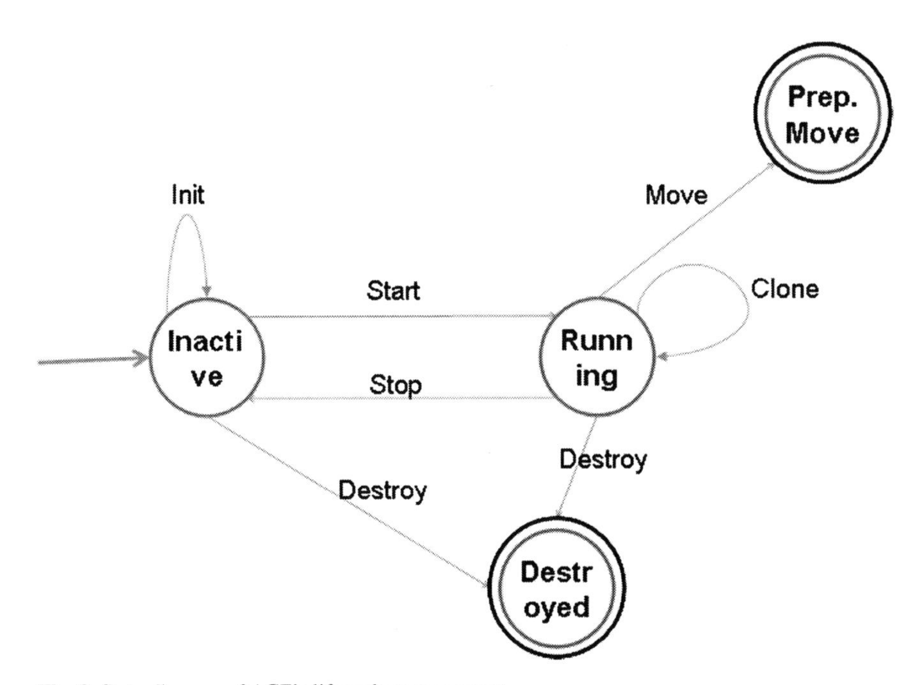

Fig. 2 State diagram of ACE's lifecycle management.

ACE. To this extent, the Executor may query the *Functionality Repository* organ to obtain them. The purpose of this repository is store the capabilities deployed.

ACEs provide services in form of functionality. These are stored in the repository, which is thus split into *Common* (i.e., guaranteed to be available in each ACE) and *Specific* (i.e., peculiar to a specific ACE) functionality. Common functionality is a major feature in terms of self-similarity. In fact, each and every ACE, regardless of its simple or aggregated nature, is capable of offering this set of capabilities albeit the set is strictly limited to the provision of basic operations. On the other hand, a service developer may create any arbitrary service and store it into the Specific Functionality Repository to equip the ACE the service is deployed in with more sophisticated and distinctive capabilities.

In addition to the organs described above, the original component model has been extended towards the provision of self-healing features through a dedicated organ aimed at interacting with a pervasive supervision framework. Activation of this organ, named *Supervision*, determines the "supervisable" nature of the ACE, and triggers monitoring activities at specific (configurable) points in the ACE aimed at verifying its own operative state. The structure of the pervasive supervision framework, as well as the description of ACEs self-healing features will be the subject of Section 6.

4 Semantic Self-Organization

The Framework supports semantic self-organization through algorithms for organized self-aggregation of ACEs in the form of autonomous *clustering*, *differentiation* and *synchronization*. The red wire among all the three self-aggregation techniques is that the choice of aggregation partners is subject to knowledge, and evaluation, of the context they are operating in, and that the self-organization will impact. This enables ACEs to conduct the aggregation according to a selection process of cognitive (i.e., semantic) nature, as deriving from the consideration of situation-aware information.

Clustering foresees an ACE initiating a "rewiring" procedure upon detection of a discrepancy in its list of required/available functionality (resulting from many different events such as the breaking of an existing collaboration link or a change in the local load due to a surge in demand). Depending on the circumstances, the initiator of the algorithm can choose one or more of its (contracted) first ACE neighbors as match-makers, and the constraint on the conservation of the total number of links can be relaxed or not. Simulation results [29] show that successful self-organization can take place, at a predictable rate, provided that well-identified conditions are met. The resulting aggregate ACEs will reflect the presence of durable complementariness between functions provided/encapsulated by individual ACEs (i.e., long-lasting GN/GA matches) and their ability to collectively identify and realize these functional clusters through local interactions. This can be generalized to any complex web of interdependencies, with individual ACEs potentially belonging to more than one functional cluster, and including "single type" aggregates designed for load-balancing rather than complementariness.

Differentiation allows ACEs to decide to "self-terminate" locally when facing an inappropriate workload. This automates the transfer of resources, between applications and according to fluctuations in the demand for a variety of co-hosted services, in a way that resources released can be re-assigned to other applications. The name *differentiation* derives from biological morphogenesis with which it shares some characteristics. Results from preliminary simulations [29] show that differentiation can help maximize the throughput of a distributed processing infrastructure while also making the system more responsive to heterogeneities in the workload when combined with the above "on demand" clustering. At the same time, results also emphasize that even the simplest set of rules tends to combine a large number of parameters to make interpretation difficult, therefore making selection of appropriate values a non-trivial problem. The algorithms integrated in the Framework enhance ACEs with capabilities to self-assess their own type according to the local observable workload, along with means for locally deciding when to change type.
Synchronization aims at finding or creating partnerships that adequately take into account the time activity pattern of individual constituting ACEs. This, in turn, involves (i) establishing a collaborative overlay that aggregates components featuring activity patterns that are *a priori* compatible starting from a random bootstrap configuration, and (ii) seeking ways of adjusting individual time-cycles so as to create opportunities for collaborations that would not exist if every individual activity pattern were "frozen" from the onset.
When the "rewiring" algorithm is employed, simulations show that the use of random time signatures allows the formation of a scale-free overlay when pruning from a complete graph and trying to secure a target number of active neighbors at any one time. In addition, they also show that distributed heuristics based on the "on-demand" clustering algorithm can be found so that near-optimal configurations obtained by pruning can be approximated.
In the economy of the Framework, and of the applications developed through it, self-organization is used as key enabler for composition of more sophisticated services, typically through aggregation of ACEs offering basic services. Furthermore, the semantic nature of the self-organization process brings a number of added values such as, for instance, the reinforcement of collaboration links in the aggregation process. As an example, consider the ACE that governs an advertisement screen in the PBA application. In order to conduct the advertisement process appropriately, it will have to aggregate to, among others, the ACE providing the multimedia contents, the ACEs providing self-healing features and, for instance, an ACE providing encryption services. Upon the rise of instabilities in the system, for instance due to a hardware failure causing a reduction of the available resources, it might enter an "emergency mode" whereby sensitive neighbors might be "clustered" to reinforce collaboration links critical to the service provision (e.g. the ACE providing multimedia content) while the self-healing infrastructure takes some action, also "synchronizing" to the "clustered" neighbors so as to optimize the aggregation's internal configuration. At the same time, it might "differentiate" with non-critical aggregations (e.g. the ACE providing encryption), to reduce the workload, thus allocating newly available resources to links retained critical. It is worth noting, finally, that

the employment of these algorithms is not automatic, but rather triggered by the situation-aware detection of well recognized instabilities.

5 Situation-Awareness

Situation-awareness refers to the ability of refining decisions according to the specific contextual situation. This capability requires models and tools for analyzing and organising pieces of contextual information into structured collections. To this end, the Framework offers a *Knowledge Network* (KN) [8] service, accessible, through aggregation, by ACEs as a system-wide service.
The KN is in charge of gathering and processing information to form a collection of *Knowledge Atoms* (KAs), which structure the latter in a data model conceived around the consideration that any bit of contextual knowledge is produced as a consequence of an event occurring in the context. Accordingly, a dedicated data model was created to represent any such fact, in a simple and expressive way, by means of a 4-tuple of the form *(Who, What, Where, When)*. This 4-tuple represents the basic unit of information in KAs and allows to account an entity (*Who*) involved in some activity (*What*) at a certain location (*Where*) at a certain moment (*When*).
The use of the data model enhances knowledge networking so as to make it possible to identify relations between atoms on the basis of content similarity (e.g. two atoms having the "who" field set to the same value corresponds to facts related to the same entity). Once relationships between atoms are identified and organized, it is possible to process them, to produce views, on the basis of equality of values in the same fields for different 4-tuples. For instance, data from a sensor network in an environment can be clustered according to the geographic closeness of sensors so that a concise perspective on an activity occurring in region larger than the one sensed by a single sensor can be generated.
Although it appears as a single ACE to the eyes of ACEs using it, the KN is internally composed by a collection of ACEs structurally organised in the architecture depicted in Figure 3. ACEs at bottom-most layer realize the concept of KA from heterogeneous data sources, from GPS devices to the Web. Data sources with very limited power, or too dynamic and ephemeral, are represented within so-called *Knowledge Repositories*. This might be the case, for instance, of data from RFID readers and sensor networks, which can be accessed via a repository rather than via individual KAs. It is worth noting that non-ACE applications may still be part of the Atom Repository through a simple interface.
On top of this layer, a number of components organize the KA so as to verify the concept of knowledge networks. The *Knowledge Organization* is in charge of organizing data in containers and exporting an interface for concept-based querying (i.e., by keyword) enabling higher level ACEs to access concept-based information. *Knowledge aggregators* allow establishing meaningful relationships among KAs, also storing data in a very expressive format and processing upon request. Results can thus be made available via a pattern-matching interface *à la Linda* [1]. The

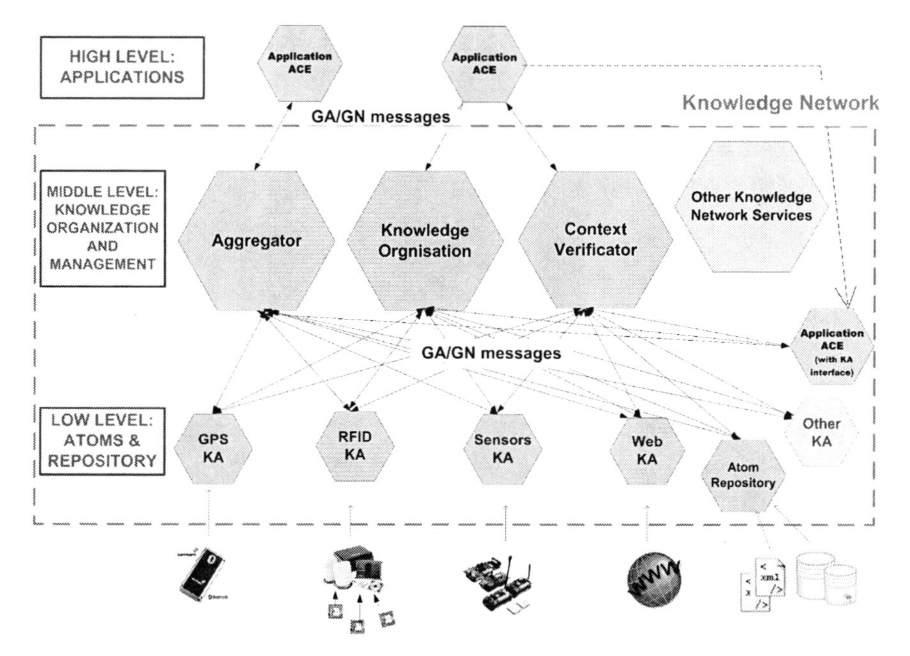

Fig. 3 Architecture of the KN.

idea behind the aggregator component is to provide specific information via pattern matching to form knowledge *views*. The *Context Verificator* verifies consistency of information in the KN according to (application- or user-) configurable parameters. This is done by accessing the KA(s) under verification and/or querying knowledge organization components to verify consistency in the knowledge. According to the outcome of such verification, then, it can notify the application about problems. An understanding of the advantages brought by the use of the KN can be obtained by looking at the PBA scenario. There, situation-awareness is the real enabler for self-adaptation in the advertisement process, as it allows the content allocation process to be driven according to contextual views of the composition of the current audience. In more detail, the ACE that takes the decision on the type of content to show, at a certain time, evaluates its appropriateness against the most recent view. Thus, by doing so it can make sure that the decision respects, in the best possible way, the nature of the interests that emerge as dominant from the view, and propose content that meets such interests. This process is realized by ACEs, representing potential content advertisers, aggregating to the KN in order to acquire situation-awareness features. By doing so, the former ones can come acquainted of the "situation by the screen", that is, the dominant interests among the audience currently in front of the screen, so as to ponder about relevance to own business, which would potentially lead to notification of interest.

6 Pervasive Supervision

Supervision refers to the ongoing observation of ACEs and the issuing of corrective measures upon detection of hazardous situations. Figure 4 shows the basic pattern for pervasive supervision, where all components are ACEs themselves. The *Sensor* links the supervision system with the ACE (configuration) under supervision in order for this pattern to be monitored internally. Each ACE exports, through the dedicated *Supervision* organ described in section 3.1. a management interface through which internal state and session objects can be accessed and verified. Monitoring data gathered in this way is aggregated by the *Correlator*, to extract meaningful indicators of current health conditions of the system under supervision, and, in turn, analyzed by *Drift Analyzers* that try to anticipate future problematic situations in the system under supervision. Additionally, information from the environment may be used to supplement the analytical process. The outcome of this analysis constitutes the input for *Assessors*, which make predictions on the current (or future) system health, on the basis of raw data or output of correlators and drift analyzers.

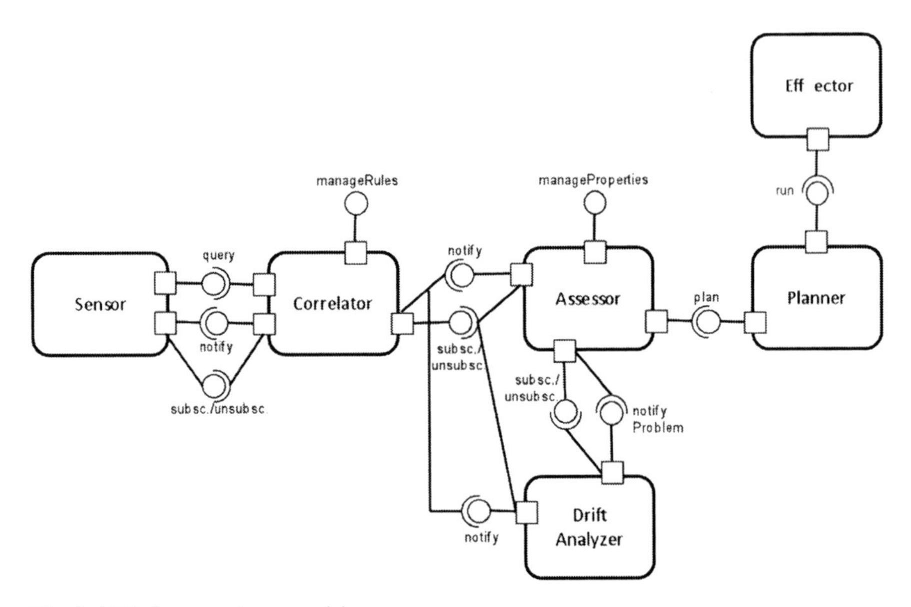

Fig. 4 ACEs for pervasive supervision

The above components are concerned with detection of potential health threats for the configuration supervised. According to the outcome of this activity, a reaction is eventually decided and put in place as follows. The *Planner* tries to compute a solution plan for a detected problem, on the basis of the assessments generated by the Assessor and the actions described in the self-model of the ACE (or ACEs) under supervision. These are finally executed, through interception and modification of internal processes of the supervised ACE(s), by *Effectors*. The actual state of

the system under supervision, its potential future behaviors and countermeasures on problem situations is derived from an analysis of the composition of the self-models of the ACEs that constitute the supervised system itself through the use of a mathematical framework for model composition, abstraction, and local refinement [11]. Automation of supervision functions to limit human intervention, referred to as *Self-management*, can benefit of self-organization techniques, already employed in the Framework, to enable highly distributed supervision of large collection of ACEs. Thus, a higher-granularity supervision system accounts for the self-adaptive nature of ACEs behavior, all of which implement the same supervision logic. Their co-operation enables for highly distributed supervision logic. Clustering and differentiation techniques are considered to foster information exchange and support of application-specific logical neighborhood to the extent of detecting and reacting to events related to ACE's behavior and their interactions through contracts.
The use of the pervasive supervision framework in the PBA scenario has the main purpose of enriching the prototype eco-system with capabilities of dynamic detection of ACEs entering faulty states. While, on one hand, this allows for a prompt heal, as per definition of self-healing, on the other hand its utilization also allows other operative ACEs to be notified of relevant anomalies as these occur. This, in turn, facilitates the design of modification rules to allow ACEs to cope with uncertain environments arising as a consequence of such anomalies. This is the case, for instance, of the ACE governing a display, which might ask allocation of an advertisement slot of time to other allocation ACEs upon notification, from the supervisor, that the one currently employed has entered a faulty state. It is worth noting that by doing so, the display ACE is also enhancing self-survivability of the display service, as the reaction guarantees the continuation of the service even though part of the aggregation is not in an operative state.

7 Security and Self-Preservation

The distinctive feature of a system built as a collection of ACEs is the absence of a centralized authority. As a consequence, *a-priori* trust relationships between ACEs belonging to different administrative domains cannot be assumed.
ACEs can show selfish, uncooperative or, in the worst-case, malicious behavior. Therefore, it becomes of paramount importance to address security issues in two distinct directions to cope with this wide range of attacks. Indeed, the behavior of ACEs needs to be monitored, and it is necessary to provide mechanisms to secure the communication infrastructure with confidentiality, integrity, and authentication. The Framework approaches these through security elements enhancing *hard-* and *soft-security* mechanisms, concerned with cryptographic mechanisms and behavioral attacks respectively. In the course of its operations, the security elements exploit the supervision infrastructure and employs semantic self-organization techniques to guarantee survivability of the system while ensuring self-preservation of (both simple and aggregate) ACEs and resources.

Hard security mechanisms aim at protecting the system from threats such as impersonation, eavesdropping, spoofing, and data modification. However, deploying cryptographic algorithms directly on ACEs would make them cumbersome, thus potentially prejudicing execution on pervasive devices. For this reason, the Framework exploits aggregation functionality and the GN/GA communication protocol to provide security in an adaptive and flexible way. Security features are thus provided through ACEs that can equip, through on demand aggregation, requesting ACEs with self-preservation and security features.

These ACEs provide a set of libraries containing basic cryptographic algorithms that can be combined on demand in order to compose more sophisticated security features. As an example consider, in the PBA scenario, the moment when the ACE governing the display communicates with the one governing the advertisement allocation service. There, the requesting ACE can be instructed to evaluate the availability of bandwidth as a parameter for deciding the type of encryption algorithms to use. The rational behind this check is that more complex algorithms will require altogether longer transmission times, which contrasts with the real-time requirements of inter-ACE communication in the PBA scenario. Then, once decided which encryption service to use, the requesting ACE might aggregate with the ACE providing the desired encryption algorithm in order to acquire the desired encryption capabilities. If at a later time the availability changes, the same ACE might want to employ a more complex encryption service, in which case the aggregation with the previous encryption ACE might be untied and another one might be undertaken with another ACE providing the service. This solution enables for flexible adaptation of ACEs to the security requirements of the application and type of service by implementing reusability of components for different purposes.

Giving ACEs responsibilities for the survivability of the system creates new challenges, whereby they are expected to react to eventual attacks, from malicious nodes targeting disruption of the system, by autonomously reconfiguring so as to exclude malicious ACEs. *Social control*, in the form of trust and reputation mechanisms, is employed in the Framework to this extent. Heuristics, or aggregation functions, can be injected so that ACEs' behavior can be captured and exclusion of malicious (selfish, or rational) entities can be enabled. Social control enables analysis of system evolution and interactions, and techniques borrowed from the game theory are used to minimize uncertainty in service provision and composition. This allows deriving conclusions on system performance in presence of selfish or uncooperative ACEs, and the cooperation level when a reputation management scheme drives interaction decisions [7]. System-wide self-preservation can also be enhanced, through a similar analysis, by exploiting semantic self-organization through the use of policies and strategies for the selection of specific partners on the basis of an evaluation of the level of involvement of an ACE. For instance, the use of such polices might enable an ACE to maximize its outcome in participating to system. This leads to an enhancement of the system survivability when the structure so formed is used to detect, and remove, malfunctioning components in order to keep the same service level[18].

The Framework targets protection against Distributed Denial-of-Service (DDoS), perhaps the most difficult type of attack to deal with. Development of feasible protection mechanisms involving both detection and reaction are under development, where two possible response mechanisms derive from generic DoS detection schemes. The first allows ACEs to self-protect by filtering detected unwanted traffic, and it is applicable to DoS attacks within the ACE communication domain. The second, exploits service migration (self-configuration) to escape malicious flows.

8 Pervasive Behavioral Advertisement Scenario

The case study scenario mentioned throughout the chapter has been fully developed as a prototype platform named *Pervasive Behavioral Advertisement* (PBA). The platform is conceived as an eco-system of ACEs developed through the Toolkit, and we believe it might represent a first step in the direction of an application with a potentially immediate industrial spin-off.

The platform is populated by a number of originally disjoint ACEs, which self-organize in *Regions* upon start as depicted in Figure 5,. Regions are formed as a result of each ACE's behavior as specified in its own self-model, and are characterized by the diversity of the service(s) offered. In other words, each region offers different services, which are made available to other regions so as to enhance on demand cooperation to the extent of delivering the promised service in the promised terms and conditions. Cooperation is then exploited through further inter-region aggregation that, in turn, fosters a system-wide self-organization optimized towards the behavioral pervasive advertisement service to be offered.

Referring to Figure 5, the *Profiling Region* makes available services for obtaining user information. This region is composed by a collection of ACEs truly interacting with badges equipped with RFID tags via RFID antennas positioned by the screen in such a way to detect the presence (and gather information) of people appearing and disappearing from the screen range. This activity is carried out on a continuous basis, as the shape of arrow denotes in the figure. Bits of information so gathered are exchanged with the *Knowledge Network Region*, whose ACEs give it the shape of structurally organized knowledge in the way previously described. To the eyes of other ones, this region appears as a single ACE providing a situation-aware information provision service. In reality, as described in previous sections, the region is composed by a number of aggregated ACEs that seamlessly provide the service in full respect of the self-similar constraints highlighted previously.

This process outcomes availability of contextual information that other regions of the system obtain and use in order to acquire high degrees of situation-awareness. The *Display Region* provides the system with displaying capabilities for the advertisement to be shown. Allocation of the actual content to be shown takes place by invoking the slot allocation service provided by the *Allocation Region*. With this respect, in the presence of a large number of screens and parties interested in buying time slots on them, solutions for allocating time slots and generating added value for interested parties must be identified. From this point of view, auctions appear

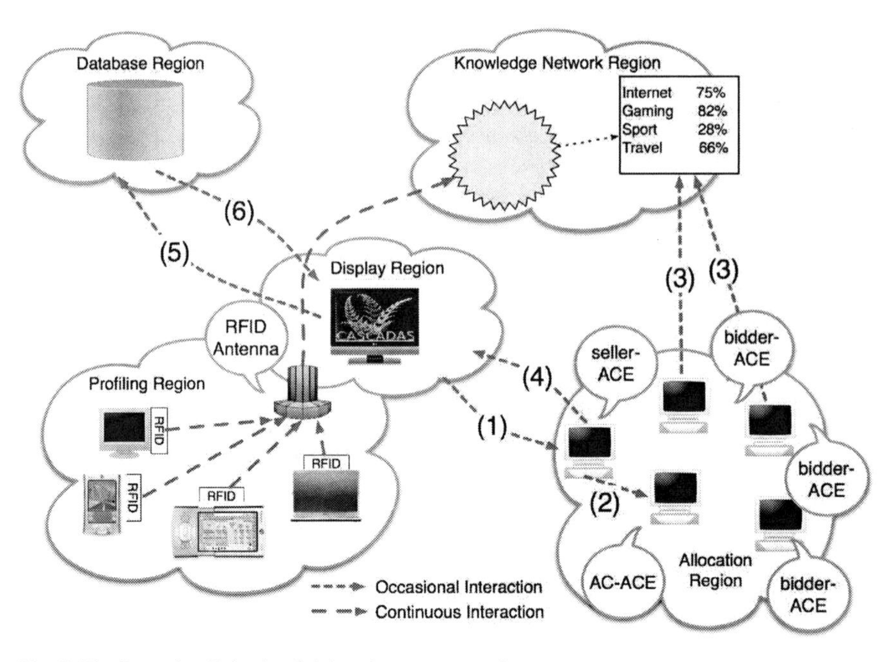

Fig. 5 The Pervasive Behavioral Advertisement scenario

an excellent solution as they prioritize allocation to advertisers who value them the most. Therefore, in our prototype the allocation of advertisement slots is decided by employing an auction-based paradigm whereby advertisers compete in a situation-aware fashion in order to acquire the rights of advertising on a specific screen at a specific time. The allocation service exported by the Allocation Region (interaction (1) in the figure) employs an iterative English auction, where ACE advertisers compete to the extent of acquiring the rights to expose own products in the slot of time under auction. Internally, the slot under auction is advertised by an *Auction Center* (interaction (2) towards the AC-ACE in the figure) through which is made available to advertisers (i.e. bidder-ACEs in figure) whose decisions on whether to submit a bid or not make intrinsic use the querying services offered by the Knowledge Network Region (interaction (3)). Thus, a bid is submitted if and only if the trends of interests reported in the (situation-aware) result of the query show sensitive relevance with the range of products the advertiser is competing to display. As in the case of the Knowledge Network Region, the Allocation Region has a single-ACE *façade* that, in turn, involves a number of aggregated, self-similar, ACEs to cooperate.

Upon auction termination, communication of the auction winner is returned back to the Display Region (interaction (4)), and is used by the ACEs there contained to select the right advertisement to display. The selection is offered as a service by the *Database Region*, which contains an advertisement database repository where advertisements are tagged based on owner and dominant relevance of interests. Thus,

selection of the right advertisement is done by simply invoking the service with auction winner and the interest the winning bid was submitted for, returned by the Auction Region, as input (interaction (5) for the query, and (6) for the response). Seamlessly from the interaction model just described, which ensures that the contents advertised on the display evolve in the same way as the interests in the audience, the pervasive supervision framework enriches the platform with self-healing features. In the current prototype, however, self-healing capabilities are used for proof of concept, and therefore the only Auction Region is put under supervision. Specifically, relevant ACEs aggregate with the supervisor, which starts its monitoring activity towards all of them. If an ACE, say, the seller, enters a faulty state, a notification is sent to all other supervised ACEs, which can thus escape the faulty state by aggregating to other sellers, and a healing action aimed at reverting the ACE back to an operational state is started.
The prototype platform has been deployed on a distributed testbed, and executions show that ACEs developed are stable and capable of supporting the interaction model described above in extensive long-term sessions. In addition, preliminary performance evaluation tests have been conducted to the extent of evaluating effectiveness of the platform in terms of impact of the advertisement currently shown on the current crowd as based on a matching of relevant interests. Results show that our system enables high impact on the current crowd, as compared with the classical "round-robin". Analysis on such data allowed inferring that the effective investment cost for the advertiser, intended as the cost per-matching-person, decreases even though the price paid for advertising results many times increased (with a consequent increase in the screen owner's revenue).

9 Conclusions

This chapter describes the distributed component-ware framework for autonomic and situation-aware communication designed and developed in the context of the CASCADAS project. The use of the toolkit as a mean for creating computer networks enables the development of these latter ones as ecologies of services, or eco-systems, according to a vision that many foresee as a natural future evolution of the current Internet.
Eco-systems are characterized by a relatively flat architecture, which makes it easy to inject new services without, or with very limited, configuration efforts, and provide an environment where services can thrive to satisfy users' needs, while accounting for the actual operative context, as these both evolve. The framework embraces this vision by enhancing the creation of dynamically adaptable services through the innovative abstraction of Autonomic Communication Elements (ACEs), which are made available through an open source development Toolkit. This latter is fully usable and capable of exploiting highly innovative networking and application services through provision of features such as situation-awareness, which enriches the system with the ability to take into consideration the context evolution in local de-

cisions by providing system-wide knowledge in a highly expressive and flexible format, and semantic self-organization, which allows aggregating and organizing ACEs on the basis of dynamic contextual changes. These features are federated into a sound component model, which, among other things, ensures uniformity of the interfaces exported by composite ACEs resulting after aggregation of ACEs offering basic services. This results in a natural tendency to show self-similar behavior.
The above features have been used to equip the framework with other, composite, features that enrich the framework itself. In detail, security and self-preservation features are provided through a dedicated set of ACEs, where confidentiality, integrity, non-repudiation and authentication mechanisms are provided in form of on-demand services. Similarly, another dedicated set of ACEs brings self-healing features through a distributed approach towards detection of changes in the health of one or more ACEs, while also anticipating reactions based on projections of the way the state will evolve; A Pervasive Behavioral Advertisement application was selected fully developed as proof of concept scenario. The structure and interactions of the collection of ACEs that form it served as a way to better illustrate the features above, and how these support such a highly dynamic application. The scenario anticipates a potential future industrial scenario, and relies on the sole CASCADAS framework in order to build a robust and effective platform in a fully distributed fashion.

Acknowledgements The material here presented is based on a research funded by the CASCADAS (*Component-ware for Autonomic Situation-aware Communications, and Dynamically Adaptable Services*) project (EU FP6-027807). Authors would like to thank Nermin Brgulja, Roberto Cascella, Peter Deussen, Elisabetta Di Nitto, Sandra Hasselhof, Ricardo Lent, Corrado Moiso and Fabrice Saffre for their help.

References

1. Ahuja S., Carriero N., Gelernter D., "Linda and Friends". In *IEEE Computer*, Vol. 19, No. 8, pp. 26-34, 1986.
2. Albert R. and Barabasi A. "Statistical Mechanics of Complex Networks". In Review of Modern Physics Vol 74(47), 2002.
3. Bicocchi N., Mamei M., Zambonelli F. "Self-organising Spatial Regions for Sensor Network Infrastructures". In *Proc. of the* 21st *International Conference on Advanced Information Networking and Applications Workshops* (AINAW'07), Niagara Falls, Sept. 2007.
4. Bigus J. P., Schlosnagle D. A., Pilgrim J. R., Mills III W. N. and Diao Y. "ABLE: a toolkit for building multiagent autonomic systems". In *Proc. of the* 1st *International Workshop on Enterprise Distributed Object Computing* (EDOC 2007), Annapolis, Oct. 2007.
5. Bouchenak S., De Palma N. and Hagimont D. "Autonomic Management of Clustered Applications". In *Proc. of the IEEE International Conference on Cluster Computing*, Barcelona, Sept. 2006.
6. The Cascadas Consortium. "CASCADAS White Paper". Available at `http://www.cascadas-project.org/Deliverable.html`.
7. Cascella R. G., "The "Value" of Reputation in Peer-to-Peer Networks". In *Proc. of the Fifth IEEE Consumer Communications & Networking Conference* (CCNC 2008), Las Vegas, Jan. 2008.

8. Castelli G., Mamei M., Zambonelli F.. "Engineering Contextual Knowledge for Pervasive Autonomic Services". To appear in *International Journal of Information and Software Technology*.
9. Cugola G. and Picco G.P. "REDS A Reconfigurable Dispatching System". In *Proc. of the 6th International Workshop on Software Engineering and Middleware* (SEM2006), Portland, Nov. 2006.
10. Deussen P. H. "Model based reactive planning and prediction for autonomic systems". In *Proc. of the INSERTech (Innovative SERvice Technologies) workshop*, Rome, Oct. 2007.
11. Deussen P. H., Valetto G., Din G., Kivimaki T., Heikkinen S., and Rocha A. "Continuous online validation for optimized service management". In *EURESCOM Summit*, Heidelberg, Oct. 2002.
12. Di Ferdinando A., Rosi A., Lent R., Zambonelli F. and Gelenbe E. "A Platform for Pervasive Combinatorial Trading With Opportunistic Self-Aggregation". In *Proc. of the 2nd IEEE WoWMoM Workshop on Autonomic and Opportunistic Communications* (AOC'08), Newport Beach, Jun. 2008.
13. Di Ferdinando A., Lent R. and Gelenbe E.. "A Framework for Autonomic Networked Auctions". In *Proc. of the 1st International Conference on Autonomic Computing and Communication Systems* (AUTONOMICS 2007), Rome, Oct. 2007.
14. S. Dill, R. Kumar, K. Mccurley, S. Rajagopalan, D. Sivakumar, A. Tomkins, "Self-Similarity in the Web". In *ACM Transactions on Internet Technology* Vol. 2(3), pp. 205-223, 2003.
15. Sourceforge Website. `http://www.sourceforge.net`.
16. Dobson S., Denazis S., Fernandez A., Gaiti D., Gelenbe E., Massacci F., Nixon P., Saffre F., Schmidt N. and Zambonelli F. "A Survey of Autonomic Communications". In *ACM Transactions on Autonomous Adaptive Systems*, Vol. 1(2), pp. 223-259, 2006.
17. Escoffier C., Hall R. S. and Lalanda P. "iPOJO: an Extensible Service-Oriented Component Framework". In *Proc. of the 2007 IEEE International Conference on Service Computing* (SCC'07), Salt Lake City, Jul. 2007.
18. Eugster P., Felber P., Guerraoui R. and Kermarrec A.-M.. "The Many Faces of Publish/Subscribe". In *ACM Computing Surveys*, 35(2):114-131, 2003.
19. Greer K., Baumgarten M., Mulvenna M., Curran K., Nugent C., "Autonomic Supervision of Stigmergic Self-Organisation for Distributed Information Retrieval". In *Proc of the Workshop on Technologies for Situated and Autonomic Communications* (SAC), Budapest, Dec. 2007.
20. Greer K., Baumgarten M., Mulvenna M., Curran K., Nugent C. "Knowledge-Based Reasoning through Stigmergic Reasoning". In *Proc. of the International Workshop on Self-Organising Systems* (IWSOS07). In: Hutchison D. and Katz R. H. (Eds.). *Lecture Notes in Computer Science*, LNCS 4725, 2007.
21. Höfig H., Benko B. K., Di NItto E., Mamei M., Mannella A., Wuest B. "On Concepts for Autonomics Communication Elements". In *Proc. of the 1st IEEE International Workshop on Modeling Autonomic Communication Environments* (MACE 2006), Dublin, Oct. 2006.
22. Huebscher M. and McCann J. "A Survey of Autonomic Computing - degrees, modelas and applications". In *ACM Computing Surveys*, December 2007.
23. Khargharia B., Hariri S., Parashar M., Ntaimo L. and Uk Kim B. "vGrid: A Framework for Building Autonomic Applications". In *Proc. of the International Workshop on Challenges of Large Applications in Distributed Environments* (CLADE2003), Seattle, June 2003.
24. Kephart J. and Chess D. "The vision of autonomic computing". In *IEEE Computer*, 36(1):4152, 2003.
25. Liu H., Parashar M. and Hariri S. "A Component Based Programming Framework for Autonomic Applications". In *Proc. of the 1st IEEE International Conference on Autonomic Computing* (ICAC-04), New York, May 2004.
26. Manzalini A., Zambonelli F, "Towards Autonomic and Situation-Aware Communication Services: the CASCADAS Vision". In *Proc. of the 2006 IEEE Workshop on Distributed Intelligent Systems* (DIS 2006), Prague, Jun. 2006.
27. Marrow P., Koubarakis M., Van Lengen R.H., Valverde-Albacete F., Bonsma E., Cid-Suerio J., Figueiras-Vidal A.R., Gallardo-Antolin A., Hoile C., Koutris T., Molina-Bulla H., Navia-Vasquez A., Raftopoulou P., Skarmeas N., Tryfonopoloulos C., Wang F. and Xiruhaki C.

"Agents in Decentralised Information Ecosystems: the DIET Approach". In *Proc. of the Artificial Intelligence and Simulation Behaviour Convention* (AISB 2001), York, Mar. 2001.

28. Melcher B. and Mitchell B. "Towards an Autonomic Framework: Self-Configuring Network Services and Developing Autonomic Applications". In *Intel Technology Journal*, Vol. 8(4), pp. 279-290, 2004.
29. Michiardi P., Marrow P., Tateson R. and Saffre F. "Aggregation Dynamics in Service Overlay Networks". In *Proc. of the* 1st *IEEE International Conference on Self-Adaptive and Self-Organizing Systems*, Boston, Jul. 2007.
30. Quitadamo, R. and Zambonelli, F. "Autonomic communication services: a new challenge for software agents". In *Journal of Autonomous Agents and Multi-Agent Systems*, 2007.
31. Sajjad A., Jameel H., Kalim U., Lee Y.-K. and Lee S. "A Component-Based Architecture for an Autonomic Middleware Enabling Mobile Access to Grid Infrastructure". In *Proc. of the Conference on Embedded And Ubiquitous Computing* (EUC 2005), Nagasali, Dec. 2005.
32. Shilit B. N., Adams N. and Recker J. "Context-aware computing applications". In *Proc. of the IEEE Workshop on Computing Systems and Applications* (WMCSA94), Santa Cruz, Dec. 1994.
33. Shilit B. N. and Theimer M.M. "Disseminating Active Map Information to Mobile Hosts". In *IEEE Network* Vol. 8(5), pp. 22-32, 1994.
34. The ACE Autonomic Toolkit. `http://sourceforge.net/projects/acetoolkit`, or `http://acetoolkit.sourceforge.net`.
35. Vassilakis C. C., Laoutaris N., and Stavrakakis I., "The impact of playout policy on the performance of P2P live streaming". In *Proc. of the* 5th *Annual Multimedia Computing and Networking* (MMCN '08), San Jose, Jan. 2008.

Autonomic Middleware for Automotive Embedded Systems

Richard Anthony[1], DeJiu Chen[2], Martin Törngren[2], Detlef Scholle[3], Martin Sanfridson[4], Achim Rettberg[5], Tahir Naseer[2], Magnus Persson[2], and Lei Feng[2]

Abstract This chapter describes DySCAS: an advanced autonomic platform-independent middleware framework for automotive embedded systems. The concepts and architecture are motivated and described in detail, focusing on the need for, and achievement of, high flexibility and automatic run-time reconfiguration. The design of the middleware is positioned with respect to the way it overcomes the specific technical, environmental, and performance challenges of the automotive domain. Self-management is achieved in terms of automatic configuration for context-aware behavior, resource-use efficiency, and self-healing to handle run-time detected faults. The self-management is governed by the use of policies distributed throughout the middleware components. The simulation techniques that have been used for extensive validation are described and some key results presented. A reference implementation is presented, illustrating the way in which the various concepts and mechanisms can be realized and orchestrated.

1 Introduction

This chapter describes the DySCAS middleware framework for automotive embedded systems.

DySCAS is motivated by the need to introduce flexible dynamic configuration into vehicular control systems. These systems are increasingly complex and this complexity impacts on the already long design and development cycles. The automotive industry needs the ability to defer some design decisions so that time-to-market is reduced without compromising the level of functionality achieved. Upgrades should be supported transparently throughout the lifetime of the vehicle; this is especially needed for those sub-components with higher rates of innovation (such as infotainment devices, GPS etc.). Vehicle owners' expectations of technology are also increasing and they demand the ability to make customization choices, and to change those choices over the vehicle's lifetime. The community also indirectly demands flexible vehicle upgrades, because vehicular legislation concerning aspects such as safety, emissions, noise

The University of Greenwich, Greenwich, London, UK. R.J.Anthony@gre.ac.uk · Royal Institute of Technology (KTH), SE-100 44 Stockholm, Sweden. {chen, martin, tnqu, magnper, leifeng}@md.kth.se · Enea, Skalholtsgatan 9, Box 1033,164 21 Kista, Sweden Detlef.Scholle@enea.com · Volvo Technology Corporation, Mechatronics & Software, SE-412 88 Gothenburg, Sweden. martin.sanfridson@volvo.com · Carl von Ossietzky Universität Oldenburg, Offis e.V., Oldenburg, Germany. achim.rettberg@informatik.uni-oldenburg.de

A.V. Vasilakos et al. (eds.), *Autonomic Communication*, DOI: 10.1007/978-0-387-09753-4_7,

etc. are constantly updated and applied differently across Europe and the world; currently such legislation only affects new vehicles as it can not be retrospectively applied. The deployment of self-management into vehicular systems has the potential to improve robustness through dynamic fault detection and handling and to improve efficiency, for example through dynamic reconfiguration to reduce power consumption.

Based on the significant and wide-ranging technical challenges and expectations of the automotive application domain we derive the key requirements of DySCAS, which include: real-time performance guarantees, flexibility, self-management, context-awareness, platform independence, resource efficiency, high robustness, run-time configuration, extensibility, and functional upgrade over the lifetime of deployed systems.

The resulting architecture specification is presented. The architecture follows a layered, service-oriented model and is designed to be able to meet stringent real-time and reliability QoS (Quality of Service) guarantees. The architecture supports autonomic behavior with specialized planner and actuator components to achieve dynamic reconfiguration to improve resource efficiency and to self-heal when faults are detected.

The component model is described; this specifies the interfaces and internal modules of middleware services, the means of execution of and interaction between services, and a strategy for component deployment. The configuration logic is distributed throughout the middleware and application components wherever deferred logic or run-time context-sensitive configuration is required. Components can be internally configured by inserting *Decision Points*, which are place holders for policies which can be loaded dynamically. The policies are intended to operate at the strategic level. To support this, a dynamic context management scheme ensures that the appropriate environment and state information is locally cached for each specific policy so that context-aware decisions can be made with very low latency. In addition to the policy evaluator, Decision Points also encapsulate a policy supervisor which detects and handles any problems arising from policy evaluation.

We go on to show how the DySCAS architecture specification can be turned into a realized system through the steps of extensive simulation to validate concepts, and the development of a reference implementation. One of the typical DySCAS use cases is described and the associated implementation issues are discussed. Policy evaluation, and the interaction between components in the reference implementation are detailed, as well as the techniques used for checkpointing, versioning of components, and scheduling. An open-source Instantiation Layer is provided to facilitate portability and to reduce the device-driver development effort. The various simulation and validation tools and techniques used are discussed and some results are presented.

Finally, open issues and on-going work areas are identified.

2 Automotive challenges and DySCAS

The general trend towards increasing use of electronics and communication technology is continuing and will have a major impact on vehicle design and operation. In this trend, system integration technologies to merge functions together, will play an important role [30].

Vehicles in series production today already contain the same amount of electronics as aircraft did two decades ago. It is predicted that the share of automotive embedded systems in respect to a vehicle's total value will reach 40% by 2015 [53], bringing in innovations and new features in vehicle control and driver assistance such as radar assisted cruise control, traffic information services, improved navigation, fuel efficiency, and many more.

Like many embedded systems in other markets, e.g. medical technology or avionics, automotive embedded systems are safety critical because of their potential effects on the environment and humans. Many automotive products are targeted towards the consumer mass market and are thus sensitive to cost, usability and reliability. Many automotive applications have real-time requirements, ranging over closed loop periodic controllers to multimedia and communication functions.

The transition from today's static system to dynamic configurations made in the DySCAS project is a large step for the automotive industry. The introduction of adaptive aspects of configuration and behaviors, and the ability to defer part of the configuration decisions, verification and validation efforts beyond the point of systems deployment, calls for enhanced support for error detection, error handling, and recovery. The intended support by the operating system and middleware is related to supervision of operations such as monitoring the system execution state of application software and devices, producing checkpoints for rollback, transferring component states, and re-flashing nodes.

A vehicle can be divided into domains, such as the chassis domain which comprises e.g. brakes, and the infotainment or telematics domain physically located in the cabin for interaction with the driver and passengers. A navigator is a typical example of an infotainment device that can be nomadic or built-in. Information interchange with the vehicle and other infotainment devices as well as a seamless human-machine interface integration into the vehicle, e.g. using the built-in display and buttons on the dashboard to control the nomadic device, are both necessary and valuable.

It is well known that technology advances drive business. The technology growth is currently very high in the infotainment domain. The business opportunities of an embedded dynamic middleware such as DySCAS, is based on the observation that the vehicle as a whole has a lifecycle three to ten times longer than its more or less tightly connected infotainment devices. This gap creates a tension and desire to upgrade both hardware and software of the infotainment devices and have them seamlessly integrated into the vehicle.

Another technology emerging on the automotive mass market is the vehicle-to-vehicle and vehicle-to-roadside communication, giving rise to a vast number of new opportunities for innovation in traffic safety and management and driving information such as road and weather conditions, which must be integrated into the vehicle to aid but not distract the driver. The external wireless communication has another deep impact – it also enables system upgrade. This is exactly what buyers expect today of modern consumer electronics such as television decoders, computer communication devices, console games, mobile telephones, etc., that regularly connect to a server, eventually performing software upgrade. To catch the business opportunities and customer satisfaction out of these two technological changes, the computer system coordinating the infotainment domain needs to be adaptable. The business prospective will be proportional to the simplicity as perceived by the customer and inversely proportional to the labor intensity of the vehicle manufacturer and service provider.

The increased use of embedded electronic systems in vehicles, however, also implies growth and change in both application complexity and system development complexity. For many advanced applications, there are apparent needs of integration of interacting data and functionalities and incorporation of behaviors with different criticalities and heterogeneous off-the-shelf components, further characterized by real-time, resource and dependability constraints. In system development, such product complexity is augmented by the involvement of multiple stakeholders and organizations, heterogeneous technologies and components, and lifecycle concerns in regards to maintenance, upgrade, variability and reuse. To cope with the technical and managerial challenges it becomes necessary to develop new technologies, tools, and methodologies [71]. The dynamic self-configuration approach of DySCAS offers here an alternative to handle the expected complexity.

On the automotive customer service side, the ability of allowing cost efficient and reliable field-upgrades of software is also considered important besides vehicle customization, personalization, and incorporation of technology innovations.

It is expected that future vehicles will provide the ability of building ad-hoc networks between vehicles and with road infrastructure access points to share information and functionality, e.g. to resolve hazardous traffic situations. These information services are based on external communication and a cornerstone is the ability of the vehicle internal architecture to handle all conceivable future information services.

The following is an example use case which we assume did not exist at the time the vehicle rolled out from the factory. The wish is to attach a device that offers a new predefined functionality to the vehicle and integrate into the vehicle the specific human machine interface: *a handheld navigation device becomes available inside the vehicle.* Its functions will become integrated into the vehicle's infotainment system e.g. via wireless bluetooth communication. Navigation directions shall be given out via the sound system of the vehicle while the current entertainment source is muted. The transition when integrating and removing the handheld navigation device must be seamless to the driver and passengers.

Current automotive embedded electronic systems adopt a static configuration scheme, in which the design assumptions on environment, system functionalities and behaviors, component compositions and resource deployment are defined during the development process and subsequently kept unaltered over the complete lifetime of the vehicle. This is, as seen from the above mentioned use cases, insufficient for many future scenarios of automotive vehicles. Further, seen from an organization perspective, the inevitable early design assumptions which cannot be withdrawn even when new information is revealed are a source of much agony and hesitation hampering the business key time-to-market issue. In contrast and extension to statically configured architectures, DySCAS will, for the dynamic operations, achieve predictability through on-line mechanisms that negotiate and reserve necessary resources in advance and provide synchronization with application conditions. When design time testability is partially lost in a dynamic self-configuring system, supervision of behavior takes over. For a networked system, it is important that decisions are made based on a consistent global view and actuated in a synchronized way. This in turn necessitates middleware support for consolidating distributed information in regards to vehicle conditions, application states, operation events, and resource availability, as well as the support for disseminating the consolidation results. For example, a software update may require re-allocation of components and thus a global view of resource utilization.

Automotive systems are often highly resource-constrained because of cost considerations in mass production. To introduce middleware solutions in automotive embedded systems, performance overheads in time and in resource utilization (e.g., bus, CPU, and memory) need to be properly handled. While overheads because of the middleware mechanisms are unavoidable, the DySCAS approach aims to keep the overhead as small as possible while making the behaviors predictable. To this end, the choice of algorithms, the instantiation, mapping, and allocation of middleware services, as well as the planning and controlling of its tasks, are all of importance.

The support for load balancing, on-line supervisory verification, validation, and error handling also in the future promise to enable a design trade-off between the costly development-time and testing effort against overhead in terms of run-time CPU capability, processing, footprint, and communication utilization. For the reasons of dependability, time-to-market and lifecycle management efficiency, scenarios of future automotive embedded systems also call for enhanced QoS (quality of service) support. This will permit post-development time optimization according to the actual resource utilizations and operating conditions.

DySCAS aims to advance today's state-of-practice technology and introduce context-aware and self-managing behaviors into automotive embedded electronic systems [24]. Targeting the above mentioned future scenarios, the DySCAS approach explicitly addresses the automotive needs in regards to configuration flexibility, quality assurance, and complexity control in particular in the infotainment and telematics domains. DySCAS has developed and proposed a middleware framework that allows automotive embedded systems to dynamically reconfigure themselves according to the environmental conditions, application states and resource deployment, to cope with unexpected events, emerging use cases and optimization needs, and external devices not known at the deployment time. The DySCAS approach aims to provide necessary run-time support for enabling a systematic and efficient implementation of QoS and dynamic configuration behaviors in automotive systems. The core is a set of middleware services that facilitate the sampling of system configuration and operation states, the computation for QoS and dynamic configuration decisions, and the actuations of such decisions. The

DySCAS targeted use cases and the mapping of these onto system requirements are discussed in [10].

3 Background and related work

3.1 Middleware for distributed computer systems

During the recent decade, many new middleware technologies for distributed computer systems have been developed and enhanced, targeting various application areas and hardware platforms. Middleware technologies for distributed software focus on the deployment and integration of independently developed software components, thus emphasizing the support for a scalable and dynamic configuration. Examples of such a middleware are the JavaBeans [67], .NET [54], the CORBA Component Model [56], Jini [68] for the plug-and-play of soft real-time services and devices, and RoSES [16] trying to achieve graceful degradation through software reconfiguration using Jini communicating over a typical automotive bus - despite nodes failing. However, not all the middleware solutions are suitable for embedded systems due to their excessive memory usage and processing overheads or their lack of support for real-time guarantees, data and state consistency, and fault-tolerance by redundancy. To improve on these important aspects, many middleware solutions specifically targeting real-time and embedded systems have been developed, including HADES [6], ARMADA [2], and the RT-CORBA [57] implementations TAO [21] and ZEN [39]. In the domains of sensor networks, ubiquitous computing and networked embedded systems, there are also middleware technologies such as RUNES [52] and 2K [42], to support advanced dynamic configuration and automated software maintenance, applying meta-object protocols and reflection for run-time adaptation of configuration and behavior [20], or QoS control for finding performance objectives when the request and availability of processing, communication and other resources change [65]. An interesting scheme is the Simplex architecture [66], designed for error recovery of runtime upgrade of experimental automatic controllers. There are also many middleware solutions aiming at particular aspects or application domains of dynamic configuration, such as the HAVi [35] software architecture for configuration and interoperation of home local area networks, and the OSGi architecture [60] for a life cycle perspective aid to coordinate development and management of network services. A more formal approach for QoS management is given by the component framework Lusceta [17], providing a QoS-aware middleware preceded by formalisms for specifying, simulating, analyzing, and run-time synthesizing QoS management.
The above cited works provide together a reference source for the design of the DySCAS architecture in regards to scalability, middleware structuring, QoS management, fault-tolerance, and execution control. In addition to this reference source, the expected defacto standard in the automotive industry AUTOSAR (AUTomotive Open System Architecture) [13], is a very important reference and metric for the DySCAS architecture. AUTOSAR provides a domain-specific approach to specification, management and integration of application software components, system services, run-time environment, and system resources. The AUTOSAR concept of a virtual functional bus provides flexibility at design time, but the runtime communication matrix is static and any adaptive behavior is confined to application level software without support from the runtime environment or any middleware.

3.2 Policy-based configuration

Autonomic behavior (self-management) in DySCAS is governed by policies which are placed throughout the middleware components. Policies provide a powerful means of representing the logic required to make decisions which is decoupled from the underlying deployed code. Policies are flexible and can be formalized by using a closed grammar described in a formal notation such as EBNF or a schema definition language. A suitably expressive language enables a wide range of behaviour to be represented at the strategic level by a relatively simple policy description. Policies can also be used at lower, mechanistic levels if required.
The simplest type of policy consists only of configuration settings that are loaded at application initialization. Typically these are Boolean flags that allow selection / de-selection of provided features. Internally there may be embedded policy logic comprising rules and actions etc., but the actual logic is fixed. behavior is thus fixed for any particular execution instance. Behavior can be modified *between* executions but the extent of the modification is limited to changes to the parameters that constitute the policy. For example [15] embeds fixed rules into agents.
A greater level of flexibility is achieved when both the configuration parameters *and* the actual decision rules are held externally to the embedded mechanism so it is possible to update the actual logic of the policy as well as its initial parameterization. An example is provided by the Policy Description Language (PDL) [48], which is an event-driven programming language. Distributed Action Plans (DAP) are used to specify distributed network management tasks. DAPs are executed by policy agents which are specified in PDL [41].
Some policy mechanisms indirectly support open-loop adaptation of the actual policy. This support is in terms of *identifying*, or *facilitating* the identification of, inefficiencies in, or conflicts between, policies. The policy behaviour remains fixed during the current execution instance. The user is notified and manually updates the policy between executions. IBM Research's Policy Management for Autonomic Computing (PMAC) [38] provides an automated policy management and deployment application to assist with the manual policy updates. Policy Schedule Advisor [49] is a utility that assists in the refining of a policy schedule to ensure efficient execution on the PMAC middleware. The Unity policy environment 'Policyscape' [64] provides a set of 'templates' that can be used as building blocks to create more-complex policies and supports automated policy creation.
The current state of the art in policy-based computing are schemes that have at least partial support for closed-loop self-adaptation, in which the policy or its support mechanisms can automatically adapt the policy's own behavior to suit contextual or environmental circumstances. The dynamic adaptation is achieved in a variety of ways and to a wide variety of extents. Ponder [22] uses meta-policies to define semantic constraints on the regular policies. A security policy implementation described in [69] uses a meta-policy to dynamically select between security policy versions, but policy updates are performed manually by administrators (although this can occur during run-time). Further examples of short-term adaptation are found in [51] in which event-trigger conditions are dynamic, and [5] in which conflicts between the obligations of security policies are automatically detected and resolved at run-time by dynamically removing conflicting obligations under certain circumstances. Similarly, the language described in [19] supports automatic detection and resolution of rule conflicts. The AGILE policy technology [8] has been purposefully designed to be highly flexible in terms of dynamic self-reconfiguration to facilitate context aware behavior in a wide variety of application domains. It achieves dynamic self-adaptation in several ways.
The policy technology used in DySCAS is AGILE-Lite; a lightweight, embedded version of AGILE [11]. A policy script can be loaded into an application at run-time to change the behavior of the application at the place where the script is inserted.
The places at which decision logic can be changed, called Decision Points (DPs), are specified at design time; [73] provides a detailed explanation of DPs and the supporting mechanisms. Policy scripts can be loaded into these points (usually when the component containing the DP is initialized) and can replaced with other policy scripts during run-time, yielding different decisions (e.g. for customization) or more advanced decisions (e.g. for functional upgrade).

The AGILE policy language has a level of flexibility more normally associated with a lower-level programming language. For example, indirect addressing is supported at the policy script level, so that all constructs can be dynamically configured by changing the parameter variables supplied. Consider 'rule' constructs which can be used to implement Boolean logic as well as simple conditional tests. It is possible to use the outcome of one rule to contextually change the behavior of another rule by changing the actual parameters (not simply the values) compared in the second rule. Further details of the language are provided in [8] and its application to the DySCAS use cases is discussed in [9, 12].
Although powerful, AGILE policies at the same time can define functionality at a high level so that developers can focus on the intended business logic and need not be experts in concepts such as autonomics and policy-based configuration. At the strategic level a policy defines desired behavior, without having to describe exactly how the behavior is achieved, (since the lower-level mechanisms are pre-built). An AGILE policy editing tool further simplifies the task of preparing policy scripts; making script editing less problematic and less error-prone, see also [7].

4 The DYSCAS Middleware Architecture

The architecture constitutes a framework where various middleware services for the intended features of dynamic architecture are defined and integrated. Figure 1 provides an outline of the DySCAS middleware architecture, including the major middleware services and the external interfaces towards the application software and target platform. The dashed blocks represent optional services allocated at the DySCAS Instantiation Interface.

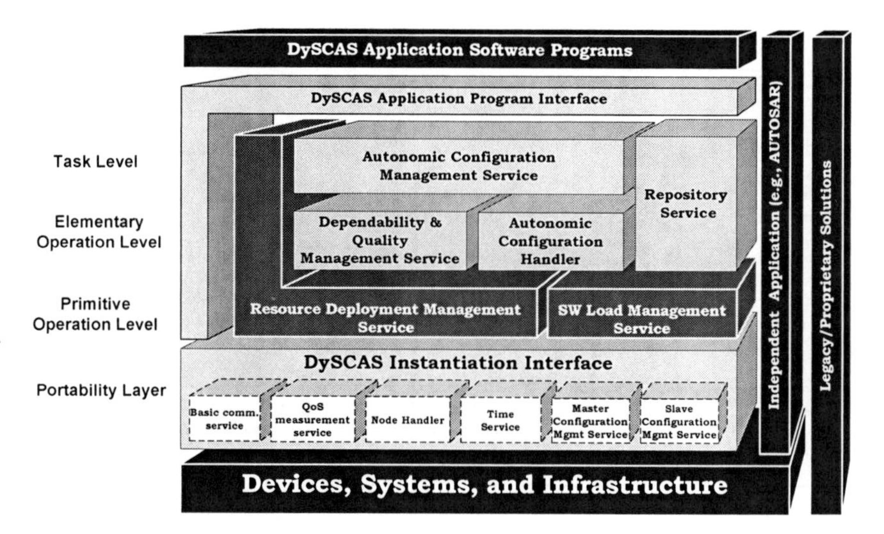

Fig. 1 A schematic view of the DySCAS architecture

The assumed context of DySCAS middleware consists of application software programs, legacy/proprietary solutions, and a target platform in terms of system software, hardware and devices, and communication infrastructure. DySCAS distinguishes independent and legacy solutions that are developed independent of the middleware, from solutions that are developed

with the middleware support in consideration. In general, the solutions that are developed or adapted for DySCAS should allow middleware interactions for information exchange and behavior control. For practical and technical reasons, the independent and legacy solutions do not need to be ported to the middleware but coexist with the middleware system and its applications. They are allowed to have direct interaction with the underlying target platform. The target platform includes the RTOS and basic I/O,as well as infrastructures of various types, such as CAN [63], LIN [47] and MOST [55]. The actual content and levels of features offered by the target platforms may differ, such as among the platforms of different OEMs. Moreover, there can also be various existing middleware solutions introduced on top of these system layers for the reasons of portability and interoperability.
A DySCAS middleware system has a layered architecture consisting of two groups of runtime services:

1. **core services**, providing the reasoning and decision-making support for the dynamic configuration management and quality control, and
2. **interface services**, providing basic support for the interactions between the middleware and its environment (i.e., the application programs and system platforms).

The core middleware services provide the support for consolidating the monitored runtime state information and for maintaining a consistent view about the overall system status in regards to configuration, resource utilization, and quality satisfaction. Such services thereby analyze the impacts of any detected or requested changes, derive necessary dynamic change tasks, and finally provide the planning and supervisory control support for performing such change tasks. The overall concept design follows the paradigms of the IBM Autonomic Computing Reference Model [37] and NASREM reference model for autonomous systems [4]. Table 1 provides an overview of the core services and their main responsibilities.

Table 1 An overview of DySCAS Core Services and some of their properties

DySCAS Core Service	Overall System Roles	Related Autonomic Features
Autonomic Configuration Management Service	**Analyzer** for the overall impacts of requested dynamic changes. **Planner** of configuration tasks.	**Self-configuring** (online configuration reasoning and work planning support). **Self-healing** (error repair and fault removal).
Repository Service	**Repository proxy** for saved files, components, policies, and diagnostics information.	Not applicable.
Dependability & Quality Management Service	**Dependability controller** **Performance Optimizer**	**Self-healing** **Self-optimizing** **Self-protecting**
Autonomic Configuration Handler	Coordinator of distributed operations of dynamic configuration work.	**Self-configuring** (on synchronized execution of dynamic configuration work).
Resource Deployment Management Service	**Monitor** of external status in resources utilization, execution behaviors, and configuration. **Executor** of dynamic configuration operations.	**Self-defining** **Context awareness** **Self-configuring**
SW Load Management Service	**Executor** of dynamic load operations.	**Self-configuring**

The interface services are responsible for enabling and facilitating the middleware interactions with application software programs as well as the underlying system and platform support. Such interactions are necessary for monitoring system states, performing the operations of dynamic changes, and obtaining necessary support for start-up and transparent interactions. An interface service can be optional as a DySCAS system can rely on external system and

platform support for the same functionalities for the reasons of performance and resource efficiency.
The fundamental benefit of a DySCAS middleware system is a layered data and control hierarchy where the configuration management problems are stepwise consolidated. Through the hierarchy, the monitored context information is handled and disseminated stepwise from low to high with increasing level of abstraction, whereas the control actuation requests are refined stepwise from high to low with more execution specific considerations.

- *Task level.* This is the highest level DySCAS service for the embedded configuration management. The main function is to process various requests for configuration changes by verifying such requests (i.e., if a request is valid), assessing the overall impacts of requested changes, and deriving a scheme of configuration tasks for performing such changes (e.g., updating a software component and other related components in sequence for a software upgrade request).
- *Elementary operation level.* Services at this level provide support for assessing qualities of the target system and for controlling the dispatches of dynamic configuration tasks. By assessing the degree of quality satisfaction, such services can derive the corresponding needs for reconfiguration, behavior change, and adaptation of effective configuration constraints. At this level, the services also provide support for synchronizing related change activities according to the status of such change activities as well as the status of the target system.
- *Primitive operation level.* The main functions at this level include the supports for consolidating monitored run-time status information and for controlling configuration setup and the deployment of platform resources (e.g., CPU and memory). Services at this level interact with the application programs and system services through the middleware interfaces. For each dynamic configuration task, the services reserve necessary resources, generate relative calls to the target platform and application programs, and perform the supervisory control for the execution (e.g., receiving the platform feedbacks and raised exceptions).

As a prerequisite for self-managed architecture dynamism, a system needs to be aware of its configuration, the contracts and constraints of its components, as well as the external conditions and internal status of applications and resources. This understanding is established through a priori available information of architectural design and variability of target systems in combination with the run-time sensed and derived context information. As the architectural information of target systems specifies how the configurations should be managed, it is referred to as the *meta-information* of configuration reasoning and management decisions by middleware. In DySCAS, it is assumed that all meta-information is derived from the offline specifications of system architecture, configuration setup, and lifecycle management. To this end, DySCAS provides an information model for formalizing various architectural concerns and status information of target systems for dynamic configuration management.
Across the architecture layers, there are three main paths of data and control, representing the main interactions of middleware services in accomplishing configuration management and QoS control. For a depiction of the basic cross-service interaction segments in these paths, see figure 2.

- *Context consolidation path.* The context consolidation path provides support for monitoring and consolidating different status information. Such status information is related to the execution of vehicle and applications, the deployment of target and external resources, as well as the satisfactions of qualities and the related effective configuration constraints. It produces a consistent view of overall system status, which constitutes the run-time context of middleware decisions. In figure 2, the basic interaction segments in this path are denoted with the symbol 1.
- *Decision path.* The decision path starts either when a request for decision is received or when a particular application or system event is detected (e.g., discovering an external device). It provides a system with the ability to reason about the correctness and efficiency of its current state, and to plan for changes without eroding the architecture or violating the

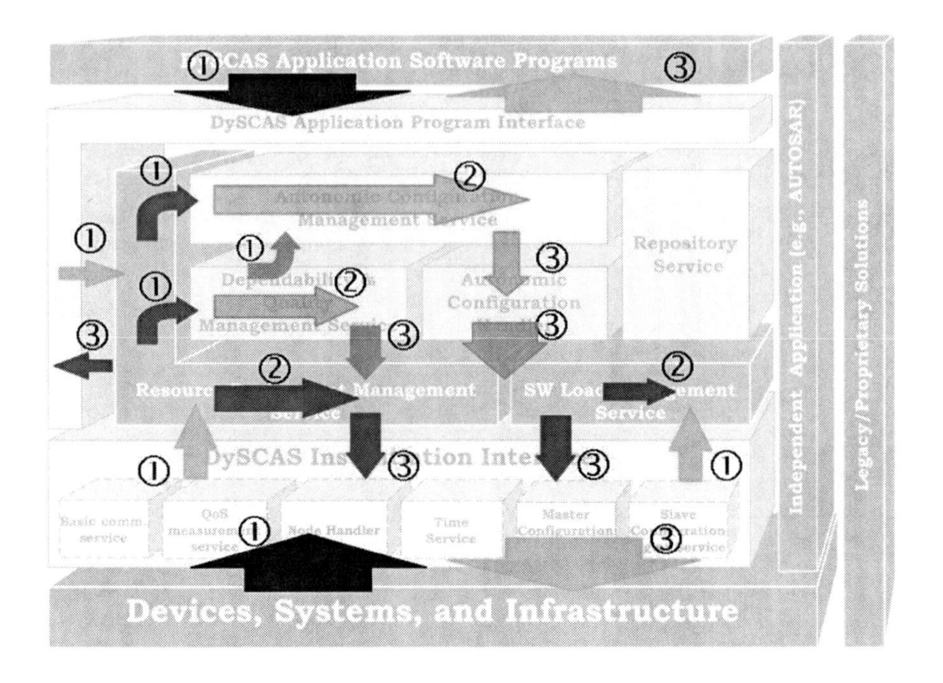

Fig. 2 The basic cross-service interaction segments in the 1. Context Consolidation Path, 2. Decision Path, and 3. Actuation Path

functionality and dependability (e.g., safety, security, and availability). The basic interaction segments in this path are denoted with the symbol 2 in figure 2, and can involve both local and global services.

- *Actuation path*. Of great importance to the DySCAS middleware system is the actuation of dynamic configuration management tasks. This path provides support for a stepwise refinement of the actuation requests and a synchronized dispatch of such requests based on the status of target systems. The basic interaction segments in this path are denoted with the symbol 3 in figure 2.

5 The Component Model for DySCAS Middleware Services

In order to achieve the intended control strategy and the desired maintainability and flexibility of middleware itself, DySCAS adopts a component-based approach. For each middleware service, there is an individual software component/process that interacts with other middleware services with signal-based communication channels via its ports. The DySCAS component model provides a common basis for packaging the middleware services, promoting the modularity, understandability, reusability, and distributed implementation. The component model specifies the interfaces and internal modules of middleware services, the means of execution and interaction, and a strategy for component deployment. The internal structuring adopts the policy-based computation pattern introduced in section 5.1. below. See figure 3 for an

overview of the concepts.
Each DySCAS service component is *active* (i.e. having its own thread-of-control implemented by the virtual execution controller). The contained modules are passive with their behaviors in regards to initialization, operation, and error handling controlled the execution controller. The design is shown in figure 4. For the implementation, the core service components can be partitioned/grouped and mapped to a real-time task/process in different styles (e.g., in one-to-one, many-to-one. etc) according to certain performance and dependability constraints.
Each DySCAS middleware core service component has a set of signal-based external ports for specifying its interaction points to/from other middleware service components. These external ports are declared as *behavior ports* [58] as the conveyed signals are handled by the component behavior (i.e., its virtual Execution Controller). Modules inside the component have internal ports for the internal interactions. Each port is connected to an *interface* that specifies the provided or required information. DySCAS differentiates four types of *ports* covering signal-based communication between components and service-based communications between modules.

1. senderPort - ports for sending signals.
2. receiverPort - ports for receiving signals.
3. clientPort - ports for forwarding calls to operations (e.g., function calls to an object).
4. serverPort - ports for accepting calls to offered operations.

For each software component that makes up the middleware, there is a virtual execution controller defining its behaviors including the start-up process and necessary error handling. Each execution controller is defined based on a hierarchically defined state machine and activity behaviours that are triggered either by external signals received through the component ports or by internal signals generated by the internal modules [58]. See figure 5.
The interfaces associated with a port specify the nature of the interactions that may occur over that port [58]. For a DySCAS middleware core service component, interfaces are used to declare the signals to be exchanged between its ports. Such messages are given either as signals for the interactions with other middleware service component, or as operation invocations for the interactions between the contained modules. A DySCAS signal is a specification of messages communicated between the service components in terms of active objects. By sending a signal in a non-blocking way, a sender can trigger a reaction in the receiver asynchronously. The attributes of the signal specify the carried data. The interfaces are shown in figure 4 and are listed in table 2.

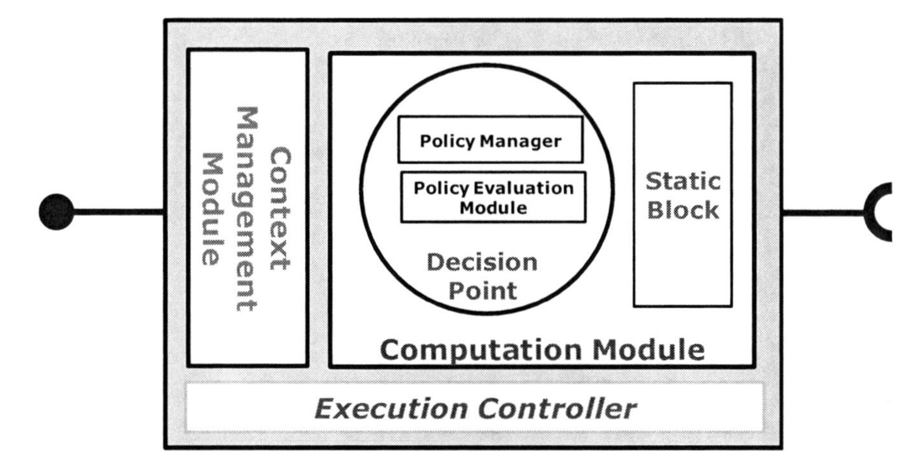

Fig. 3 Key concepts in the DySCAS MW component model

Table 2 definition of component interfaces

Interface	Purpose
I_ContextInfoNotification	Declaration of signals for notifying the decision context and other data.
I_ContextInfoPubSub	Declaration of signals for (un)publishing and/or (un)subscribing the decision context and other data.
I_ServiceFeedback	Declaration of signals for providing the feedbacks of requested middleware services.
I_ServiceRequest	Declaration of signals for requesting middleware services.
I_dyscasEvent	Declaration of signals for disseminating synchronisation events.

DySCAS service components communicate by asynchronous message passing with buffering queues. Referring to the state machine model of the execution controller, a component always reads and removes all the messages from all the input message queues, and carries out the computation using these messages. After the computation, the component writes messages to its output message queues if necessary. Mandatory attributes of these message queues are length and data type. DySCAS provides some predefined attributes of middleware data for specifying how multiple instances of such data should be handled during message sending and receiving, and queue handling. It distinguishes the persistent data that should not be overwritten with newly generated instances until it is used/read (e.g., for the requests of middleware services), from the transient data that can be overwritten with newly generated instances (e.g., for the monitored context data of internal states and external conditions). In the DySCAS component

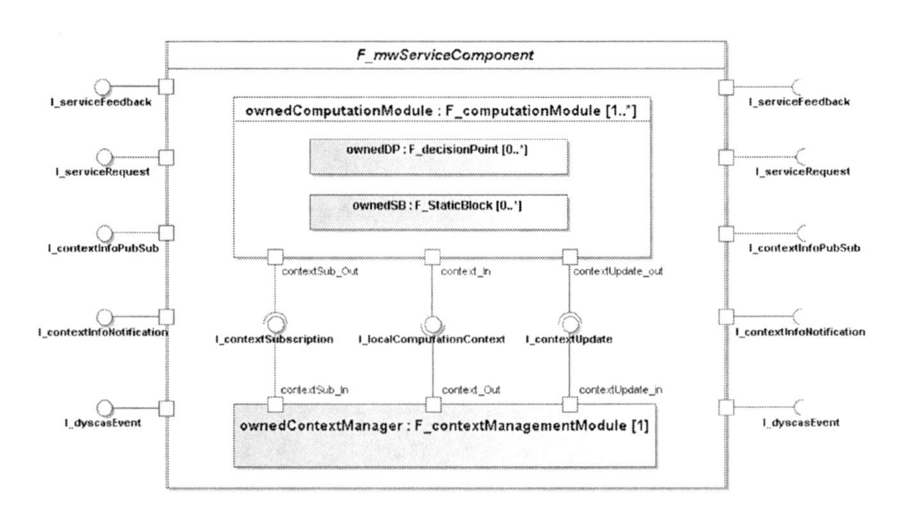

Fig. 4 Composite description of the DySCAS core service component model

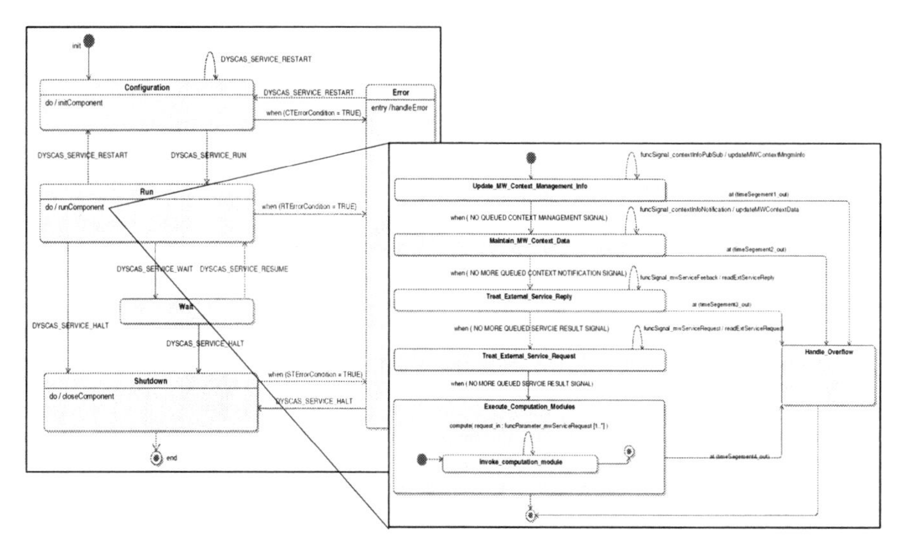

Fig. 5 Hierarchical state machine behavior description for middleware service components through the execution controller

model, a message queue can be either nonblocking on both reading and writing, or nonblocking on reading and blocking on writing. If a message queue is empty and a component attempts to read it, the reading component is not blocked. It may either perform the remaining computation using the old values or abort the current job, depending on the algorithmic design of the component. If a message queue is full and a component still wants to write into it, the writing component may be blocked if the content of the queue is history sensitive (i.e., with not overwritable persistent data), like the requests to integrate a new device or the task migration request. If these service requests were lost, a perceivable deterioration of system performance could be noticed by the user. In such a case, the writing component should be synchronized either through the events indicating the occurrence of a queue overflow situation or through careful scheduling. If the content of the queue is not history sensitive, the writing component may immediately receive a status report of the queue or overwrite the old messages in the queue.

5.1 Policy-based configuration in the DySCAS component model

The configurational flexibility in a DySCAS system is realized using techniques inspired by the policy-based computing paradigm. In contrast to those approaches that only provide change from one entire system configuration to another, this method allows incremental changes to occur independently at various points throughout the system. This enables future proofing of vehicles at the point of development, facilitating upgrades throughout their lifetime.

Rather than a centralized, monolithic intelligence system; in a future vehicle control system, the configuration 'intelligence' needs to be distributed across components for reasons of flexibility, maintainability and scalability. DySCAS achieves this at three levels:

1. A flexible, designed-for-purpose middleware, which incorporates -
2. A versatile component model which supports dynamic mapping of components' context information requirements, and embeds -
3. Policy-based configuration, in which each component can include a number of policies which can be easily upgraded without changes to the deployed code.

The DySCAS architecture has been designed with dynamic self-configuration as a core feature. The fundamental concept is that each software component that makes up the middleware may embed one or more Decision Points (DP), which are place-holders for policies. Each DP can be dynamically configured by loading a policy at initialization. Within a DP, the Policy Manager is responsible for requesting the appropriate policy from the on-system repository, whilst run-time evaluation of the policy is handled by the Policy Evaluation Module (see figure 3). If a component has multiple DPs, each operates independently having its own policy and context requirements.
The DySCAS component model facilitates the use of policies in a robust run-time framework comprising three main policy-related features: DPs; a dynamic wrapper (DW) which decouples policies from their host components, handling faults and making the integration of policies transparent to the design of host components; and dynamic context management which decouples context producers from context consumers and greatly enhances the flexibility of configuration changes and the ease by which these change can be effected. The DP, DW and dynamic context management are described in later sub-sections. The model also specifies the following policy-related functions:

- A method for design-time embedding of DPs into software components;
- Run-time support for the operation of DPs;
- The ability to specify default behavior per DP, which is actioned if for example, a policy is not loaded;
- A mechanism to dynamically load the appropriate policy into a DP from an on-system repository;
- A mechanism to dynamically replace a policy with a new version;
- A mechanism to automatically map the required context information to each DP.

5.1.1 Decision Points

A basic characteristic of computing systems in the automotive domain is the difficulty in making changes post-deployment during a long vehicle lifecycle. Under current practice this can only be achieved by directly servicing each vehicle by suitability qualified personnel with specific equipment. In contrast DPs allow flexible run-time configuration of software components. Policies are developed and validated off-line, and can then be loaded to the in-vehicle system by a variety of means, depending on the circumstances requiring the new policy: at vehicle commissioning; during routine service; on-line purchase of functionality upgrades; automatically loaded to the vehicle from a manufacturer's system via a wireless network hotspot, perhaps to achieve a field upgrade remotely; or from an attached USB memory stick, enabling driver / owner customization.
The software developer identifies, at design time, places in the software where dynamically changeable behavior is appropriate. At each of such places a DP is embedded into the compiled code, marking out the possibilities for reconfiguration after deployment. The way in which a decision is made (the logic) is not statically compiled into the DP; it is specified in a policy which is loaded (in the form of a data file, via an API method) into the DP at runtime. This separation of decision logic from the compiled code is key to post-deployment incremental upgrade in DySCAS.
When inserting a DP it is only necessary that the developer identify the number of possible decision outcomes of the DP, and also that a special 'Default' behavior is defined (which can

coincide with one of the regular outputs). If any problem should occur during the evaluation of the DP at run-time, the policy library mechanisms will sidestep the policy evaluation and automatically return the default outcome. In this way dynamic policy logic collapses down to statically defined, predictable behavior if any problems are detected. The DP acts as a sandbox for policy evaluation. This behavior is discussed in more detail later in section 5.1.2 'The Dynamic Wrapper'.
The actual program code that constitutes the software component is provided as static function blocks and apart from policy logic a component can only be upgraded by recompilation and redeployment. The use of DPs is illustrated in figure 6.

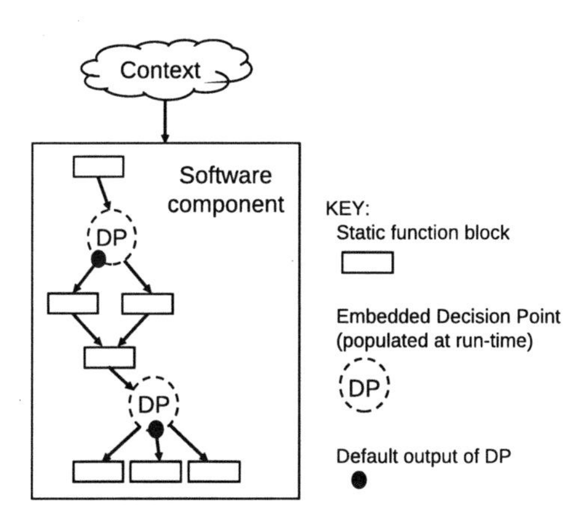

Fig. 6 A software component with a combination of multiple DPs and statically compiled functional blocks

The DP concept also future-proofs systems. There are circumstances where a developer is aware that future enhancements to behaviour will be necessary, but is not aware of the details at the time of application deployment. In such cases DPs can be embedded at the appropriate places in the logic and very simple policies can be provided initially which can be replaced with more-sophisticated logic when necessary.

5.1.2 The Dynamic Wrapper

The AGILE policy library is encapsulated into the Decision Point and is referred to as the 'Policy Evaluation Module' (see figure 3). The Dynamic Wrapper (DW) is actually implemented as part of the Policy Evaluation Module but is logically separate as it oversees the evaluation of policies. This provides DP-internal localized self-management which is transparent to the rest of the system, providing dynamically configured interfacing and automatic fault-hiding.
The DW monitors the whole run-time policy lifecycle. This lifecycle includes the following steps:

1. A policy is loaded into the Policy Evaluation Module.
2. The DW monitors the parsing of the policy scripts and from this extracts a list of 'environment variables' used by the policy (each environment variable relates directly to one item of context information needed during policy evaluation).

3. The DW automatically requests subscription to the necessary context items identified from the list of environment variables.
4. The DW monitors the provision of context items from the middleware resource manager service.
5. When invoked, the DP initiates evaluation of its policy. The DW checks the outcome of this evaluation, checking for internal errors which have to be handled transparently (details of these errors are provided below).
6. If for any reason the policy cannot be correctly evaluated, the DW returns the design-time developer selected 'default' return value on behalf of the policy.
7. If any internal faults have occurred the DW sets internal error flags. The value of these flags can be retrieved into the host component's operating space via a call on the DP API.

By automatically and transparently handling any problems arising at run-time related to the evaluation of policies, the DW ensures that robustness is not compromised by the addition of the policy logic mechanism. The 'dynamic' behavior aspects automatically downgrade to statically-specified behavior when necessary.
Run-time errors that are detected and handled by the DW include: No Policy loaded; No appropriate policy found in repository; Policy had parse errors, or was not referentially self-consistent; Policy outcomes set does not match the DP configuration; and Required context items not available.
Through its silent error handling (i.e. by trapping errors generated in the Policy Evaluation Module, and returning a predefined 'legal' return value to the component), the DW makes a significant contribution to system robustness. The implementation of dynamically configurable components has been achieved in a manner which can be only advantageous in comparison to static components and will never decrease the system stability and integrity. From a component developer point of view, the DW is entirely transparent, because it works silently and cannot cause component failure. The only way to detect its intervention is to check whether error flags were set or not. In general, a developer can simply use the decision result produced by the DP in their code.

5.1.3 Dynamic context management

The achievement of context-awareness depends on the ability to map the necessary context information from context providers to context consumers. Any form of direct mapping increases the complexity of the system, inhibits change because of inter-dependencies and also impacts on scalability in terms of the number of components and the amount of context information they use. Design-time mapping restricts the extent of dynamic configuration that can be achieved, for example new polices could be added but would be limited by the predetermined context information provided to the particular DP.
Thus DySCAS uses a dynamic context mapping, in which components subscribe to the particular context information they need, via a context manager service within the middleware. Context providers and context consumers are de-coupled allowing flexibility for incremental upgrade of individual components and dynamic upgrade of policies.
The dynamic context management operates on a publish-subscribe basis. When a policy is loaded into a DP the associated DW automatically subscribes to the required context items (using the Context Management Module shown in figure 3 as a local agent for the required communication). The context management service then pushes out the context items to DPs only when the value changes. The value is cached within the Policy Evaluation Module, in the form of an 'environment variable' within the policy logic. This approach has the following key benefits:

1. Depending on the rate of change of context values, this method has the potential to keep communication overheads low and likewise the number of context switches when components have to handle received messages.

2. The latest context value is already held in the policy logic at the point of policy evaluation. There is no need to request a context update and suffer the latency of retrieving the value from a different component.
3. Context requirements are policy specific. When a particular policy is upgraded the associated DW simply changes its context subscriptions; no other components or policies are affected.
4. A completely new policy-enabled component can be added to the system, which either publishes or subscribes new context information. In both cases the addition of the new component can be transparent to the rest of the system.

A further key benefit of dynamic context management is that it provides transparency to the developers of software components in that a context-producing component does not need any design time consideration of consumer components, and a consumer component does not need any design time consideration of what context will be used or where it is generated.
Managing the context information in this way also supports reconfiguration of the location of running software. For instance, an important software component may be shifted from one node to another due to resource availability. The context information required by this component's DPs can be routed to the new location dynamically.

6 Autonomic reconfiguration

Automatic reconfiguration is triggered by the middleware itself or by external events. A configuration change can for example cause reallocation of tasks to processing nodes, of which addition or removal of tasks is a special case, and switching QoS modes for applications, requiring other resource amounts for that specific application. Even though defining the exact algorithms to be used for the reconfiguration is not the core aim of DySCAS, having a substantial understanding of them is vital to build interfaces and structures that make it possible to incorporate different relevant approaches.
In this section we treat the following aspects of automatic reconfiguration:

- *Task migration*, where tasks are moved between ECUs, where one specific purpose includes *dynamic load balancing* where allocation is performed during runtime.
- *Quality of Service* techniques where parameters affecting the performance and resource usage of an application are adjusted during runtime to control overall resource usage in the system.
- *Admission control*, where some applications are denied to start execution due to an insufficient amount of available resources.

As briefly discussed in [62], there is an important trade-off between these reconfiguration techniques and it is difficult to decide which will be most effective without taking application characteristics into account. A further difficulty is that there do not exist unified frameworks even when considering a single of these techniques [61]. This is illustrated by the large number of different abstractions available for example in terms of task models, component models and machine models. While, unification would be highly desirable and would have been of great value in architecture design, this has not been within the scope of DySCAS. Instead we have attempted to develop an architecture that will allow incorporation of a large class of approaches. The following sections describe autonomic reconfiguration schemes that we developed. These schemes show how reconfiguration algorithms can be integrated into the architecture.
Two additional key concerns in developing an architecture that supports autonomic reconfiguration are algorithm complexity and system verification. As shown in e.g. [28], finding an optimal solution to many of these problems is intractable. This is the case even for static systems [70]. It is however seldom necessary to find an optimal solution, hence the focus of the

work has been on efficient and effective heuristic solutions which can be applied to resource-constrained systems (in terms of both memory and computation power). System verification is treated further in this section and also in section 8 (A Framework for modeling, designing and analyzing dynamically configurable systems).

6.1 Task migration as an actuation mechanism

When a reconfiguration has been decided upon, it needs to be actuated. Reconfiguration might imply task migration to another node during runtime, something which is further discussed in [40]. This also implies that the new node will have to be able to start a new activity (e.g. OS process) for the migrated application at runtime. This could be implemented using dynamic loading [45] or through pre-allocation of programs. Although most resource-efficient OS implementations don't have such capability, there exist a few examples of very compact operating systems that do [23]. There is even an implementation of an automotive style system, capable of performing task migration in the DySCAS sense, if given an external trigger to do so [40].
Finally, to accomplish migration, one also has to consider the typical hardware heterogeneity in an automotive environment. If tasks are to migrate between nodes with different types of processors, they will have to be written in an interpreted language or run in a virtual machine (like Java), or be provided in several compiled binaries. Also, checkpointing [34] is a technique that can be used to provide hardware-independent storage of application state to be transferred to the new node.

6.2 Using policies for flexible reconfiguration mechanisms

DySCAS reconfiguration is implicitly dynamic, but can be triggered in a variety of different ways. As described above, there are a large number of possible approaches for reconfiguration, and which one is best for each individual reconfiguration problem can in general only be decided on a case-by-case basis. This means that an implementation of the DySCAS middleware will preferably need to be shipped with several of these mechanisms. Policies will then be able to decide which mechanisms should be used for each individual task, and what tuning parameters to use in utility functions etc. Using policies for this purpose also makes it possible to improve the reconfiguration support "in-field" by downloading new policies for the Autonomic Configuration Management Service, which is responsible for reconfiguration planning, along with normal software upgrades.

6.3 Algorithms and an approach for Dependability and Quality Management and Autonomic Configuration Management

This section describes developed algorithms and a concrete realization of configuration management, focusing on two DySCAS core services; the dependability and quality management service (DQMS) and the autonomic configuration management service (ACMS). In the models we make the following assumptions:

- There are local DQMS services, allocated to each node of the embedded system, and a central (coordinating) DQMS. The ACMS is only realized as a central service.

- Applications have QoS levels that represent different qualities and re-source usages (e.g. by providing different algorithms/programs).
- Three types of resources are considered: CPU execution, memory and network bandwidth.
- A major assumption in this implementation is that a task may execute only if all its required input signals are available in the network.

Each node in a network thus hosts a local DQMS, which performs admission control and optimizes the resource utilization at that node. If the admission control rejects certain tasks from running at that node, the local DQMS submits a request for load balancing to the global DQMS, located at a master node of the network. The global DQMS chooses other nodes, if possible, to redeploy the tasks. The ACMS enforces the interdependency among tasks, such as task communication and precedence constraints.
While the DQMS manages the physical resource constraints and ACMS maintains logic constraints, the two are closely coupled, as shown in Figure 7. The admission control algorithm of DQM first determines a set of schedulable tasks according to the estimated resource usage of tasks and the resource limit of the system. ACMS further chooses, from the schedulable task set, the tasks that satisfy the logic dependency, namely the active task set. Only the active tasks will run under the scheduling of RTOS. During the run-time, the actual resource usage of tasks and the resource limit vary dynamically owing to the changing workload, the emergence of new tasks, and the hardware failure. DQMS, therefore, dynamically allocates resources to the tasks and maintains their minimal QoS. If some tasks must be rejected owing to resource overload, or new tasks can be allowed owing to low resource utilization, DQMS invokes the admission control to decide a new schedulable task set. These decisions form a loop and must be carefully designed to avoid instability.

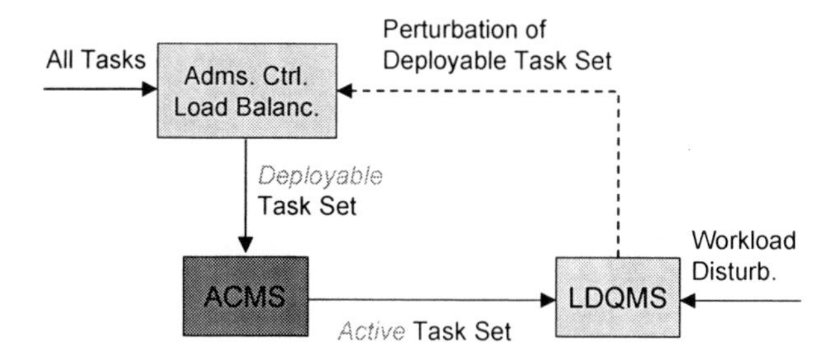

Fig. 7 Interaction of dependability and quality manager and automatic configuration manager

The algorithm design of DQMS and ACMS aims to solve the following self-configuration problem [27]: given an ECU network with limited resources and a group of dependent tasks with estimated resource usage and benefits, find an optimal task set that maximizes the total benefit. This problem is NP-hard [33]. Given the limited computational power of automobile ECUs and the requirement for timely response, we opt for approximate answers rather than optimal solutions, so that acceptable decisions can be decided quickly. The following two sections briefly preview the design of DQMS and ACMS, respectively.

6.3.1 Dependability and Quality Management Service

The DQMS dynamically manages the physical resource constraint over the network and allocates proper resources (CPU, memory and network bandwidth) to each application task during runtime. To realize dynamic resource allocation, we assume that each task can run in discrete

QoS levels. Each level is associated with the estimates on resource usage and the benefit to the system. With this information and the given resource limitation, the example implementation aims to (1) provide quick response on resource deployment, (2) maximize the overall benefit of the system, and (3) support as many application tasks as possible.

Every node incorporates a local DQMS which realizes two services: local admission control and QoS adjustment. The former service decides if a group of application tasks is schedulable at an ECU without violating resource limitation. The latter contains a suite of algorithms to maximize the system benefit by allocating more resource to more important tasks. These algorithms provide not only a suboptimal solution based on resource estimates, but also reject unknown disturbances using a feedback mechanism. The main functional modules of local DQMS are shown in Figure 8.

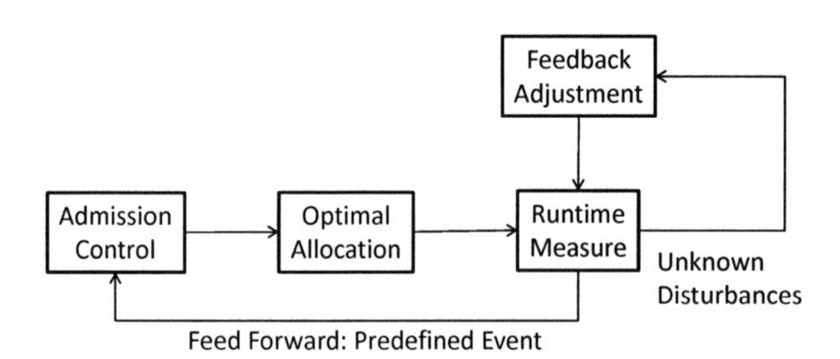

Fig. 8 Functional Modules of Local DQMS

Following the DySCAS architecture, this implementation includes a global DQMS to increase the number of deployable tasks in the system. The global DQMS decides how to migrate tasks over the network when a node suffers re-source overload. When this happens, the local DQMS selects the tasks to be relocated and reports the request to the global DQMS. Receiving the request, the global DQMS tries to find a candidate node that could support the task and reaches the highest system benefit. In the situation that no other node has enough idle resource to accept the task, the global DQMS finds a node that is least influenced by the task migration, i.e., receives the lowest rejection penalty.

The local DQM function has been realized in a Matlab/Simulink/TrueTime [59] environment (See Section 8 for more details on DySCAS simulation environments). In one simulation, four tasks running at one ECU are considered. Owing to the resource constraint, they cannot all run at their highest possible QoS levels. The simulation shows the scenario that at 3 seconds, a failure occurs and the CPU loses 50% of the processing power. At 7 seconds, the failure is recovered and the CPU resumes its full power. Figure 9 shows the change of QoS levels of the tasks in 12 seconds. In this graph, two tasks always keep the lowest QoS level 1, because they provide small benefits to the system.

6.3.2 Autonomic Configuration Management Service

An acceptable configuration is a combination of middleware components and application programs satisfying logical dependencies. An application program requires certain preconditions to run and provides services to others. If all pre-conditions are not satisfied in the network, the task may not execute. This type of logical dependency also encompasses precedence constraints.

According to the DySCAS architecture, the configuration of the network is maintained by the global ACMS. The ACMS takes as inputs from every node their intended application tasks,

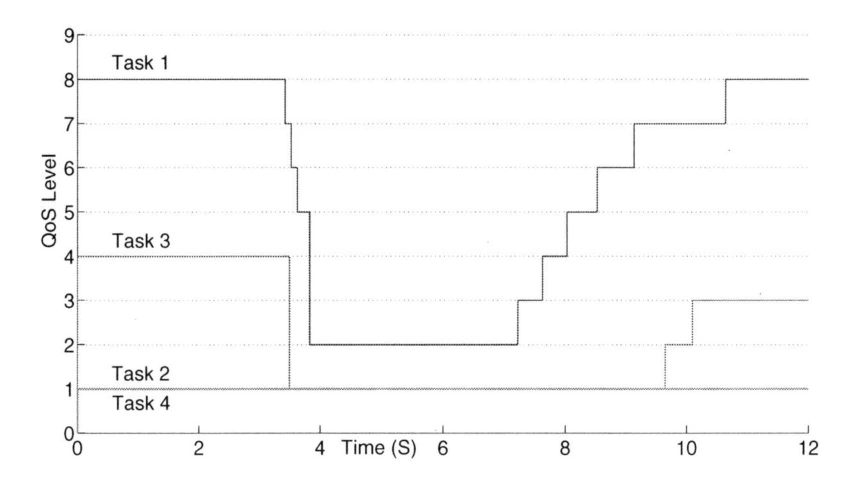

Fig. 9 Change of QoS Levels of four tasks in the example system

and their provided/required communications including preconditions. The ACMS dynamically maintains these data for the entire network and decides the possible configuration. The dependency of all application tasks is modeled as a bipartite graph and new configurations are inferred based on this graph.
To find an acceptable configuration, the concept of a *stable* task set is defined. Each application task in this set has all its preconditions satisfied by the services provided by other tasks in the set or the sensor inputs. To respect the logical dependency, the output of ACMS is always a stable task set. Under mild assumptions, in any set of networked applications, the greatest stable task set always exists, which is the optimal configuration under the given logical dependency and which can be computed in polynomial time.
Also this proposed method has been implemented in Matlab/Simulink/TrueTime and verified via simulation [28]. A simulation assumes the simple imaginary distributed system depicted in Figure 10. The system consists of two ECUs connected by a Controller Area Network (CAN). The two big squares represent two ECUs and the small rounded squares represent application tasks.
All tasks, except the highlighted task dev, are permanently allocated to the target ECUs. Task dev serves for a new device that might be attached to ECU1 in the future; hence it is normally absent in the system. The inputs of a task represent the preconditions and the outputs the provided services. When the device is connected / disconnected to ECU1, it will trigger an external interrupt, which loads / removes the application task associated to the device in the ECU and calls upon a system configuration.
In the simulation the device is attached at 3 seconds and removed at 7 seconds. The two plots in Figure 11 show the execution schedules of all tasks at the two ECUs. The simulation agrees with the theoretical prediction.

6.4 An Approach for Load Balancing

The *Load Balancer* spreads tasks between the vehicle's ECUs in order to get optimal resource utilization and decrease computing time. It evaluates possible migration of tasks based

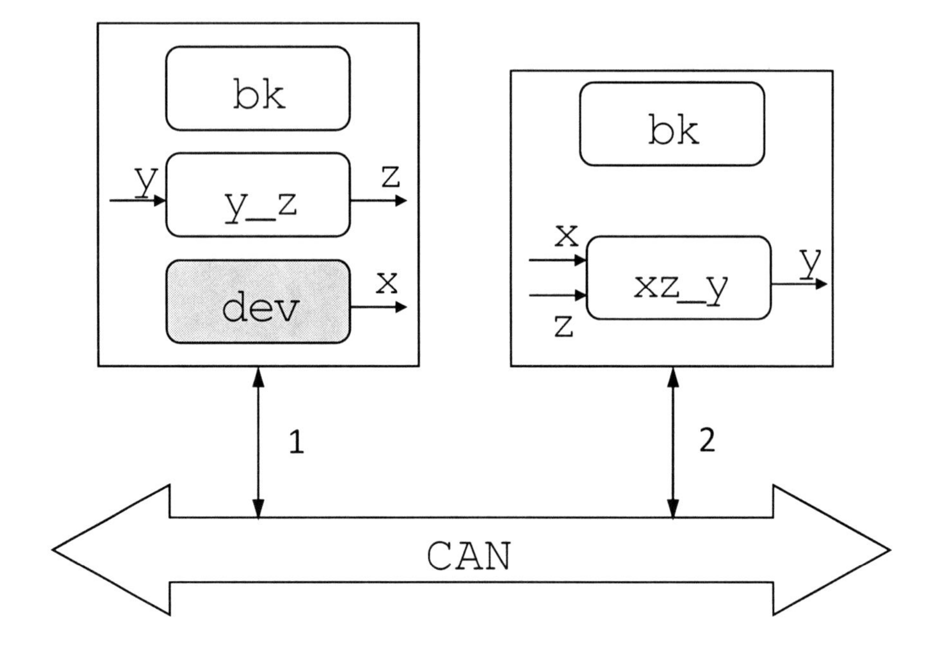

Fig. 10 A Simple Networked Embedded System

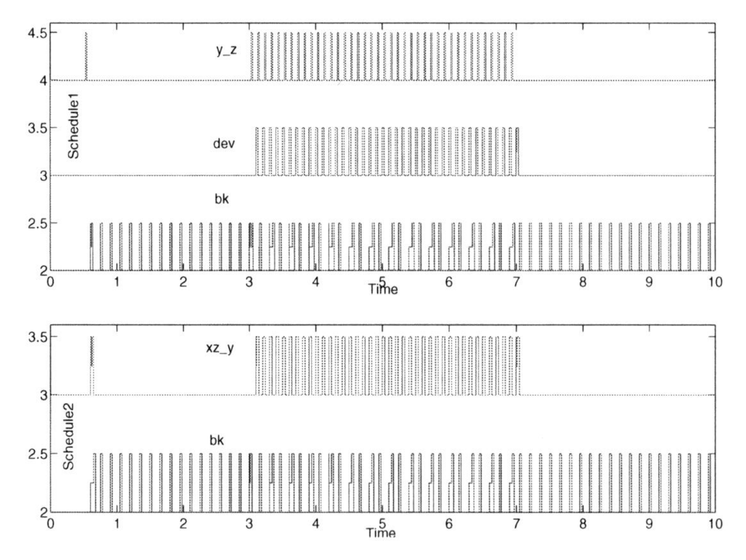

Fig. 11 Task Schedules of the Configuration Process

on different load balancing strategies. To guarantee a suitable migration the *Load Balancer* considers the current resource situation on the ECUs with aid of the *Resource Manager*. If a failure has occurred the *Load Balancer* tries to find a feasible migration, based on the characteristics of the tasks and ECUs. If, after a migration has occurred, the host device is detached, the migrated process will be re-started on the original ECU. In this case the *Event Manager* is responsible to inform the *Load Balancer* to initiate this re-start.
The *Resource Manager* supervises the resources of the local ECU. To be aware of the complete network resource situation all *Resource Managers* synchronize with each other. Thus the *Load Balancer* gets the current resource situation of the complete vehicle infrastructure with aid of its local *Resource Manager*.
In our approach, the middleware is located on each ECU in the vehicle. Every ECU has a unique ID. The ECU with the lowest ID is the master. Thus it is responsible for the control of the entire vehicle network. Newly connected devices are discovered by its *Event Manager*, device information is registered by its *Registry*, and its *Load Balancer* is responsible for the evaluation of the possible migration with the aid of the local *Resource Manager*. If the master ECU fails a new master will be chosen with the aid of the Bully-Algorithm.
ECU failure detection will be handled by a hardware interrupt. It initiates an error correction in our middleware. That means, to correct the error, tasks of the omitted node are migrated to other ones, which are able to execute them. In this chapter we will not focus on the failure detection but on error correction. Therefore, our middleware must be able to migrate tasks. A detailed knowledge of the task characteristics is needed. It is important to know if it is a real-time task or not.

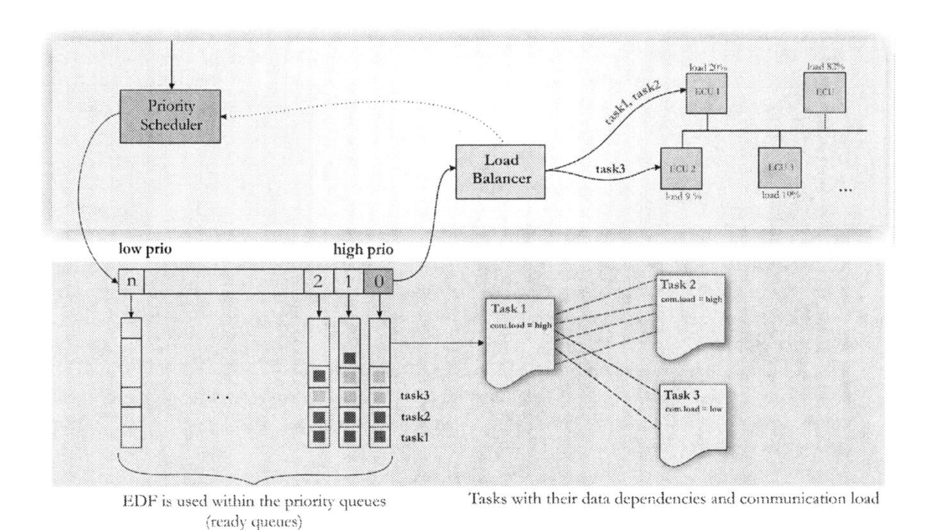

Fig. 12 Failure correction handling - the task migration mechanism

Figure 12 presents our approach for task migration. We assume that each task has a priority and we have a detailed knowledge about their hardware requirements. Additionally, the data dependencies between the tasks are known. As we can see from figure 12 we start with a priority scheduler. This will schedule the tasks according their priority in priority queues. That means each priority has its own task queue. Within the queues the tasks are scheduled by a simple earliest deadline first (EDF) scheduler to ensure a flexible schedule. Real-time (RT) tasks have a high priority. The *Load Balancer* works on the priority queues beginning from the queue with the highest down to the lowest priority. For each selected task a possible set of

ECUs which are able to execute the task is evaluated. After that a data dependency check will be done. That means, we look at those tasks that interact with the inspected one. In the case that interaction is weak the *Load Balancer* selects an ECU from the previously evaluated set of ECUs and finally migrates the task and deletes the task in the priority queue. In case of a strong interaction the *Load Balancer* will try to avoid unnecessary bus load, by selecting an ECU from the ECU set that is able to execute both tasks. Afterwards both tasks will be deleted in the priority queue. If the *Load Balancer* could not find a possible ECU for migration the task will be deleted from the queue with the outcome that a migration is not possible.

The previous paragraph gives an overview of the migration, but there are still some open issues we will discuss in the following. If an ECU with more than one task running on it fails, we will migrate the tasks to one or more ECUs according the classification of the tasks (see figure 12). That means tasks with high priority will migrate first, followed by the other ones. During the migration phase, the timing of the tasks are taken into account. After a task migration, we have to decide to start the task "from scratch" or from the state it had before the ECU failed, but how is this state recognized? For this, we need the context of the task. Our solution is the following; if we have a context available (e.g. stored in an external flash memory of the ECU and still available) we will invoke the task with the context, otherwise not. This gives a brief overview how our middleware migrate tasks. Finally the decision as to which tasks are migrated is done by the *Load Balancer*.

Figure 13 shows a sequence diagram where a failure occurred in the radio system. We assume the tasks from the radio system can migrated to the navigation system.

As we can see in Figure 13 the *Event Manager* detects the failure of the radio system, this is done by the function `failure_detection(error_code)`. Afterwards the *Event Manager* triggers the *Load Balancer* with the `initialize()` function. The *Load Balancer* asks for all device information from the *Registry* `req_loads(*device[0..n])`. Then the *Resource Manager* runs the `schedule()` function to calculate all possible schedules. The *Load Balancer* will get the device information back from the *Resource Manager* with `ack_loads(*device[0..n])`. Finally the *Load Balancer* runs `initiate_load_balancer()` that calculates which tasks could be moved from the device with the failure to another one based on the information of the schedules, the load of each processing element in the car network, the communication costs and regarding the feasibility. In our case it will decide to move tasks from the radio to the navigation system.

In the last paragraph we described the interactions between the four tasks, which are necessary to support load balancing. Now we will discuss the internal data structure of our middleware. The *Event Manager* triggers the Registry and initializes the *Load Balancer*. The *Registry* itself interacts with the *Resource Manager* and the *Load Balancer*. The *Resource Manager* hands over the actual status of the entire system to the *Load Balancer*.

To perform the scheduling in the *Resource Manager*, we can select between different scheduling strategies. They are instantiated within the scheduling mechanism class of the internal data structure.

The *Registry*, as well as the scheduling mechanism, needs information about all tasks and devices. This is handled by the so called `list` class. It contains linked lists of devices and tasks and offers functions to manage the lists. As described before list offers all functions to manage the task list, but additional functions to set the status of the tasks are needed. The status of the task can be running, waiting or sleeping. Besides this, the task manager is able to create a new task. The information of a task is stored in the data structure provided by the task control block. The parameters of the generated structure are set by the task manager with functions from the list class. The list class uses the functions from the task control block to get task information.

For the devices we have the same functions available as for the tasks. This is realized in the device control block. Each device has a list containing the task-ids that are running on the device. By setting the global variables of our middleware we can initialize the system and can set it in running mode.

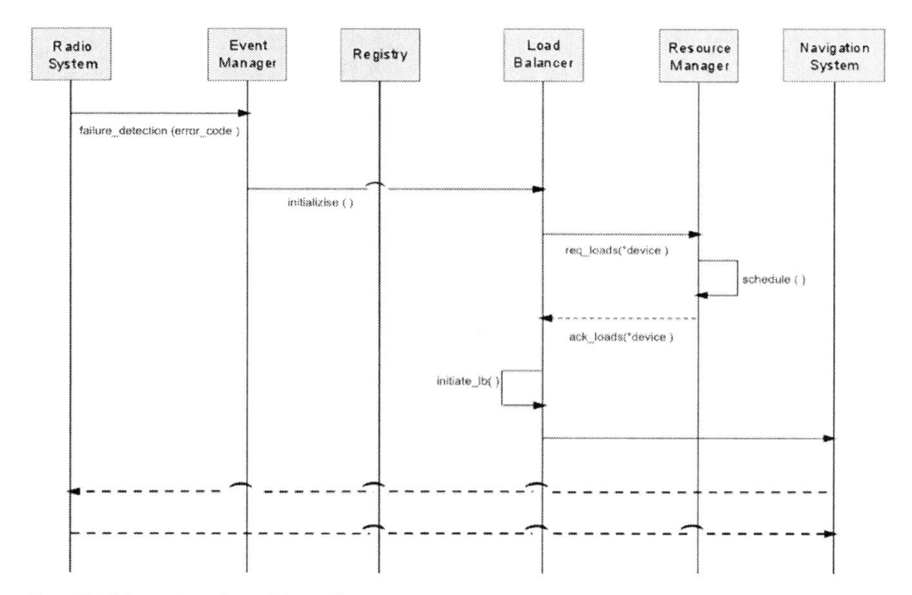

Fig. 13 Failure detection of the radio system

6.4.1 Load Balancing Strategy

There are several possibilities to balance the load after an error happened inside the vehicle infotainment network. Initiated by the *Load Balancer* component the new resources can be used and applications or tasks can be migrated to the additional device.
In the following the cost-based load balancing strategy is briefly described. Within the cost based strategy the *Load Balancer* evaluates possible migration of tasks from one ECU to another. It evaluates a set of ECUs where the task could be migrated. Hence the migration is only a useful option if:

- the cost of migrating is lower than the cost of keeping tasks with their original device and
- it is feasible to migrate a task or a set of tasks from one ECU to another one.

The cost benefit ratio for tasks of busy devices is computed which helps the *Load Balancer* to form the decision of whether to migrate or not. The calculation of the migration costs of a task is realized according to the priority list of the Most Loaded strategy. Most Loaded generates a priority list which ranks the tasks from the busiest processor. In that way the tasks with the highest priority will be migrated to the resources of the additional device.
Let us assume we have tasks t_i with $i = 1$ to n, and the utilization of the task running on an ECU is u_i. Additionally, let U_j the maximum utilization of ECU e_j with $j = 1$ to m. Then the upper bound for the utilization of an ECU e_j is:

$$\sum_{i=1}^{n} u_i \leq U_j$$

For the communication we can make the following assumptions. Let c_k with $k = 1$ to r the communication channels in the vehicle and C_k the maximum costs a channel c_k supports. Furthermore, let $m_{i,k}$ be the cost task t_i produces on channel c_k. Then we can define the following bound for the *communication cost* of a channel c_k:

$$\sum_{i=1}^{n} m_{i,k} \leq C_k$$

Now our *Load Balancer* has to find an optimal balancing for all tasks within in the vehicle network regarding the utilization, communication cost and the feasibility. This can be done with integer linear programming (ILP) or other optimization methods.

7 A reference implementation of DySCAS

The architecture description provides freedom for the system designer to optimize the middleware for specific deployments as it does not specify the actual implementation in detail. Section 7.1 places the DySCAS concepts and architecture into the context of an example implementation called *SHAPE*, highlighting many of the DySCAS features that contribute to the dynamic self-management and high robustness.

7.1 Implementation of the DySCAS architecture

The application layer is the highest layer in the system. This is where the system developer will deploy applications, using the many systems calls that are available via the SHAPE API. The calls are normally only available as signals for inter process communication. Each call enables middleware services of varying types. The application has to register and subscribe system services. In addition, there is negotiation to decide which service characteristics are of interest. The next section describes a typical DySCAS use scenario.

7.1.1 Attaching a new device - a use scenario

When a new device is available within the network (wireless or wired), the device announces its presence. In addition, the system announces its existence. The device is registered and set up in the system as a system member. The new device may offer service to selected parts or to the entire network. In addition, the new device may also want subscribe to services offered by the system.
Services are offered in terms of characteristics rather than physical requirements. It is now up to the system to match the offered services with required services where possible.
Example: offered service = hardware device, type = data storage, size = 512kb. If the required service's minimum storage size constraint is 256kb, then this offered service is acceptable.
With this constraint satisfaction matching instead of hard matching the system is able to choose and optimize during runtime for the current situation.

7.1.2 Application policy management

The application layer provides calls for applications to use policies to increase runtime flexibility. It is possible to add Decision Points into an application and let the middleware execute them. This implies that the application needs context information for decisions. Context may include system-level state, and possibly some application-specific state, which can be optionally handled by the system context management. One reason for choosing this option is lack of local memory (e.g. on the new device where the application runs), another reason is to enable sharing of the context among several applications.

Context Management

The Context Manager is an application-supporting tool for managing the application-local context. Instead of storing the context within the application, the application is able to store the context in a shared storage managed by the Context Manager within the middleware. This storage is still local for fast access. All applications are now able to share the context as if it is a system context. Besides the mentioned sharing, the Context Manager also is able to maintain the storage. The application may then focus on its main mission. The context is made available globally via the network of local Context Managers. An algorithm for optimal data managing, data constancy and high performance is employed within the context management sub-system.

Application Policies

Every application is able to take advantage of the SHAPE built in policy management, thus running the same policy engine as the system itself. The Policy Manager adds the policies the application wants to use. The Policy Manager operates with both local and global scope, in similar manner as the Context Management.

7.1.3 Checkpoints

Checkpointing saves the state of a defined set of processes at certain times or events. The system may then be able to restart the set of processes with a corresponding state the processes had at a certain event or time. Checkpointing also facilitates process migration, which is key functionality necessary for load balancing or process re-configuration. This in turn enables down-sizing the vehicle hardware requirements whilst still keeping the system robust.
Checkpointing also supports the possibility to run hibernation. In a vehicle, a device is perhaps detached from the network and therefore loses power. When reattached, it will resume relatively quickly and be in service. DySCAS incorporates QoS support. For instance, in the case of power management, a signal starts the hibernation of a specific device. This may be a controlled or non-controlled power down action.
Checkpointing is a useful tool for debugging running systems, reducing the development cycle time. The built in test library is able to store debugging information that makes it possible to load and run analyses in a simulator. Simple debugging is able to operate on line directly in the running system. In SHAPE the system is a peer-to-peer system. This means that each node and each process has to maintain its own checkpoints internally.

7.1.4 Versioning

In the case where some upgrade action failed and a roll-back is the best option, the system will do this. This is a typical risk when updating and changing software. The newly loaded module may be corrupt but passed all security checks; subsequently the system realizes the problem and suggests the decision to remove the faulty module and roll-back to a previously working configuration.
Versioning not only provides the possibility of loading new versions of software components into the system; it also provides extended flexibility.

It is possible load several more or less complete policy schemes for a given application into the repository. During runtime the system may swap the version of policy schemes (i.e. load different policies into the various DPs) and thereby change the application's behavior.
Also it is possible to store sets of different versions of loadable software components, ready to download to RAM or flash. A final example of versioning is the possibility to store different sets of configurations, so that the configuration of a specific vehicle can be changed back and forth very quickly. Also it is possible that only certain configurations are possible. The complex part of is that the system has to keep track of dependencies, and prevent invalid configurations.

7.1.5 Scheduling resource usage

Scheduling is a common issue in real-time systems. One task is that analysis is difficult with growing systems and design work tends to increase complexity. For real-time systems there is no other option than hard work, but for the kind of system that DySCAS represents the boundaries are not set. Here the system does not have to be defined completely from the start. Some processes will be added later and some may be removed. Similarly with other resources like communication channels, memory, and system services etc. These have limited capacity and their utilization changes during runtime, and during the lifetime of the system due to configuration changes.
To build an event-based system is then a typical choice, but not optimal. A possibility is to design the aperiodic activities as periodic, i.e. event triggered and time-triggered. Anyhow, to ensure soft real-time behavior will require advanced design. The DySCAS project has considered these constraints, and created a mix of several methods to ensure optimal performance.
For instance, there are different schedulers for scheduling different resources. There is one adaptive controller for scheduling processes and for load balancing. There are two other schedulers for communication. It is possible that the scheduler will be improved and merged into one in the future.
A communication system in automotive systems should support different configurations of nodes that change over time. Adding or removing nodes during execution should not affect the actual mission of operation of the specific network. Such flexibility is crucial in a DySCAS system, the typical complexity is illustrated in figure 14.

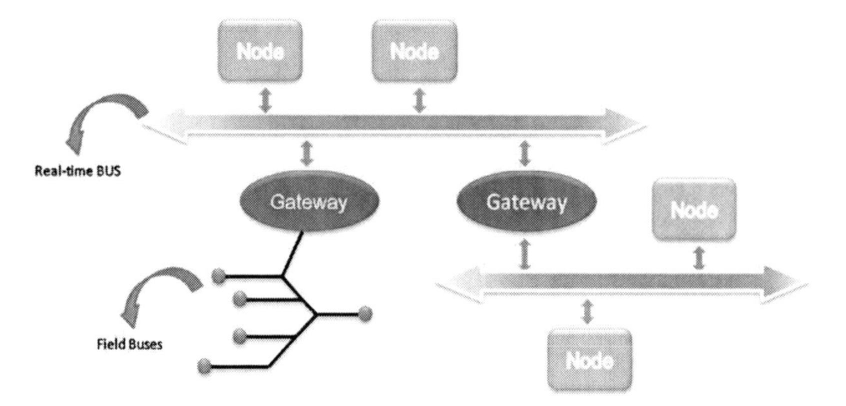

Fig. 14 Real-Time communication architecture [46]

Scheduling Communication in DySCAS

Optimal scheduling of communication in a distributed ad-hoc network is complex. There are several aspects to handle. Nodes and processes are asynchronous; therefore the corresponding IPC is also asynchronous. Without any control there is an obvious risk of overloading the network media or starving a node. In the DySCAS concept it is necessary to balance the needs of high responsiveness by handling dynamic changes as fast as possible, whilst on the other hand not disturbing the ongoing activities. To guarantee these two opposing demands the scheduling support is a difficult task. The SHAPE example implementation of DySCAS incorporates a scheduler which addresses this issue. The scheduler depends on several preconditions available in SHAPE. The System has to provide features for controlling processes and a network media such similar to Controller Area Network (CAN). It is also expected that the communication mechanism for IPC will use a link handler concept or similar, such as LINX [26]. Timing support is also required.

7.1.6 Communication Services in a distributed DySCAS system with LINX

The DySCAS project is a distributed platform system providing mechanisms to configure heterogeneous systems during run-time, with multi-vendor supplied hardware in the same network. Also the running applications are probably as complex as the hardware. Of course the system may be static and only accept the default configured system or it may take advantage of part or all of the dynamics that the DySCAS system enables e.g. through policies.
Nevertheless, a system that changes its configuration during run-time because a new device (or processing node) is attached to the system requires sophisticated communication support. In addition to high performance and reliability there is also a need for full transparency and support for distributed systems using potentially multiple operating systems (because of the connection of heterogeneous devices). For this reason the SHAPE project adopted the advanced and open source protocol, LINX [1, 25, 26]. This is a transparent, system-wide interprocess communications service with high performance because of its direct message passing technology. LINX connects multiple operating systems in a seamless fashion and is suitable for the complete range of processors including high-end CPUs, as well as DSPs, and scales well to large systems with any system topology.
The DySCAS architecture has a strong advantage due its platform independent middleware design. This implies that all fundamental code is non system dependant. It neither contains any calls to the operating system nor calls to the hardware. Instead the platform depends on an interface layer, the *Instantiation Layer*. This layer is divided into the *System Level* and the *portability level*. This design ensures that porting is an easy task and the middleware will always ensure the correct behavior irrespective of what platform it is running on. The Instantiation Layer is open source, to encourage fast driver development.

8 A framework for modelling, designing and analysing dynamically configurable systems

This section describes our work and results towards a methodology and tools supporting design and analysis of DySCAS systems.
In order to develop a system such as DySCAS, various types of tools are required covering a wide range of design activities from architecture modeling and design, verification/validation

through simulation and formal analysis, to prototype implementation. Clearly, when DySCAS-type systems make it to the market, there will also be a need for tools supporting the entire system life cycle including tasks such as configuration, deployment and maintenance, such tools are however not treated here. Figure 15 illustrates currently available tools (solid line boxes and connections) and future possible tools (dashed line boxes and connections) and their interconnections. Current support for the design includes simulation, safety analysis using FMEA and formal verification with the purpose of providing feedback to the architecture modeling and specification work which is carried out using UML. In general, modeling and analysis of these different aspects plays the important role of improving our understanding of DySCAS systems, in communicating DySCAS concepts by means of models, and in verifying and validating designs. In the following, current support tools and experiences are described (solid line boxes). A reference implementation is described in section 7.

The architecture has been captured by UML models encompassing structural and behavioral aspects. A main emphasis has been placed on defining architectural function blocks and their interfaces (including signals and data). For this purpose UML composite structure and class diagrams are used. As a prerequisite, a component model is defined with derived classes representing different types of components (application services, core services and instantiation services). With respect to behavior a generic execution controller for DySCAS core components is captured with a state machine that describes execution modes (start-up, normal operation and fault handling). Sequence diagrams are used to describe interactions (protocols) between components.

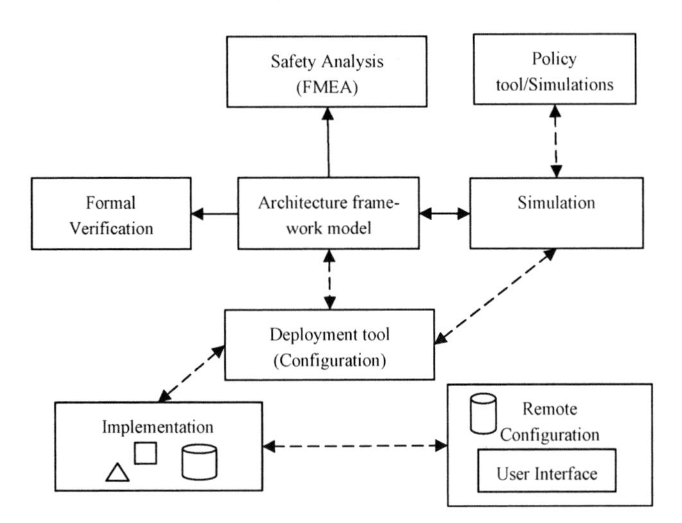

Fig. 15 Existing and desired design activities (methods/tools) used in the development of the DySCAS architecture. The solid lines illustrate where information has been transferred by manual model transformation. In the future we wish to automate these transformations. The dashed lines illustrate desired new tools and connections, for example where a configuration tool can be used as a basis for defining a concrete DySCAS implementation and for configuring simulations.

8.1 Simulation

In developing dynamically configurable systems, it would clearly be useful to have a simulation platform which is able to simulate structural and behavioral changes such as addition or removal of devices, policy behavior, and system and application performance. In particular, the dynamics part of a DySCAS system (structural changes, varying loads and partial failures) constitute major concerns for modeling and simulating a DySCAS system. The choice of level of abstraction is also based on these requirements. A very low level of abstraction is beneficial for detailed analysis but at the same overly detailed and time-consuming for many of the aspects of interest to evaluate. On the other hand, a very high level of abstraction will give rise to results which are obvious and may not be useful.
Simulations in the DySCAS project are classified as logical, base and policy simulations.

- *Logical simulations*. The main purpose of these models and simulations is to verify structural/architectural properties (component model, interfaces and signals) and logical behavioral properties (state-machines and activity diagrams). The modeling emphasizes platform independent modeling and supports both interface verification as well as validation through simulation of the system behavior. We have chosen Matlab/Simulink/SimEvents as tools due to their ability to support simulation of discrete event and continuous time systems simultaneously.
- *Base simulations*. The main purpose of these models and simulation is used to evaluate system behaviors, including algorithms for configuration management, quality of service and load balancing. The modeling has included explicit platforms abstractions in order to incorporate aspects such as allocation to processors, platform performance as well as application performance. For these simulations we have chosen the TrueTime [59] toolbox due to its support for modeling logical as well as real-time operating systems and network protocols.
- *Policy simulation*. For simulating the policy-based configuration aspects, a separate platform i.e. AGILE is used [8].

The logical simulation models are directly based on the architecture design model, as captured in UML. The UML model is mapped to SimEvents manually to test the logical correctness of the DySCAS architecture. The table below summarizes the mapping from UML to Simulink. For a detailed description reader is referred to [28]. The work on logical simulations has focused on capturing the specifications in more details. We have also explored the incorporation of platform behavioral abstractions into the logical simulations.

Table 3 Transformation scheme between Architecture design model in UML and Matlab/Simulink/Simevents

UML	SimEvents/Simulink
Component model structure	Subsystem
Ports	SimEvents Connection port
External Signals	Entities
Signal attributes	Attributes
Decision functions	Attribute function
Execution controller	Stateflow
Computational modules	Combination of gates, queues and servers
Activity diagrams	Combination of Stateflow, gates, queues and servers

The base simulations emphasize behavioral aspects (with simplified architectural aspects). The simulation set-up makes it possible to integrate and test various adaptive management algorithms (see e.g. [3, 18, 43, 50, 72]) including our own algorithms for DQMS and ACM as illustrated in Section 6.

8.1.1 Policy simulation

The DySCAS architecture incorporates the AGILE policy technology [8]. An implementation library, suitable for integration with the DySCAS middleware, and deployment on the embedded platforms that DySCAS targets, has been developed within the project. There are many functional and non-functional requirements of the policy library for which support has been incrementally added and evaluated throughout the development process.
For pragmatic reasons the policy library was initially developed in parallel with the middleware. This meant that the policy library functionality could not be tested on the real platform, with real sensor data and real applications, during the early stages of the project.
To allow the policy development to proceed, a series of simulations were developed to stress-test the policy grammar and the policy library mechanisms. The in-development policy library was embedded into mocked-up middleware and application environments with simulated sensor and system inputs (which provide context to the policy decisions).
The simulations have been carried out throughout the lifetime of the project and in addition to feeding into continuous improvement of the policy library itself, the results and findings have also helped drive the design of the middleware and the various mechanisms and services within. Particular aspects of the middleware design and operation that have been influenced by these outcomes including the policy-load path and the repository that stores the policies within the system; and the dynamic context management service which provides the appropriate context information required for a particular policy to operate. In addition the policy simulations have helped identify the requirements of meta-data attached to policies to guide the policy loading and tracking.
As only the environment and middleware has been simulated, with the real policy library in-place, the functionalities of the policy library mechanisms (including those of the Dynamic Wrapper and the Decision Point API) have been tested; whilst a number of concepts relating to the wider use of policies within the middleware and in the context of the DySCAS use cases have been validated.
Concepts and functionalities validated and tested respectively, include: Policy load, Automatic subscription to required context information, Run-time change of policy, Decision point concept, Support for multiple decision points per component, Multiple policy evaluation (when multiple decision points are used simultaneously, each with its own policy), and Error handling behavior for the following fault conditions: no policy available in repository; policy not loaded into decision point, or parse error occurred; and needed context information not available. Figure 16 illustrates the way in which existing and mocked up parts were combined to enable validation of policy-related concepts, testing of policy-library functionalities and also feasibility of DySCAS use cases.
The simulations have also helped to test the policy language grammar for completeness and closed qualities. Various policies with collectively a wide-range of behaviors have been written for test purposes. This policy-writing exercise in turn led to the design and development of a custom policy editing tool to simplify the policy writing process and to ensure that syntax errors are avoided (the editor supports visual development of policies and automatically generates the policy script).

8.2 Safety analysis and formal verification

When introducing systems that include run-time configurations it is of utmost importance that the autonomous computer is conceived as trustworthy and robust. While the DySCAS architecture mainly addresses non safety critical domains such as infotainment and telematics, malfunctioning applications and middleware may still disturb users, thereby potentially creating hazardous events. In addition, some of the scenarios also relate to more critical operations such as attaching a trailer to a truck and failures in a DySCAS system could in the worst

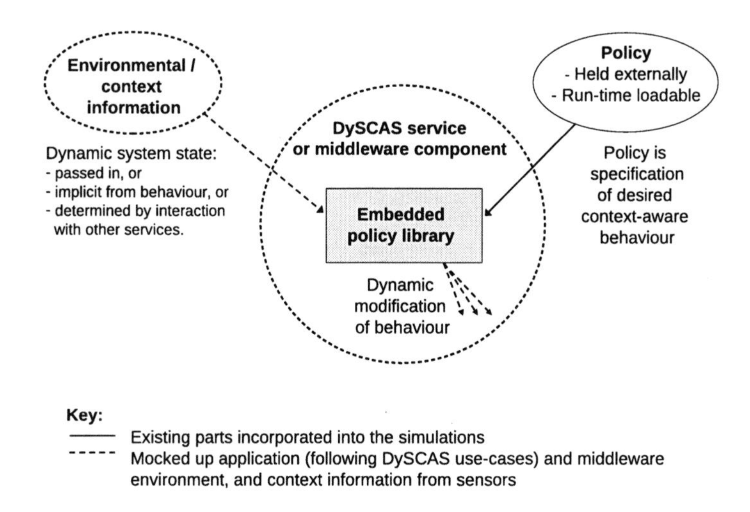

Fig. 16 Combining real and mocked-up parts of the system to validate a wide range of policy-related concepts

case propagate from non critical to critical parts. For these reasons, several means for addressing both validation and verification have been and are being addressed. In the following we describe safety analysis and formal verification of component interactions.

8.2.1 Safety Analysis

A safety analysis has been performed iteratively involving three steps (constituting traditional ingredients in safety analysis [44]:

- Hazard analysis. In hazard analysis it is studied what can go wrong at the system interfaces, and more precisely to identify system states that under certain circumstances can lead to system failure that may harm humans and cause financial loss. This is supported by considering the use cases of DySCAS, defining central interactions across the interfaces.
- Functional failure analysis (FFA). In FFA, the possible failures of key functionalities and their consequences are investigated together with means to avoid or handle such failures. FFA could be seen as a failure mode effects analysis (FMEA) carried out at a very high abstraction level.
- Architecture (re)design. Weak or vulnerable points of the architecture are re-visited in order to avoid or mitigate critical failures.

In the following we give a brief overview of these steps (for more details see [29]).
The analysis concerns the DySCAS middleware software, which is interacting with application components and the supporting embedded systems platform. The hazard analysis is however performed at the level of the super-system since this is where hazards appear. The DySCAS super-system includes the DySCAS middleware, applications and the underlying platform. This system is thus delimited by human machine interfaces, internal communication with other vehicle embedded systems (e.g. body electronics and dynamics control), external communication with external devices, and by power supply interfaces.

Failures at interfaces can occur in the time-, value- and energy domain, leading to special classes of failures such as omission and commission. A DySCAS system together with its applications may thus cause the following hazards:

(1) Driver distraction. A distracted driver may in the worst case result in accidents. An interface causing various distractions will also reduce the confidence in the system and may cause complaints (costs).
(2) Energy failure, leading to reduction or loss of power in the vehicle. Such a failure may impact critical functions in the vehicle.
(3) Erroneous communication (in data and/or time) to other vehicle embedded systems. Such a failure may lead to failure of critical functions.
(4) Erroneous communication (in data and/or time, input or output) to vehicle external embedded systems. Such a failure may lead to disclosure of confidential information, erroneous charging, or intrusion. Intrusion may in turn lead to other failures and hazards.

In our investigation we have chosen not to treat (2). Power management has not so far been included into DySCAS. When included, failures due to inappropriate power management also have to be considered.
Potential causes of (1) include inappropriate or unexpected HMI outputs (e.g. sound, light and misleading information on displays). While HMI interfaces are not part of the DySCAS system, it is important in the DySCAS design to consider the degree of autonomy (for example when a human should acknowledge a certain action such as to download software) and the level at which error handling is carried out (DySCAS vs. application level). The goals are to provide intuitive and robust behavior where individual tailoring is possible through the use of policies.
It should be noted that hazards belonging to (3) are handled either by a DySCAS gateway or through appropriate communication and security protocols which we assume to be in place.
A functional failure analysis (FFA) was carried out for the DySCAS middleware with the purpose to identify the possible failures of key functionalities, their consequences, and ways to avoid or handle such failures. The failures were classified as catastrophic, critical, marginal and insignificant. A number of possible failures were detected and assessed. Many of these were deemed as critical, for details see [28]. In general, critical functional failures are those that involve faulty decisions, expose faulty information at the interfaces, and represent security risks.
A general conclusion regarding the hazard and FFA analysis is that it is essential for DySCAS (and similar) systems to consider and handle the following aspects:

- Address security for relevant functions/services.
- Ensure that critical data (internal sensors and state) can be provided reliably or detected as faulty.
- Design robust interactions and verify these through formal verification, simulation and testing.
- Pay special attention to robust decision making (sensing, decision algorithms and policies).

It is clear that especially decision making is critical.
A number of measures have accordingly been taken including specific features of the DySCAS architecture and in verification and validation through simulation and formal verification. Nevertheless it must be pointed out that further work is required, for example in managing the transition from design to implementation.

8.2.2 Formal Verification

As part of the DySCAS project formal verification has been explored as one part of the verification work. Formal verification has been applied to algorithm development (verifying con-

figuration algorithms - see [28] and [27]) and for verifying certain interactions between architecture components. The work and experiences indicate the usefulness and relevance of the approach in which simulation is combined with formal verification. Simulation is not adequate alone for ensuring the correctness of the architecture model, as it cannot explore all possible activity traces resulting from complicated communication, interaction, and resource contention of multiple processes.
Formal methods have been applied to one of the DySCAS use cases, "Attach Device". This use case is in the following referred to as GUC1 (DySCAS General Use Case 1). The verification concerns the communication as specified by the architecture model (in this case through a UML sequence diagram).
The investigation improved a faulty design in the behavioral model of the *localResourceManager*, and confirmed the reversibility and responsiveness of the behavioral model.
A UML sequence diagram [31] describes a group of independent objects transmitting synchronous and/or asynchronous messages. In this investigation, the dynamic behavior of each participating object is modeled by a finite state automaton [36]. A synchronous message between objects represents a direct function call and is represented as a joint event shared by the corresponding automata, forcing a rendezvous of the related automata. An asynchronous message, however, is transmitted via a message queue, which is also modelled as an automaton. Unlike the former case, the automata of objects participating in an asynchronous message call need not synchronize for the message. Rather they each synchronize with the message queue. From a system perspective, the message queues are limited resources used by participating objects: They have limited capacities and cannot overflow or underflow.
Having obtained the corresponding automaton models of a sequence diagram, we also formalize desired properties as automata and submit all them to a computation tool. These desired properties typically include systemic requirements e.g., freedom of deadlock, reachability, reversibility, and user defined correctness specifications. The computation tool adopted in this investigation is XPTCT[1]. While this software tool is intended for supervisory control design of discrete-event systems modelled as finite automata, we use it only for automaton computation.
The remainder of the section shows a few snapshots of the verification process. Sequence diagram "GUC1_deviceAttach" contains 11 independent objects, among which the automaton model of *localResourceManager* is shown in Figure 17. The events are named after the message calls in the sequence diagram. If the object sends out an asynchronous message, the message name is immediately followed by "!". If the object receives a message, the message name is followed by "?".

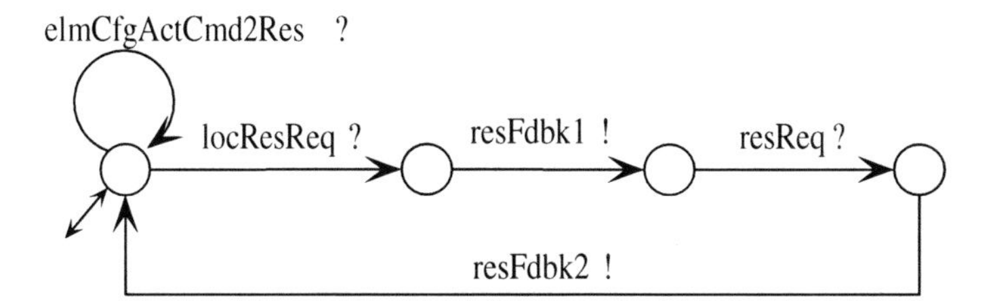

Fig. 17 Model of the *localResourceManager*

Going through the sequence diagram "GUC1_deviceAttach" and the automata of objects, we define a message queue for every pair of events with the same name but ending with the complementary symbols "?" and "!". Altogether there are 23 message queues. Message queues have the similar properties: They have limited lengths. When full, senders cannot send more

[1] Free to download from `http://www.control.utoronto.ca/DES`

messages; when empty, receivers cannot receive. In one message call, one piece of message is inserted into or taken out from the queues. The model of a generic queue is shown in Figure 18. The states are numbered from 0 to $n \geq 1$, where the integer number at each state indicates the number of messages in the queue and n the length of the queue. At state 0, the queue is empty. So it can only receive a message from a sender via event *fromSender*! but cannot provide any message to a receiver. At state n, the queue is full. So it can only be read by the receiver via event *2Receiver*? but cannot be written by the sender.

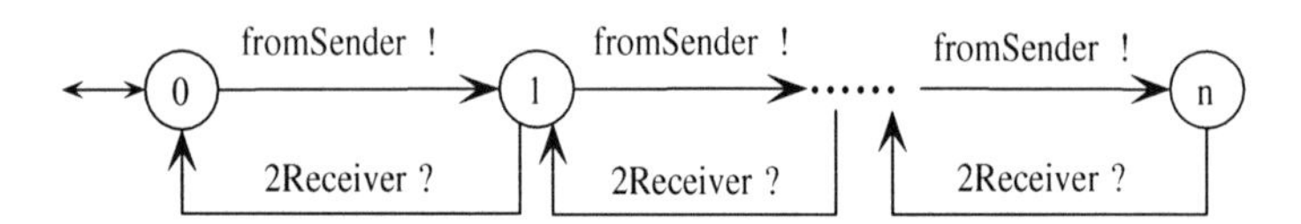

Fig. 18 Generic Model of a Message Queue

Since XPTCT, as well as most other model checkers, e.g., SPIN and SMV, does not support indeterminate parameter n, we, for conceptual evaluation, set the lengths n of most message queues as 1. Therefore the models have 2 states. Events *fromSender*! and *2Receiver*? are replaced by corresponding labels for each message queue.
We first check if the GUC1 process is *reversible*, namely the communication process can repeat in the same way for ever. Evidently, this property is stronger than deadlock freeness. The reversibility property is implied by the *nonblocking* property of the automaton model of GUC1. An automaton is *nonblocking* if and only if every reachable state can further reach a marker state via the state transition relation of the automaton. Note that we choose the *marker* states identical to the initial states for all previous automata.
Computing the *synchronous product* (concurrent composition) [36] of the automaton models of all objects and message queues, we obtain an automaton describing all possible activities pertaining to GUC1. This automaton has a deadlock, which implies that the GUC1 process, as modeled above, is *not* reversible.
The direct reason for this deadlock is the faulty design of the model of *localResourceManager* of Figure 17. According to this model, *localResourceManager* periodically serves requests *locResReq*? and *resReq*? in the given order. This model, however, overlooks the possibility that in one cycle, the second request may not happen at all owing to decisions made between the two requests. Thus the model of Figure 1 waits for request *resReq*? for ever and a deadlock arises. To eliminate this deadlock, we change the model of *localResourceManager* to the new automaton in Figure 19. Now the object can answer requests without any sequential constraint.

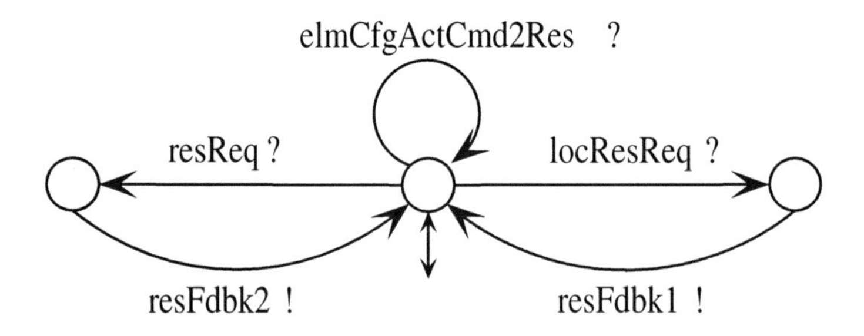

Fig. 19 New Model of the *localResourceManager*

We repeat the previous computation with this new model. We obtain a new automaton with 4644 states and 16395 transitions, and it is nonblocking. Therefore the entire GUC1 process, after the modification on a behavioral model, is reversible.
The other property that we verified is the *responsiveness* of the model. When there is a new device attached, the system, in particular *localDeviceManager*, always generates a response. Either the device is not recognized, or there is not sufficient resource, or the device is not authorized, or the system receives a dynamic configuration feedback on the new device. Our computation confirmed this property of GUC1.

9 Open issues and ongoing work

Several issues are still open in the DySCAS project. This is not strange since the DySCAS middleware touches upon several scientific challenges and involves extensive engineering work - all of which are not possible to cover in one project.
Solving the DySCAS challenge requires competences and results from many scientific disciplines to be integrated and further developed, including scheduling, feedback control, artificial intelligence, fault-tolerance and software engineering. Continued work is necessary in finding common and suitable abstractions. Both theoretical combinations as well as clever engineering will be required.
It can be noted that the configuration problem itself, even within traditional static design of embedded real-time systems, is a big challenge. Extending it to do dynamic configurations requires increased emphasis on design-time techniques as well as deployable run-time techniques. Robustness and verification are treated in this chapter, integration with legacy systems and implementation on resource-constrained systems are issues that are still subject to work. The last two are further elaborated in the following.

9.1 Integration with a legacy statically reconfigurable platform

Statically configured systems play a key role in the automotive industry today. This is illustrated by the effort to develop a standardized middleware platform, the AUTOSAR initiative [13]. DySCAS-type systems will probably first be introduced for less critical and quick changing functions, represented by the infotainment and telematics domains. This means that a DySCAS system must be able to interface to the other domains inside a vehicle. In addition, a migration strategy from statically to dynamically configurable systems has to be considered. The main approach taken within DySCAS has been to investigate the use of a gateway to separate the two networks [32]. A gateway approach is a natural solution since gateways are frequently used today to separate the existing domains. This approach has several advantages and allows the two networks to coexist. Integrating the two types of middlewares into the same domain would be a very difficult challenge.

9.2 Implementation on a resource-constrained platform

One of the main restrictions for automotive systems is that of resource-constrained platforms. Practical deployment of DySCAS implies that all the used algorithms need to execute efficiently, even on a constrained embedded platform. The automotive computing environment

poses specific limitations in terms of cost-constraints (large series production), size and energy. Already today, the increasing power usage of automotive electric architectures is a design challenge in itself.
As a way of approaching these issues, a minimalistic implementation of the DySCAS concepts is under development. During the DySCAS project, a first proof-of-concept prototype implementation has been built for the Coldfire MCF5213 microcontroller and the Movimento Puma ECU. This implementation has received the project name DyLite. The main aim is to further evaluate the concepts for QoS and reconfiguration. Several delimitations of the work have been decided upon; the implementation will only run on two previously well known hardware and software platforms, it will use a lightweight protocol based on concepts and experiences from both the SHAPE reference implementation as well as a previous project at KTH, SAINT [14]. The DySCAS-Lite/QoS implementation can also be seen as a design exploration in exactly how small an implementation of a self-aware middleware can be made.

10 Conclusions

Vehicles are increasingly technology rich, which has an obvious benefit in terms of potential functionality but also has a negative impact on the duration of the design phase with many technical decisions to make; there is considerable risk of committing to a wrong approach or missing opportunities through lack or foresight. A large fraction of the technology within a modern vehicle is made up of embedded computing systems which form a distributed control system. This implies the capability to support dynamic reconfiguration although current systems are static. The introduction of dynamic configuration and context-aware behavior into vehicular control systems, as targeted by DySCAS, opens up whole new realms of future use cases. However, to achieve the DySCAS goals there are a large number of very demanding technical requirements which must be met.
This chapter has examined the DySCAS concept, putting the benefits and the technical requirements and challenges into perspective, and has introduced the DySCAS architectural specification for future vehicular systems.
The technical descriptions have focused on the achievement of autonomic behaviors within the DySCAS architecture: self-configuration for responsiveness to resource requirements and to optimize services, and self-healing to automatically detect, analyze and deal with run-time hardware and software faults. Policies provide flexible configuration and customization, and allow deferment of some design decisions potentially shortening the time-to-market and reducing manufacturing cost and risk.
A dynamic context mapping service decouples components so that component upgrades can be performed in isolation. The strong obligations of robustness, validation and verification are met by wrapping the dynamic configuration mechanism with an automatic fault-handling mechanism which silently downgrades a problem component to statically defined default behavior.
Extensive simulation and validation activities have been reported, as well as a reference implementation. Various partners in the project are also developing a range of demonstrators with a view to disseminating and promoting the DySCAS concept within the automotive community. Whilst very detailed, this chapter does not do full justice to the scope, sophistication and capabilities of DySCAS, or to the full extent of the design, development and validation effort that have been necessary. However, the chapter does provide a valuable insight into the techniques used, and the challenges and long-term benefits associated with developing a complete autonomic system for automotive control systems.

Acknowledgments

The DySCAS project is funded within the 6th framework program "Information Society Technologies" of the European Commission. Project number: FP6-IST-2006-034904. Project partners are: Volvo Technology AB, Daimler AG, Enea Services AB, Robert Bosch GmbH, The University of Greenwich, The University of Paderborn, The University of Oldenburg, The Royal Institute of Technology (KTH), Systemite AB, and Movimento. The project started in June 2006 and runs until February 2009. Further details are available at the project website [24].

References

1. LINX download page at Sourceforge.net. http://sourceforge.net/projects/linx/
2. Abdelzaher, T., Bjorklund, M., Dawson, S., Feng, W.C., Jahanian, F., Johnson, S., Mehra, A., Mitton, T., Shaikh, A., Shin, K., Wang, Z., Zou, H.: Armada middleware and communication services. Journal of Real-Time Systems **16**, 127–153 (1999)
3. Abdelzahler, T.F., Atkins, E.M., Shin, K.G.: QoS negotiation in real-time systems and its application to automated flight control. IEEE Transactions on Computers **49**(11), 1170–1183 (2000)
4. Albus, J., Proctor, F.: A reference model architecture for intelligent hybrid control systems. In: Proceedings of the International Federation of Automatic Control. San Fransisco, CA (1996)
5. Ananthanarayanan, R., Mohania, M., Gupta, A.: Management of conflicting obligations in self-protecting policy-based systems. In: Proc. 2nd Intl. Conf. on Autonomic Computing (ICAC), pp. 274–285. IEEE, Seattle (2005)
6. Anceaume, E., Cabillic, G., Chevochot, P., Puant, I.: Hades: a middleware support for distributed safety-critical real-time applications. In: Proceedings of International Conference on Distributed Computing Systems, pp. 344–351 (1998). URL http://ieeexplore.ieee.org/xpls/abs_all.jsp?arnumber=679736
7. Anthony, R.: Policy autonomics website. http://www.policyautonomics.net/
8. Anthony, R.: A versatile policy toolkit supporting run-time policy reconfiguration. Cluster Computing: The Journal of Networks, Software Tools and Applications **11**, 287–298 (2008)
9. Anthony, R., Ekelin, C.: Policy-driven self-management for an automotive middleware. In: Proceedings of First International Workshop on Policy-Based Autonomic Computing (PBAC 2007), at the Fourth IEEE International Conference on Autonomic Computing, in Jacksonville, Florida, USA, Jun. 11 – 15 (2007)
10. Anthony, R., Rettberg, A., Jahnich, I., Törngren, M., Chen, D., Ekelin, C.: Towards a dynamically reconfigurable automotive control system architecture. In: Proceedings of International Embedded Systems Symposium. IFIP, Irvine, CA, USA (2007)
11. Anthony, R.J.: The Agile policy expression language for autonomic systems. ITSSA **4**(1), 381–398 (2006)
12. Anthony, R.J.: Policy-centric integration and dynamic composition of autonomic computing techniques. In: ICAC '07: Proc. 4th Intl. Conf. on Autonomic Computing. IEEE Computer Society, Jacksonville, Florida, USA (2007)
13. AUTOSAR Consortium: http://www.autosar.org
14. Axelsson, M., Eriksson, M., Francke, T., Hammarstrand, F., Lindell, A., Nyqvist, O., Persson, E., Strömberg, C., Svensson, M., Thrönqvist, N.: An automotive embedded systems demonstrator; the Saint truck - Saint3: mechanics and EE platform enhancements, intelligent model supported configuration and reverse steering. Tech.

Rep. TRITA MMK 2008:01, ISSN 1400-1179, ISRN/KTH/MMK/R-08/01-SE, Mechatronics Lab, Department of Machine Design, Royal Institute of Technology (KTH) (2008). URL http://www.md.kth.se/saint/publications/Saint3/SAINT_3_FinalReport_MMK_KTH.pdf
15. Basra, R., Lu, K., Rzevski, G., Skobelev, P.: Resolving scheduling issues of the London underground using a multi-agent system. In: 2nd Intl. Conf. on Industrial Applications of Holonic and Multi-Agent Systems (HoloMAS), LNAI 3593, pp. 188–196. Springer Verlag, Copenhagen, Denmark (2005)
16. Beveridge, M., Koopman, P.: Jini meets embedded control networking: A case study in portability failure. In: Proceedings of the The Seventh IEEE International Workshop on Object-Oriented Real-Time Dependable Systems (WORDS 2002). IEEE Computer Society, Washington, DC, USA (2002)
17. Blair, L., Blair, G., Andersen, A., Jones, T.: Formal support for dynamic QoS management in the development of open component-based distributed systems. In: IEE Proceedings Software, vol. 148 (2001)
18. Brandt, S.A., Nutt, G.J.: Flexible soft real-time processing in middleware. Real-Time Systems **22**(1-2), 77–118 (2002)
19. Chomicki, J., Lobo, J.: Monitors for history-based policies. Policies for Distributed Systems and Networks, pp. 57–72. Springer (2001)
20. Cointe, P. (ed.): Meta-Level Architectures and Reflection: 2nd International Confer-ence, Reflection '99, St. Malo, France, *Lecture Notes in Comp. Science*, vol. 1616. Springer (1999)
21. Corman, D., Loyall, J.P., Schantz, R.E., Schmidt, D.C.: Integrated adaptive QoS management in middleware: A case study. In: RTAS '04: Proceedings of the 10th IEEE Real-Time and Embedded Technology and Applications Symposium, p. 276. IEEE Computer Society, Washington, DC, USA (2004)
22. Damianou, N., Dulay, N., Lupu, E., Sloman, M.: The Ponder policy specification language. Policies for Distributed Systems and Networks, pp. 18–38. Springer, Berlin (2001)
23. Dunkels, A., Finne, N., Eriksson, J., Voigt, T.: Run-time dynamic linking for reprogramming wireless sensors. In: Proceedings of the 4th international conference on Embedded networked sensor systems, pp. 15–28. Boulder, Colorado, USA (2006)
24. DySCAS Consortium: DySCAS project website. http://www.DySCAS.org
25. Enea AB: LINX performance test. Tech. rep. http://www.enea.com/EPiBrowser/Literature%20(pdf)/Pdf/Not%20leadgenerating/Datasheets%20and%20Brochures/LINX%20DS%20Final.pdf
26. Enea AB: LINX protocols. Tech. rep. http://www.enea.com/EPiBrowser/Literature%20(pdf)/LINX/LINX%20Protocols.pdf
27. Feng, L., Chen, D., Persson, M., Naseer Qureshi, T., Törngren, M.: Dynamic configuration and quality of service in autonomic embedded systems. Tech. Rep. TRITA-MMK 2008:12, ISSN 1400-1179, ISRN/KTH/MMK/R-07/12-SE, Department of Machine Design, KTH, Stockholm, Sweden (2008:12)
28. Feng, L., Chen, D., Törngren, M.: Self configuration of dependent tasks for dynamically reconfigurable automotive embedded systems. In: Proceedings of 47th IEEE Conference on Decision and Control, Cancun, Mexico, Dec. 9 – 11 (2008)
29. Feng, L., Törngren, M., Chen, D.: Safety analysis of dynamically self-configuring automotive systems. Tech. Rep. TRITA-MMK 2008:13, ISSN 1400-1179, ISRN/KTH/MMK/R-07/12-SE, Department of Machine Design, KTH, Stockholm, Sweden (2008:13)
30. Forsight Vehicle: Foresight vehicle technology roadmap, technology and research directions for future road vehicles. Tech. rep., Society of Motor Manufacturers and Traders Ltd (2004)
31. Fowler, M.: UML Distilled: A Brief Guide to the Standard Object Modeling Language. Addison-Wesley Longman Publishing Co., Inc., Boston, MA, USA (2004)

32. García, J.: A gateway for interconnecting statically and dynamically configurable embedded systems. Master's thesis, Department of Machine Design, KTH, Stockholm, Sweden (2008). Report number MMK2008:76 MDA330
33. Garey, M.R., Johnson, D.S.: Computers and Intractability: A Guide to the Theory of NP-Completeness. Freeman and Company (1979)
34. Hägglund, J.: Analysis and design of application policies and checkpointing in a distributed automotive middleware. Master's thesis, Uppsala University (2008)
35. HAVi: Home audio video interoperability. http://www.havi.org
36. Hopcroft, J.E.: Introduction to Automata Theory, Languages, and Computation. Pearson Addison Wesley (2007)
37. IBM: An architectural blueprint for autonomic computing. ibm and autonomic computing. Tech. rep. (2003). http://www-306.ibm.com/autonomic/pdfs/ACwpFinal.pdf
38. IBM Research: Policy technologies. http://www.research.ibm.com/policytechnologies/. IBMResearchPolicyTechnologies
39. Klefstad, R., Schmidt, D.C., O'Ryan, C.: The design of a real-time CORBA ORB using Real-Time Java. In: Proc. IEEE Intl. Symp. Object-Oriented Real-Time Dist. Comput-ing (2002)
40. Kluge, F., Mische, J., Uhrig, S., Ungerer, T.: Building adaptive embedded systems by-monitoring and dynamic loading of application modules. In: Workshop on Adaptive and Reconfigurable Embedded Systems (APRES'08). St. Louis, MO, USA (2008)
41. Kohli, M., Lobo, J.: Realizing network control policies using distributed action plans. Journal of Network and Systems Management **3**(11), 305–327 (2003)
42. Kon, F., Marques, J.R., Yamane, T., Campbell, R.H., Mickunas., M.D.: Design, implementation, and performance of an automatic configuration service for distributed component systems. Software: Practice and Experience (2005)
43. Lee, C., Lehoczky, J., Siewiorek, D., Rajkumar, R., Hansen, J.: A scalable solution to the multi-resource QoS problem. In: Proceedings of the 20th IEEE Real-Time Systems Symposium, pp. 315–326. Phoenix, AZ, USA (1999)
44. Leveson, N.G.: Safeware: System safety and computers. Addison-Wesley Publishing Company (1995)
45. Levine, J.R.: Linkers and Loaders. Morgan-Kauffman (1999)
46. Li, Y.: Real-time analysis of managed communication for the DySCAS middleware. Master's thesis, Linköping University (2008)
47. LIN: http://www.lin-subbus.org/
48. Lobo, J., Bhatia, R., Naqvi, S.: A policy description language. In: Proc. AAAI, pp. 291–298. Orlando, USA (1999)
49. Lotlikar, R., Vatsavai, R., Mohania, M., Chakravarthy, S.: Policy schedule advisor for performance management. In: Proc. of the 2nd Intl. Conf. on Autonomic Computing (ICAC), pp. 183–192. IEEE, Seattle (2005)
50. Lu, C., Wang, X., Koutsoukos, X.D.: Feedback utilization control in distrib-uted real-time systems with end-to-end task. IEEE Transactions on Parallel and Distributed Systems **16**(6), 550–561 (2005)
51. Lymberopoulos, L., Lupu, E., Sloman, M.: An adaptive policy based management framework for differentiated services networks. In: Workshop on policies for distributed systems and networks, pp. 147–158. California (2002)
52. Mascolo, C., Zachariadis, S., Pietro Picco, G., Costa, P., Blair, G., Bencomo, N., Coulson, G., Okanda, P., Sivaharan, T.: RUNES middleware architecture. d5.2.1. Tech. rep., RUNES Project. RUNES/D5.2.1/PU1/v1.7, FP6. Inf. Society Technologies. EC (2005)
53. McKinsey & Company: Auto catalog (2000)
54. Microsoft: .NET framework. http://www.microsoft.com/net/default.mspx
55. MOST: Most. www.mostcooperation.com
56. Object Management Group: CORBA 3.0. http://www.omg.org/
57. Object Management Group: Real-time CORBA joint revised submission. Tech. Rep. OMG Document orbos/99-02-12 ed. (1999)

58. Object Management Group: Unified modeling language: Superstructure, version 2.1.1 formal/2007-02-03. Tech. rep. (2007)
59. Ohlin, M., Henriksson, D., Cervin, A.: TrueTime 1.5 – Reference Manual. Department of Automatic Control, Lund University, Sweden. http://www.control.lth.se/truetime
60. OSGi Alliance: The OSGi service platform – dynamic services and networked devices. http://www.osgi.org. OSGi
61. Persson, M., Naseer Qureshi, T.: Survey on dynamic load balancing in distributed computer systems. Tech. Rep. TRITA-MMK 2008:11, ISSN 1400-1179, ISRN/KTH/MMK/R-07/12-SE, Mechatronics Lab, Department of Machine Design, KTH, Stockholm, Sweden (2008)
62. Persson, M., Naseer Qureshi, T., Törngren, M.: Suitability of dynamic load balancing in resource-constrained embedded systems: An overview of challenges and limitations. In: Proceedings of Workshop on Adaptive and Reconfigurable Embedded Systems (APRES), Apr. 21, 2008, part of the 14th IEEE Real-Time and Embedded Technology and Applications Symposium (RTAS), St. Louis, MO, USA (2008)
63. Robert Bosch GmbH: CAN specification, version 2 (1991)
64. Ronen, O., Allen, R.: Autonomic policy creation with singlestep unity. In: Proc. of the 2nd Intl. Conf. on Autonomic Computing (ICAC), pp. 353–355. IEEE, Seattle (2005)
65. Schmidt, D.C.: Adaptive middleware: Middleware for real-time and embedded sys-tems. Communications of the ACM **45**(6) (2002)
66. Sha, L., Rajkumar, R., Gagliardi., M.: Evolving dependable real-time systems. In: Proc. IEEE Aerospace Conference (1996)
67. Sun Microsystems: Enterprise Java Beans specification. http://java.sun.com/products/ejb/docs.html
68. Sun Microsystems: Jini network technology. http://www.sun.com/software/jini/
69. Tan, J., Poslad, S.: Dynamic security reconfiguration for the semantic web. Engi-neering Applications of Artificial Intelligence **17**, 783–797 (2004)
70. Tindell, K., Burns, A., Wellings, A.: Allocating hard real time tasks. an np-hard problem made easy. Journal of Real-Time Systems **4**, 145–165 (1992)
71. Törngren, M., Chen, D., Malvius, D., Axelsson, J.: Model based development of automotive embedded systems. Automotive Embedded Systems Handbook. Industrial Information Technology. Taylor and Francis, CRC Press (2008)
72. Wang, X., Jia, D., Lu, C., Koutsoukos, X.D.: DEUCON: Decentralized end-to-end utilization control for distributed real-time systems. IEEE Transactions on Parallel and Distributed Systems **18**(7), 996–1009 (2007)
73. Ward, P., Pelc, M., Hawthorne, J., Anthony, R.: Embedding dynamic behaviour into a self-configuring software system. In: Proceedings of 5th International Conference on Autonomic and Trusted Computing (ATC-08), Stavanger, Norway, Jun. 23 – 25 (2008)

Social Opportunistic Computing: Design for Autonomic User-Centric Systems

Iacopo Carreras[1], David Tacconi[1], and Arianna Bassoli[2]

Abstract The proliferation of mobile devices equipped with short-range wireless connectivity allows users to produce, access and share digital resources in a wide number of everyday occasions. In this chapter, we consider a content distribution application scenario, aimed at the diffusion of data in autonomic computing environments, and investigate the way the social attitudes of mobile users impact the design of an autonomic opportunistic communication system. We analyze the results of a simulation which combines both a real-world pattern of proximity-based encounters, as measured in an office environment, with a series of user-defined preferences regarding content. Results show how the system design space varies according to these social parameters, and the importance of designing systems which are build taking into account the user and its social habits and preferences.

Key words: Mobile computing, opportunistic communication systems, user preferences

1 Introduction

The proliferation of mobile technologies (such as mobile phones, gaming consoles and MP3 players) equipped with short-range wireless connectivity (such as Bluetooth and WiFi) has encouraged the development of applications that allow users to produce, access and share digital resources in a wide number of everyday occasions and without the support of a fixed infrastructure [1, 2, 7]. The development of such applications presents challenges in terms of both user interaction and technical feasibility, as users' behaviors needs to be taken into account, especially the mobility of users and the variability of contexts traversed, and technological limitations, especially in terms of battery/processing power and wireless bandwidth, cannot be underestimated.

From an user-interaction perspective, applications have addressed the possibility for users to access data from certain locations (location-based) or to share data with other users in proximity (mobile peer-to-peer).

London School of Economics
Science, Houghton Street
London WC2A 2AE, UK
e-mail: a.bassoli@lse.ac.uk

A.V. Vasilakos et al. (eds.), *Autonomic Communication*, DOI: 10.1007/978-0-387-09753-4_8,

From a technical perspective, the use of opportunistic communication systems [10] to support mobile peer-to-peer applications [12, 13], exploiting the proximity of mobile nodes for exchanging personal or contextual information, is becoming an increasingly popular research topic. Originally, such a paradigm emerged as a way to provide connectivity in intermittently connected scenarios, such as the cases of Interplanetary Internet and developing regions [3]. In cases where connectivity could not be taken for granted, nodes were able to temporarily buffer data, and forward it at the next communication opportunity. The paradigm applied was therefore a "store-and-forward" one, where nodes were expected to first store, and then forward any information destined outside the local network.

Recently, significant attention has been devoted to mobile application scenarios, where the carriers of information are represented by people with their personal handheld devices. In this case, the characteristics of the social network in which the data is being diffused is extremely important and can significantly influence the performance of such systems. This consideration has led to the studies on how to exploit social interactions in mobile systems. In [6], the community structure behind the social interactions driving the data diffusion process has been studied in order to improve the performance of forwarding algorithms. Starting from experimental data sets collected in real-world experiments, it was shown that it is possible to identify a limited set of nodes that were much more active than the others. Such nodes represent *hubs* of the network. Further, the authors show how it is possible to divide the network into overlapping communities of nodes, and how these communities depend on the social relations of the nodes themselves. Finally, it was shown that by incorporating the knowledge of both the social activity of the nodes and the communities to which they belonged, an extremely efficient trade-off between resources and performance could be achieved.
The relevance of users' behavior in opportunistic networks was further assessed in [8], where the impacts of different social-based forwarding schemes were evaluated in the case of a DTN routing protocol scenario, a worm infection scenario, and a mobile P2P file-sharing system application scenario. Also, in this case, the evaluation was based on real world mobility patterns, obtained from bluetooth proximity measurements. In particular, by applying a threshold on the periodicity of meetings it was possible to classify encounters in terms of strangers or friends, and analyze the properties of these 2 classes of meetings separately. Starting from this classification, the authors conducted a statistical analysis of these 2 categories of meetings, concluding that (i) while most of the encounters were between strangers, meetings among friends account for almost two-thirds of the overall meetings, (ii) networks of both strangers and friends were scale-free and (iii) the network of friends had a very high-clustering coefficient. Finally, the authors showed that incorporating this friend/strangers distinction in the forwarding policies can be beneficial in different application scenarios, such as P2P file sharing or in the prevention of the spreading of worms.
What appears to be missing from current studies on opportunistic communication systems is the realization that while mobility and proximal encounters might provide opportunities for improving the performance of the network, these mobile encounters are inherently wed to the social world in which they take place. Therefore, opportunistic communication systems must take into account the types of applications which they are meant to support, and to recognize the significance of the users' preferences regarding the data that they are interested in accessing and sharing through these systems. Further, with respect to real-world implementations of such systems, we believe it is also important to acknowledge the effects of current technical limitations (such as the bandwidth provided by short-range wireless technologies like bluetooth) on the potential performance of an opportunistic communication based application scenario.
In this chapter we present a study which is a first step towards addressing such concerns; we take into account both technical limitations of the specific technology adopted, as well as the preferences expressed by users. We believe that a proper user-centric approach for opportunistic communication research is one that assesses not only the opportunities that social behavior presents, but also the constraints that it carries on for the success new technological advance-

ments. Such approach has to begin with considering real world situations and the spontaneous flow of people's activities, behaviors, sociality patterns and everyday choices. From this starting point, by better understanding the social context chosen for the analysis and conducting experiments within this context, researchers have the opportunity to understand the potential, the limitations and the applications of new opportunistic communication systems.
In order to test this user-centric approach, we have conducted a study within an office environment, in which we have first analyzed the connectivity patterns emerging between people working there; such patterns have been recorded by means of off-the-shelf technologies. This initial step provided insights into the real-world performance level of opportunistic services running over commercially available devices such as smartphones. Next, we have gathered actual user-expressed preferences through the means of a questionnaire. Third, we have combined these technical and social parameters to develop a simulator, capable of reproducing the measured pattern of inter-user contact, and of emulating the content distribution network based on the preferences of the users. This allowed us to see not only when data *could* be distributed, but when it actually *would* be, and we were able, to investigate how the size and format of different types of content significantly impacted the diffusion process. Finally, starting from the assumption that people tend to share information within communities of similar interests, we have develop the concept of *affinity*, which is a metric measuring the similarity of users' preferences. We have analyze how such an affinity metric influences the structure of the network, and the circulation of the content. By combining user constraints and realistic contact patterns, we were able to accurately simulate how content is distributes over the network, thus providing valuable insights into the design and dimensioning of opportunistic content distribution systems.

2 The Study

The aim of our study is to investigate the technological and user-defined constraints which pertain to the design of an opportunistic content distribution application scenario, that allows users to access and distribute digital content such as music, videos or news over Bluetooth-based Smartphones in a given social environment. We have assumed that people always bring with them their smartphone with Bluetooth always on. With a good approximation networking interactions occurred through Bluetooth can be assumed as social interaction among people.
The methodology we applied in the study is briefly depicted in Fig. 1. First, we conducted a real-world experiment within an office environment, monitoring the encounters between co-workers equipped with Bluetooth-enabled mobile phones thanks to a simple Java application. Such encounters were then analyzed in order to understand the network of interactions generated by the encounters between co-workers. At the same time, we conducted a survey to assess people's preferences, in terms of preferred news topics and data formats.
In the second phase, we tested the performance of the network by taking into account both user preferences and current technical limitations of the Smartphones, and ran a series of simulations in which we measured the consequences of introducing different data formats (such as text, audio and video) over a Bluetooth-based network. In addition to this, we evaluated the impact of distributing content over the network depending on the preferences selected by users.
In the last phase, we combined technical constraints and user preferences, looking at how the performance of the network is affected by users transmitting heterogeneous data formats and only accepting the data they might be interested in.

The social environment we have selected for our real-world investigation is a work environment, which has been chosen for several reasons. First, an office represents a relatively enclosed environment where people are likely to interact with each other frequently. Within this

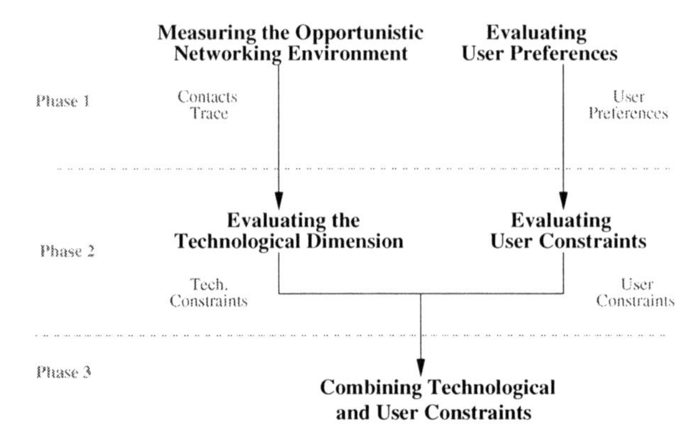

Fig. 1 The methodology applied in the study consisted of 3 parts: in the first phase, we evaluated a specific office environment in terms of opportunistic networking characteristics, and user preferences; in the second phase, we evaluated how technological constraints and the user constraints influence the design of opportunistic communication systems; in the third part, we evaluated the combination of both constraints.

setting, then, we monitored people's natural behavior without having to impose an artificial set of rules for interactions. Second, this environment allowed us to monitor social interactions between the same group of people occurring over an extended period of time. Finally, an office represents a socially-rich environment where encounters often occur not only for work but also for socializing purposes.

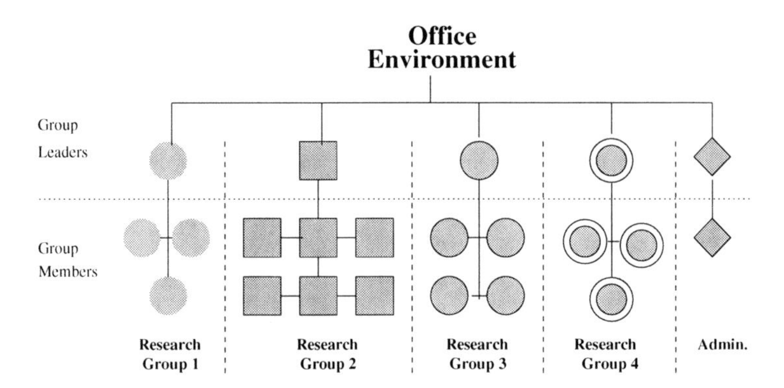

Fig. 2 Working environment organizational structure.

Fig. 2 presents the organizational structure of the office environment under consideration. It is comprised of 21 people organized into 5 groups, with each group composed by a group leader and a variable number of staff members. The workers chosen for the study had different roles within the organization and were working on different floors in the building. With respect to the social characterization of the people involved in the experimentation, 23% were women, and the ages of participants ranged from 25 to 56, with 28% of the participants being younger than 30.

3 First Phase: Understanding The Technological and User constraints

3.1 Assessing contact opportunities of an office environment

Opportunistic communication systems are typically characterized by the contact patterns of the nodes of the network [4, 5]. A "contact" is defined as the communication opportunity deriving by the physical proximity of two nodes. Clearly, such contact patterns depend from several technological aspects, i.e., the communication technology adopted, the noisiness of the environment, the mobility of nodes, and represent an extremely relevant aspect to study. We have then run a set of experiments to measure the connectivity pattern of nodes in the office environment under consideration.
Similarly to the experiments conducted in [5, 11], people's encounters have been monitored by tracing their proximity for a 4 weeks period. During the experiment, 21 workers - with different roles within the organization and working on different floors of the same building - were equipped with a mobile phone running a Java application discovering and tracing neighboring peers approximately every 60 seconds. Whenever the proximity of another device was detected, its bluetooth address, together with the meeting timestamp [1], was saved in the permanent storage of the device for a later processing. In order not to overload the devices memory, a bluetooth enabled laptop acted as a gateway toward a centralized database, gathering the stored information from any smartphone in proximity and transmitting such information to a remote repository. The result of this experiment is a trace which includes a series of contacts, with each contact fully characterized by a timestamp, the *ID*s of the met nodes and the duration of the meeting. Tab. 1 presents a summary of the experimentation settings.

Table 1 Summary of the experimentation settings. Totally, 21 people were equipped with a bluetooth-enabled phone searching for nearby peers once every 60 s. This resulted in 179332 meetings over a 4 weeks time period. The effective number of contacts is obtained by aggregating consecutive contacts into a longer one.

Participants	21
Experiment Duration	4 weeks
Registered Contacts	179332
Effective Contacts	14100

The meetings duration is inferred from consecutive positive peer discovery inquiries performed by nodes. In fact, the java application is simply tracing the proximity of nodes. This means that a 5 minutes contact is identified by 5 consecutive mutual discoveries (nodes are running a discovering phase approximately every 60 sec.). A post-processing routine has then been applied to the collected raw data in order to aggregate multiple contacts into a single one,

[1] Nodes are using the phone's internal clock for determining the timestamp. Since SIM cards are inserted in the phones, it possible to synchronize such clock to the GSM network. This ensures a sufficient level of precision, especially when compared with the granularity of the peer discoveries.

with duration equals to the sum of consecutive peer discovery inquiries. The effective number of contacts (Tab. 1) is the result of this operation.
Fig. 3 depicts the distribution of the contacts duration. It can be easily observed that most of the meetings are relatively short in duration, lasting no more than 100 sec., although few contacts persist for a very long time. This is due users not carrying the mobile phone with them, and leaving it on their desk.

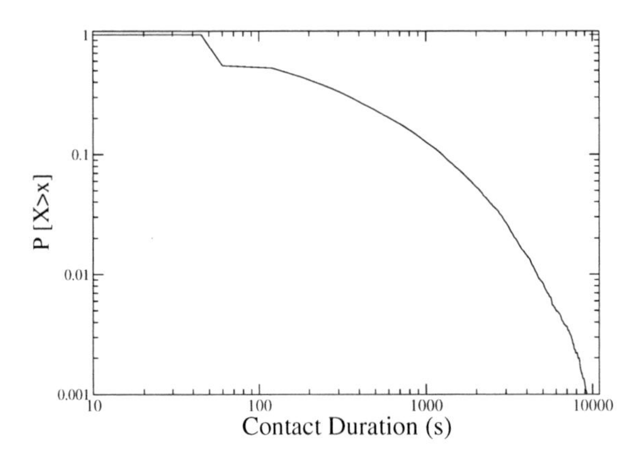

Fig. 3 Contacts duration distribution.

As in any real socially environment, there are couple of nodes meeting very often (i.e., people of the same research group), and people meeting rarely (i.e., people of different groups). This behavior is captured by the intermeeting time among nodes, which measures the time elapsed between the consecutive meeting of nodes couples. The statistical properties of the intermeeting time is represented in Fig. 4.

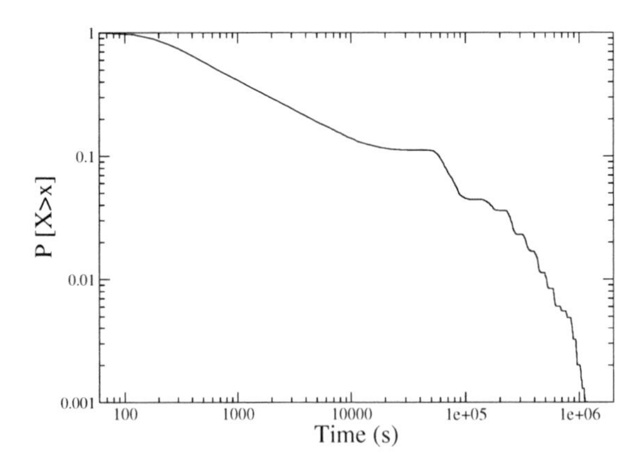

Fig. 4 Intermeeting time distribution.

Tab. 2 summarizes the overall properties of contacts duration and intermeeting time.

Table 2 Intermeetings and contacts duration.

Metric	Mean	Variance	Min.	Max.
Inter- meet-ings	24156	101588	90	1812085
Contacts Duration	473	1047	45	55717

We have then considered the network's *Contact Graph* (CG) [9], which is a graph-based representation of the network of contacts, as obtained from the experimentation. In such graph, vertexes represent the nodes of the network, and edges a contact (or a series of contacts) between a couple of nodes. The presence of each edge is regulated by some metric such as, e.g., the cumulative contact duration between nodes couples, or periodicity of such contact. In our work, the edges of the *CG* are regulated by the cumulative contacts duration, which measures the overall time that 2 nodes i and j have been in contact during the entire duration of the experimentation. In other words, an edge exists between any 2 nodes if they have been close to each other, over the entire duration of the experimentation, for a period of time longer than a predefined threshold d_{thr}. As such, the *CG* shows the stronger relations existing among the people in the office environment.
Fig. 5 presents the *CG* in the case of a cumulative duration threshold d_{thr} equals to 40000 s., which corresponds to nodes being in proximity for approximately 30 minutes per day. Given this threshold, Fig. 5 shows strong relations among nodes such as, i.e., people working in the same group or going regularly at lunch together. This is confirmed by the different shapes of nodes [2] correspond to different groups within the office environment. As expected, people working together tend to have stronger relations and to be somehow isolated with respect to other colleagues. A few nodes (e.g. nodes 6 and 18) guarantees the connectedness of the network, and represent those people working in collaboration with different groups.

3.2 Assessing users expectations

As in any socially-rich environment, people tend to behave differently, depending on their specific role played in the work environment, their preferences, their attitude toward socialization, and many other factors. In order to assess realistic user preferences, we have conducted a survey with the users involved in the experimentation. The questionnaire consisted of 10 questions addressing their preferences in terms of content (e.g.. music, cinema and news), formats (e.g., text, audio and video) and granularity (i.e., more or less focused) of the information they could access and distribute over such network.
The first part of the questionnaire regarded the users preferences in terms of content. In particular, we asked users to rank the different music, cinema and news categories according to their preferences. Further, we asked them to provide a list of their favorite 5 music bands and 5 actors. This provides us a first feedback on common interests.
The second part of the questionnaire addressed the expectations of users in terms of content potentially deriving from the proposed opportunistic content distribution application scenario. As an example, users were asked to rank the information format they would expect to receive, or the level of granularity of the information circulating over the network. A brief summary of

[2] In Fig. 2 the different shapes are explained in correspondence of the office environment organizational structure.

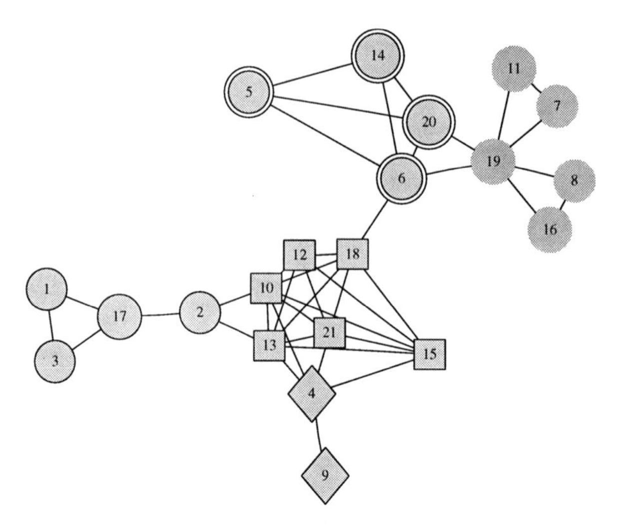

Fig. 5 Graph-based representation of the network of contacts, as obtained from the real-world experimentation. An edge exists between any 2 vertexes if the corresponding persons have been in proximity for approximately 30 minutes per day.

Table 3 Users preferences expressed for the cinema and music question categories.

Cinema

Thriller	Fiction	Dramatic	Romantic	Comedy	Horror	Docum.	Italian
19%	14%	14%	10%	15%	9%	11%	7%

Music

Etnic	Rock	Pop	Disco	Hip-Hop	Jazz	Classic
13%	18%	16%	12%	10%	15%	14%

News

Meteo	Politics	Chronicle	Economy	Sport	Culture
11%	21%	18%	16%	14%	15%

Data Format			Content Level of Detail		
Video	Audio	Textual	Focused	Medium	High Level
28.95%	36.85%	28.95%	45%	21%	34%

the questionnaires results is presented in Tab. 3. As it can be seen, there is a wide heterogeneity in the preferences expressed by users. As an example, for the music category, no specific genre can be excluded a priori. Similar conclusions apply for the other categories.

4 Opportunistic Content Distribution Application

In an office environment people already use various digital technologies to communicate and exchange data. Compared to established networked technologies, opportunistic communications and mobile peer-to-peer applications add to the richness of this environment as they allow people to share resources only when in physical proximity and often without their full awareness of any data sharing happening. Various applications can be supported by an oppor-

tunistic communication paradigm, and - as we already claimed - these need to be taken into account when assessing the potential performance of the network.
For this study we chose to consider a quite simple application scenario where users download (from an unspecified destination that could be the Internet or a particular location) daily content such as, e.g., music, news, on their mobile phones depending on what they usually are interested in. Such information can be then exchanged between users when they happen to be in proximity. When an encounter occurs, from a networking perspective, users can exchange all the data available on their phone, or keep only the one that matches their interests and forward the other to other users that might be interested, or they can only exchange the data they are interested in.
In order to fully understand the performance of such a content distribution network, we have followed a two steps approach: first we analyzed the technological constraints imposed over such network, initially abstracting users preferences and interests. To this extent, we have developed a trace-based emulator of the diffusion process, able to replay the nodes contact pattern measured during the real-world experimentation, and to simulate a data diffusion process. Second, starting from the preferences expressed by users in the questionnaires, we have reproduced a simplified behavior of users, taking into account their preferences, and the content they could be potentially be interested in. We introduced the concept of *affinity*, which is is a measure of the likelihood they will exchange content of mutual interest, and we evaluated the impact of affinity over the opportunistic diffusion of data.

4.1 The Technological Dimension

Given the described office environment, we have simulated an epidemic data diffusion process over the collected contacts trace. Epidemic-style forwarding [14] is based on a "store-carry-forward" paradigm: a node receiving a message buffers and carries that message as it moves, passing it on to new nodes upon encounter. Alike the spread of infectious diseases, each time a message-carrying node encounters a new node not having a copy thereof, the carrier may decide to infect this new node by passing on a message copy; newly infected nodes, in turn, behave similarly. Epidemic diffusion, while being far from optimal in terms of utilized resources [15], is the only viable approach for those application scenarios where information is not delivered from a source to a destination, but rather seamlessly diffused among users.
In order to evaluate the impact of different application constraints over the diffusion process, we have build a simulator of the content distribution process. The simulator replays the contacts trace gathered during the experiments, and integrates several realistic parameters such as, e.g., bluetooth service discovery time, data transfer rate [3]. With the simulator, we have investigated how the content diffusion varies, depending on different system parameters, as imposed by the opportunistic content distribution application scenario.
Since the aim of the considered application is to distribute heterogeneous content to the interested users, we assumed different content formats to be injected into the network. In particular, as summarized in Tab. 4, we assumed three type of data: text, music and video. Each is characterized by a specific size, and a corresponding transfer time, which is determined by the data rate of the bluetooth.
As a first step, a randomly chosen node injecting a message at a random time in the network. From that instant the epidemic diffusion starts, and is stopped when the 90% of the nodes have been reached by that message. We assumed all nodes to be equally interested in the content

[3] We have run separate measurements for evaluating the bluetooth performance that is possible to experience through the $JSR82$ Java APIs for bluetooth. Measurements show that is possible to obtain up to 600 Kb/sec for sufficiently long data transfers, and an average of 25 s. service discovery time.

Table 4 We assumed the opportunistic content distribution application to diffuse different formats of contents: text, music and videos. Each format is characterized by a specific size, and by the time needed for transferring it over the bluetooth communication medium.

Data Format	Size	Transfer Time
Text	$100Kb$	$12s$
Audio	$5Mb$	$90s$
Video	$20Mb$	$300s$

circulating in the network, and we measured the time that is needed in order to infect different fraction of nodes (*Network Infection Ratio*). For each node we run a set of 150 simulations, and we repeated the simulations for all the 21 nodes of the network. Fig. 6 presents the results of this first experiment, together with the 98% confidence interval obtained over the 21×150 runs. As it can be observed, the infection proceeds fast up to the 80%, and slows down above this value. This is due to few nodes missing, from time to time, for some days from the office. Video diffusion is slower, since shorts contacts are not enough to exchange $20Mb$ of data over bluetooth, but, in any case, is fast enough to reach most of the network nodes in 1 day (80000 s.).

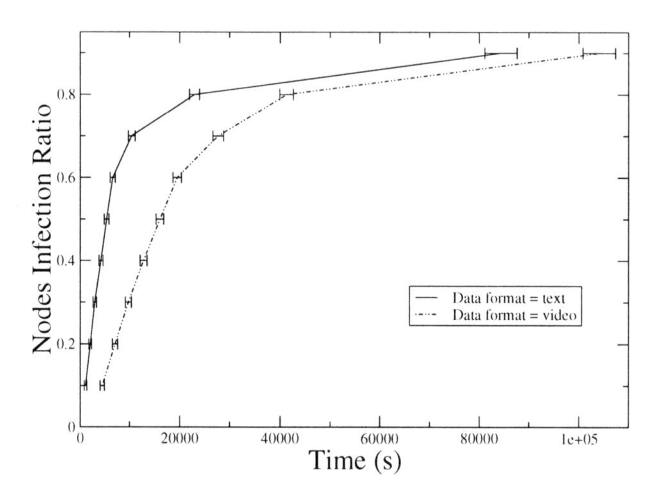

Fig. 6 Network Infection Ratio over time in the case of text and video content.

In the second experiment, we moved one step forward towards the addressed opportunistic content distribution application scenario. We assumed that contents are regularly injected into the system, as required by any content distribution application. Further, contents are fully characterized by their format, e.g., text, music or video, and by the Time To Live TTL, which represents the validity in time of the specific content. The TTL, while being useful for avoiding users to exchange useless information, is also extremely important in mobile networks for avoiding the system to collapse due to the exponential growth of data circulating in the network. The specific format of contents is chosen in accordance with the preferences expressed by the users in the questionnaires (Tab. 3).
Contents are sorted in the local storage of nodes according to their generation time, i.e. from the freshest to the oldest, and this order is respected when selecting the data to be sent. At

each meeting, two nodes merge the respective storage, avoiding duplication of data[4] and with respect to the contact duration time. In particular, we assume that the contact duration T is equally shared between the 2 nodes meeting, and that each node disposes of $T/2$ seconds of time for sending data to the encountered peer [5].
We have then run a second set of experiments, where a limited set of messages are injected in the network according to a predefined *Message Injection Rates* (MIR). Each simulation is stopped when all messages injected have expired, due to their TTL constraint. We measured the NIR in the case of different and of different TTLs, and different MIRs. In order to obtain a sufficiently small confidence interval, for each setting considered we run 10 simulations, varying the instant at which messages are injected in the network, and averaging results over all the messages injected in the simulation. In Fig. 7, the results of this experiments are presented. Increasing the MIR corresponds to increase the network load, and this is reflected, independently from the specific TTL considered, in a significant decrease of the experienced NIR for high MIR values. Also the value of the contents TTL significantly impacts the diffusion of contents in the network: the lower its value, the smaller the time available for a content to diffuse. For the considered office environment, where the dynamism of nodes is relatively small, it is necessary to adopt high TTL values, e.g., greater than 12 hours, in order to reach a sufficiently large number of users.

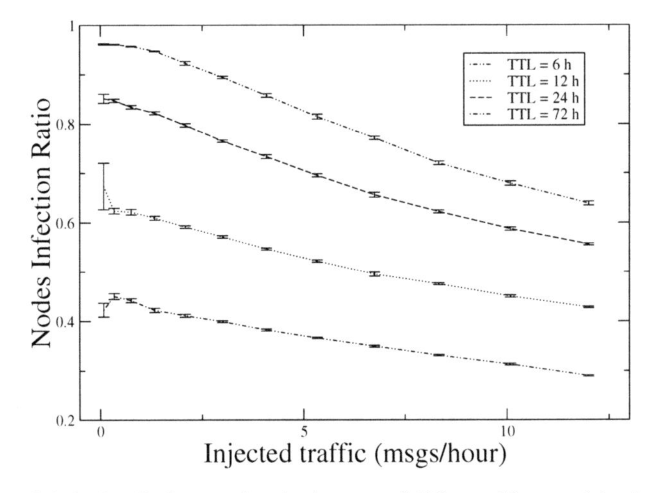

Fig. 7 *Network Infection Ratio* over time in the case of different *Message Injection Rates*, and different TTLs.

Finally, we investigated how well the different content formats diffuse, when varying the MIR. Fig. 8 presents the NIR in the case of a TTL equals 24 hours, for a MIR spanning from 1 to 15 messages per hour. As it can be observed, textual information, due to the limited content size, is insensitive to increases of the MIRs, and also in the case of 15 messages injected per

[4] In reality, nodes can avoid duplication of data by first running an information discovery protocol. This would introduce an additional overhead that is in any case negligible, with respect to the size of the data being exchanged.

[5] This assumption can be easily removed by assuming nodes to alternatively transmit predefined chunks of data. If the size of the chunk is sufficiently small, this corresponds to equally share the available contact duration T between the 2 nodes.

hour is able to reach the 85% of the nodes. Differently, when a high number of video contents are pushed into the system, the opportunistic network saturates, as the limited duration of the contacts is not enough for the nodes to exchange all data. This is clear from Fig. 8, where the video MIR decreases down to 50% very soon.

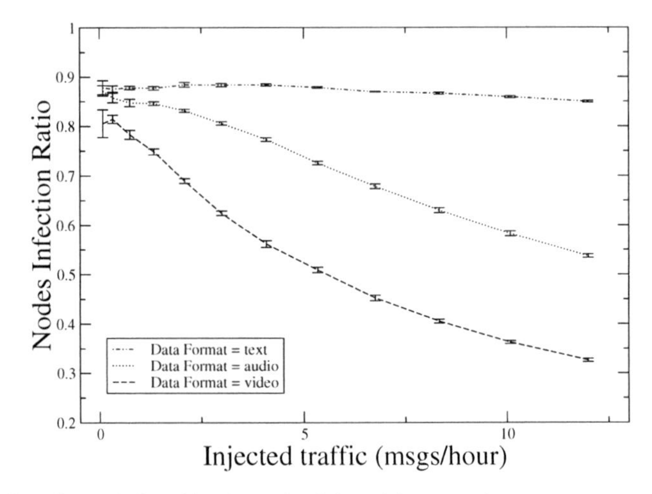

Fig. 8 Number of users infected by the packet injected the network vs message generation rate in the case of textual, audio and video data format, $TTL = 24$ hours, 10 runs for each point, 98%confidence interval.

From this first analysis, we can easily conclude that, for the office environment under consideration, the opportunistic content distribution application can safely support textual information, but not video. Further, given the limited users dynamism, content TTL should be at least 24 hours in order not to expire before having reached a sufficiently high number of users.

4.2 Evaluating User Preferences

Starting from the assumption that people tend to share information within communities of similar interests, we reproduced a simplified behavior of users, starting from the preferences expressed by users in the questionnaires. We evaluated then how his impacts the considered application scenario.

4.2.1 User Interests and Affinity Measure

We have assumed each user to be characterized by a set of interests, consisting of the union of different content categories (i.e., music, cinema, news). More formally, to each user i we associate the interests $I_i \bigcup \{I_{i,0}, \ldots, I_{i,N}\}$, where $I_{i,k}$ is the k_{th} interest of user i. Each content category I_k is fully characterized by a finite hierarchical set of nested subcategories (i.e., rock, pop, etc.). We have then associated a weighted tree data structure to each content category,

with weights representing the relevance given by a specific user to that sub-category. Each weight is relative to specific sub-categories level, and detail of each sub-category increases while navigating the tree.
An example of such structure is reported in Fig. 9. In this case, the user is interested in three distinct content categories: sport, cinema and music. For the case of music, the user is interested in rock music, and his favorite bands are *U*2 and *Moby*.

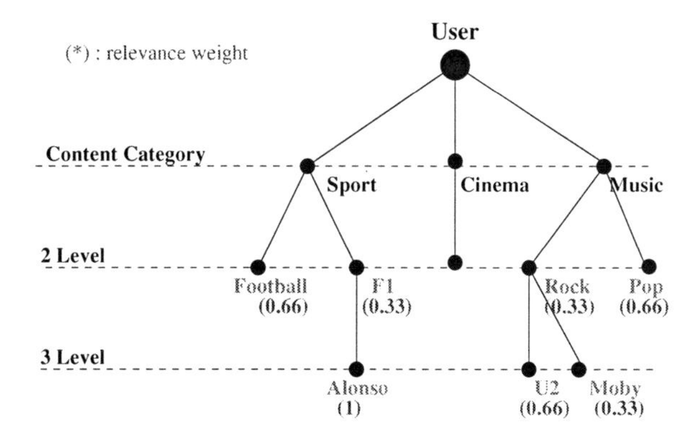

Fig. 9 User interests tree representation.

For each content category, we have then defined the *affinity* between 2 users as the similarity between the trees representing the corresponding interests for that specific category. The affinity between 2 users i and j is then defined as follows:

$$a_{i,j} = \sum_{k=1}^{N} \sum_{l=1}^{\min(h_{I_{i,k}}, h_{I_{j,k}})} w_l r_l(I_{i,k}, I_{j,k}),$$

where $r_l(I_{i,k}, I_{j,k})$ is the Pearson coefficient [6] of the subset of elements of trees $I_{i,k}$ and $I_{j,k}$ at depth l, $h_{I_{i,k}}$ represents the depth of the content tree k for user i, and w_l is the weight given by users to depth l of the content tree. Clearly, the affinity ranges from 0 to 1, and reflects the similarity between the trees representing the preferences of two users.
Fig. 10 presents an example of how such affinity metric is evaluated. In this case:

$$a_{1,2} = 0.2r_1(I_{1,1}, I_{2,1}) + 0.8r_2(I_{1,2}, I_{2,2}) = 0.50,$$
$$a_{2,1} = 0.8r_1(I_{1,1}, I_{2,1}) + 0.2r_2(I_{1,2}, I_{2,2}) = 0.15,$$

and the difference is due to the fact that user 1 retains more relevant less focused information, for which there is a higher match.
Starting from the user preferences expressed in the questionnaires, we evaluated the affinity among users for the different content categories considered. The results are summarized in Tab. 5, where the last case is the total affinity between 2 users. As it can easily observed, users interests are very close with respect to news, while the music content category leads to larger differences.
We have then introduced the *Affinity Graph* (AG), in which an edge is drawn between two vertexes if the corresponding nodes presents an affinity above a given threshold λ_{thr}. Differently

[6] Pearson's product moment correlation coefficient is defined as $r = \frac{\sum(x_i-\bar{x}_i)(y_i-\bar{y}_i)}{\sqrt{\sum(x_i-\bar{x}_i)^2 \sum(y_i-\bar{y}_i)^2}}$, and measures the linear correlation between two variables.

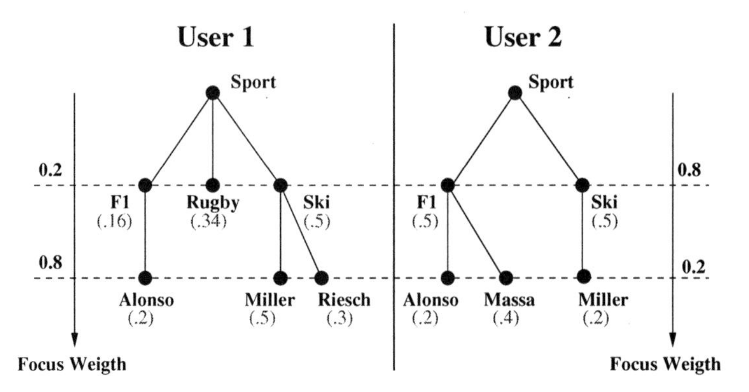

Fig. 10 Example of evaluation of users affinity.

Table 5 The affinity among users, for the content categories considered, music, cinema and news, and for the total category.

Content Cat.	Min	Mean	Max	Variance
Music	0.004	0.36	0.98	0.26
Cinema	0.003	0.39	1	0.23
News	0.003	0.54	1	0.23
Total	0.38	0.36	0.41	0.02

from Fig. 5, in this case constraints are derived solely by the affinity among users, and not from the strength of the "contact" links.
Fig. 11 shows the AG for the different content categories, and an affinity threshold $aff_{thr} = 0.75$. As it can be easily observed, depending on the specific content category and affinity threshold, the topology of the graph changes significantly. For an affinity of 0.75, all the graphs are already partitioned, with many isolated nodes. This is particularly true for the $Total$ case where, a part for a subgraph of 3 nodes, the remaining ones are completely isolated from the rest of the network. The reason for this is that the affinities for different categories tend to compensate each other, and users showing similar preferences in terms of, e.g., music have opposite ones with respect to cinema. This situation is mapped into a lower value of the total affinity, and heavily influences the resulting network structure.

5 Phase 3: combining users and technological constraints

In this last part of the study, we have evaluated how the combination of both the technological and user-defined constraints impact the opportunistic content distribution application.
Let us now assume users to exchange data only if they share common interests. At the system level, this requirement is mapped to users exchanging data only if their affinity exceeds a certain threshold aff_{thr}, otherwise not. Clearly, the higher the threshold, the closer must users interests have to be for a data exchange to occur. We have then simulated the opportunistic content diffusion over the collected traces taking into consideration the interests expressed by users in the questionnaires. Differently from the previous evaluation, messages are now injected in the system and tagged with a specific category (belonging to the set of those present in questionnaires). When two nodes start an information exchange, data is organized in the

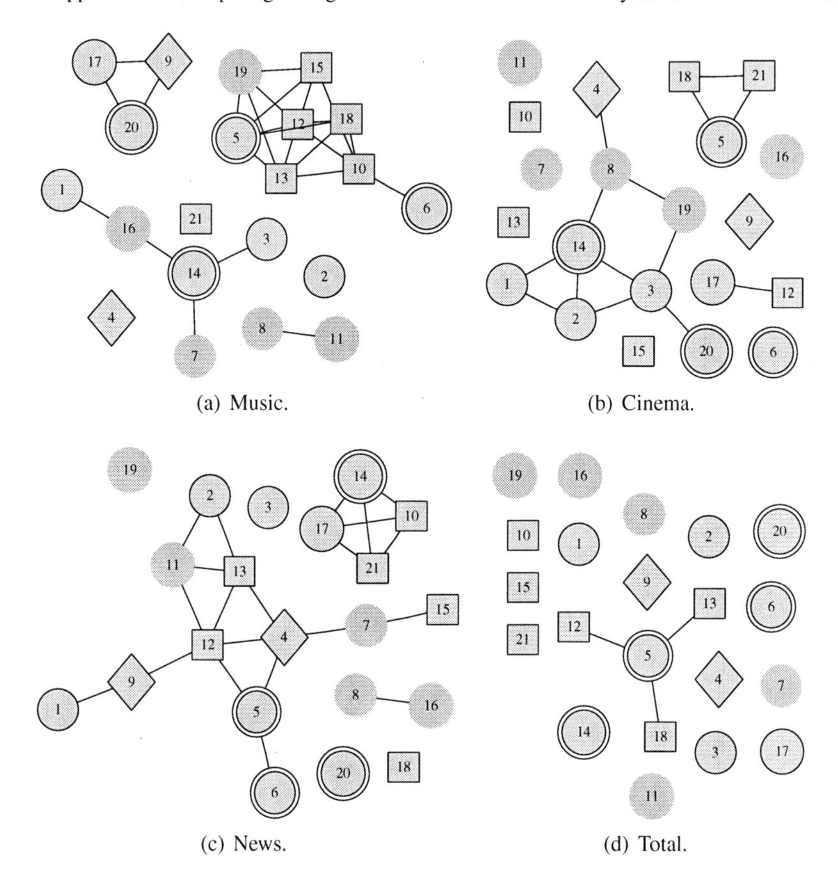

(a) Music. (b) Cinema.

(c) News. (d) Total.

Fig. 11 Affinity Graph for the music, cinema, news and total categories. The affinity graph is build by drawing an edge between two vertexes if the corresponding nodes presents an affinity above a given threshold $aff_{thr} = 0.75$.

storage from the most to the less relevant for the other node, so that in the available meeting time the most interesting information are delivered first.

Fig.12 shows, for different $TTLs$ and for a fixed infection ratio of 1 message per hour, how the node infection ratio varies in correspondence of different affinity thresholds. Again, each point is evaluated as the average of 10 different runs and with a confidence interval of 98% and simulations are stopped only when all the contents are expired. As intuitively clear, for lower values of aff_{thr} a loose filter is applied to the data exchange and the information is influenced by the technological dimension only. Results change when increasing aff_{thr} up to 0.7, above which no sufficient match is present among users preferences for a data exchange to occur. We can then conclude that the opportunistic content distribution performance consistently degrade as the aff_{thr} increases. In fact, for high values of aff_{thr} (i.e. higher that 0.7) users collect only content they are really interested in, without storing contents that could be of interest for other nodes in the network, showing in such a way a selfish behaviour.

We have then evaluated the behavior of the network if data of a single content category are injected and the affinity among users is evaluated only with regards to that single category.

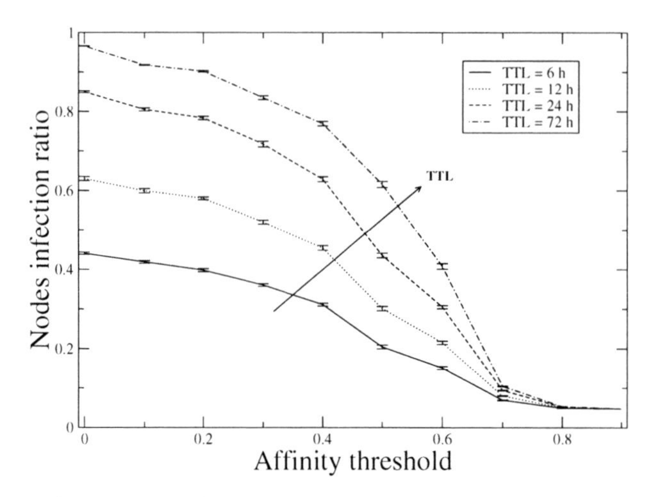

Fig. 12 Nodes infection ratio vs affinity threshold in the case of heterogeneous data formats, TTL of 6, 12, 24 and 72 hours and injection rate of 1 message per hour.

The simulation settings are the same as before. Fig. 13 depicts network performance (i.e. node infection ratio) for the three content categories we have assumed throughout this paper: it can be observed that the behavior is the same independently from the specific category, but that nodes still exchange some contents for the highest values of aff_{thr}. This is due to the fact that, as shown in the previous section, the affinity among users regarding a single content category is obviously higher than the total one, suggesting that opportunistic content distribution applications performs better with more specific contents.

Finally, we have analyzed a particular case of content distribution in order to show the importance of combining the technological and users constraints in the considered application. We have preliminarily isolated 3 users in the network, whose interests were very close in the content category "News" (Fig. 11). These users correspond to nodes '2', '4' and '5'. As a consequence, an information generated by '5' will almost certainly interest also node '2' and '4'. At the same time, we have observed that in Fig. 5 these nodes belong to different groups, and with a high probability a message delivery among them would occur only if the content distribution network performs well, e.g. all the nodes encountered by '5' store the message and forward it finally to '2'.

In a highly loaded network (i.e. injection rate of 10 messages per hour) with a TTL of 24 hours we have evaluated for increasing values of aff_{thr} the message delivery ratio for the above mentioned nodes, i.e. the number of times one specific message injected by '5' reaches '2' and '4'. In particular, we have monitored a single message injected by '5' and stopped the simulation either when it reached both '2' and '4', or when it expires. The separated results for each nodes couple ('5'-'2' and '5'-'4') are reported in Fig. 14 for the text and video formats. Each point in figure has been averaged over 50 different runs with a confidence interval of 98%. A value of 1 for the message delivery ratio means that all the messages generated by '5' have been forwarded to '2' or '4', while a value of 0 means that none of the considered information have been delivered before expiration. We can observe the different behaviour between the two formats for each source-destination couple: it is interesting to notice how, for the same simulation settings, a textual information can be easily delivered even for high values of aff_{thr} (e.g. up to 0.7 for '2') while a video not. At the same time we notice that two nodes

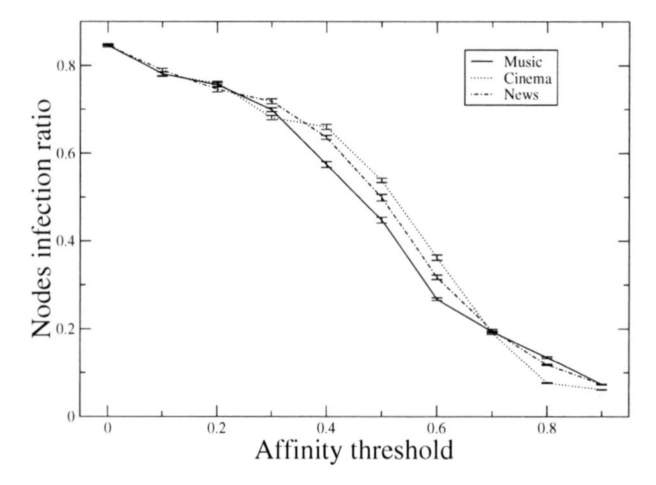

Fig. 13 Nodes infection ratio vs. affinity threshold, for the three content categories considered, TTL of 24 hours and injection rate of 1 message per hour.

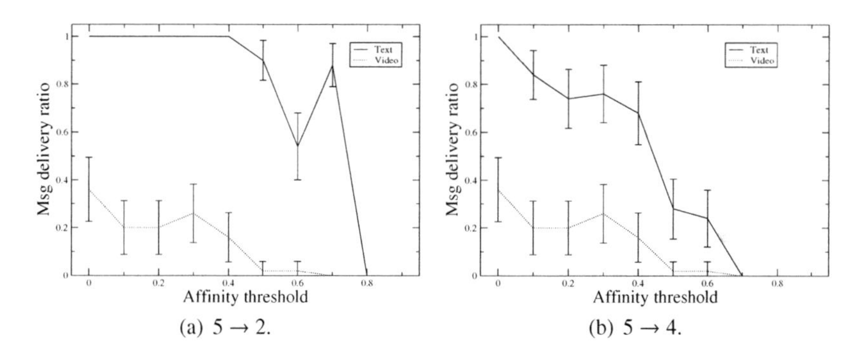

Fig. 14 Graph resulting from an affinity threshold of 0.6, in the case of music and news categories, respectively.

that should apparently behave similarly, results with a different infection. This is due to the fact that when the network is congested, contents diffusion paths differs significantly among each other.

6 Discussion

Reflecting on these results, we begin to see that considering technical constraints in isolation from expressed user preferences, or vice versa, does not paint a complete picture. When we begin to consider not merely inter-user encounters, but the duration of these meetings, coupled

with the format and size of data which the users might want to share, a network graph like that of (insert ref) Figure 5 begins to emerge. The length of an encounter, then, becomes extremely important when we consider not only *if* people were in proximity with one another, but how regularly, or how briefly they met. In recognizing that data transmission is, as of yet, not instantaneous, the graph of potential data exchange becomes less complete. Further, when we incorporate user-expressed preferences about the types of data they might wish to exchange, the graphs based on these affinities become even more fragmented. From these results we do not mean to say that opportunistic content distribution systems are not feasible, but rather to motivate the importance of both current technological limitations as well as user desires.
In future work we plan to investigate how these important factors might be effectively incorporated into the design of future systems. One interesting avenue for exploration we have considered is to attempt to identify hubs of these social networks and to lower their level of affinity in order to aid in the diffusion of data. Alternatively, we also propose to make the inner-workings of the network more transparent to users, encouraging an awareness of the trade-off between requesting only very specific types of data, or content which is of considerable size (e.g. video), and the amount of data that user is likely to receive. By attempting to incorporate social factors into the parameters of the applications performance, and, on the other hand, by making users more aware of the technical implications of their expressed choices, we hope to be able to create an opportunistic content distribution system which achieves an optimal level of real-world performance.

7 Closing Remarks

In this paper we presented a study of an opportunistic content distribution system, which enables users to seamlessly download and diffuse multi-media content such as music, videos or news. We evaluated a specific use case in an office environment, taking into consideration the realistic contact patterns of users, as measured in a 4 week period, as well as users' preferences, as obtained from questionnaires. We then evaluated how data diffusion varies, depending on these two system design dimensions. Results show how the design of such systems is a complex, non-trivial tasks and needs to account for many factors, both technological and user-defined, in order for the system to support the desired amount of traffic, and for users the effectively receive the content they are interested in. Future work will be devoted to the implementation and evaluation of the entire system architecture, with the aim of understanding the broader social impacts of the everyday use of such a system.

References

1. Bedd: Bringing people together. http://www.bedd.com/.
2. Juicecaster: Know first ... show first. http://www.juicecaster.com/.
3. S. Burleigh, L. Torgerson, K. Fall, V. Cerf, B. Durst, K. Scott, and H. Weiss. Delay-tolerant networking: an approach to interplanetary internet. *IEEE Comm. Mag.*, 41(6):128–136, 2003.
4. I. Carreras, I. Chlamtac, F. D. Pellegrini, and D. Miorandi. Bionets: Bio-inspired networking for pervasive communication environments. *IEEE Trans. on Vehicular Technology*, 56(1):218–229, Jan. 2007.
5. A. Chaintreau, P. Hui, J. Crowcroft, C. Diot, R. Gass, and J. Scott. Impact of human mobility on the design of opportunistic forwarding algorithms. In *Proc. of INFOCOM*, Barcelona, Spain, April 23–29, 2006.

6. P. Hui and J. Crowcroft. Bubble rap: Forwarding in small world dtns in ever decreasing circles. Technical Report UCAM-CL-TR-684, Univ. of Cambridge, Computer Laboratory, May 2007.
7. imity: Social situations to go. http://www.imity.com/.
8. A. G. Miklas, K. K. Gollu, S. Saroiu, K. P. Gummadi, and E. de Lara. Exploiting social interactions in mobile systems. In *Proc. of UBICOMP*, Innsbruck, Austria, Sept 2007.
9. M. Musolesi and C. Mascolo. Designing mobility models based on social network theory. *ACM SIGMOBILE Mobile Computing and Communication Review*, 11(3), July 2007.
10. L. Pelusi, A. Passarella, and M. Conti. Opportunistic networking: data forwarding in disconnected mobile ad hoc networks. *IEEE Comm. Mag.*, 44(11), Nov. 2006.
11. Reality Mining. Machine perception and learning of complex social systems. http://reality.media.mit.edu/.
12. O. Riva and S. Toivonen. The dynamos approach to support context-aware service provisioning in mobile environments. *Elsevier Journal of Systems and Software*, 80(12):1956–1972, 2007.
13. J. Su, J. Scott, P. Hui, J. Crowcroft, E. de Lara, C. Diot, A. Goel, M. Lim, and E. Upton. Haggle: Seamless networking for mobile applications. In *Proc. of Ubicomp*, Sept. 16-19 2007.
14. A. Vahdat and D. Becker. Epidemic routing for partially connected ad hoc networks, 2000.
15. X. Zhang, G. Neglia, J. Kurose, and D. Towsley. Performance modeling of epidemic routing. *Comput. Netw.*, 51(10):2867–2891, 2007.

Programming and Validation Techniques for Reliable Goal-driven Autonomic Software

Damian Dechev, Nicolas Rouquette, Peter Pirkelbauer and Bjarne Stroustrup

Abstract Future space missions such as the Mars Science Laboratory demand the engineering of some of the most complex man-rated autonomous software systems. According to some recent estimates, the certification cost for mission-critical software exceeds its development cost. The current process-oriented methodologies do not reach the level of detail of providing guidelines for the development and validation of concurrent software. Time and concurrency are the most critical notions in an autonomous space system. In this work we present the design and implementation of a first concurrency and time centered framework for verification and semantic parallelization of real-time C++ within the JPL Mission Data System Framework (MDS). The end goal of the industrial project that motivated our work is to provide certification artifacts and accelerated testing of the complex software interactions in autonomous flight systems. As a case study we demonstrate the verification and semantic parallelization of the MDS Goal Networks.

1 Introduction

In this work we describe the design, implementation, and application of a first *concurrency* and *time* centered framework for verification and semantic parallelization of real-time C++ within the JPL Mission Data System Framework (MDS). MDS provides an experimental goal- and state- based platform for testing and development of autonomous real-time flight applications [22]. The end goal of the industrial project that motivated our work is to provide certification artifacts and accelerated testing of the complex software interactions in autonomous flight systems. The process of software certification establishes the level of confidence in a

Damian Dechev
Texas A&M University, College Station, TX 77843-3112, e-mail: dechev@tamu.edu

Nicolas Rouquette
Jet Propulsion Laboratory, NASA/California Institute of Technology, e-mail: nicolas.rouquette@jpl.nasa.gov

Peter Pirkelbauer
Texas A&M University, College Station, TX 77843-3112, e-mail: pirkelp@tamu.edu

Bjarne Stroustrup
Texas A&M University, College Station, TX 77843-3112, e-mail: bs@cs.tamu.edu

A.V. Vasilakos et al. (eds.), *Autonomic Communication*, DOI: 10.1007/978-0-387-09753-4_9,

software system in the context of its *functional* and *safety* requirements. A software certificate contains the evidence required for the system's independent assessment by an authority having minimal knowledge and trust in the technology and tools employed [6]. Providing such certification evidence may require the application of a number of software development, analysis, verification, and validation techniques [20]. The dominant paradigms for software development, assurance, and management at NASA rely on the principle "test-what-you-fly and fly-what-you-test". This methodology had been applied in a large number of robotic space missions at the Jet Propulsion Laboratory. For such missions, it has proven suitable in achieving adherence to some of the most stringent standards of man-rated certification such as the DO-178B [25], the Federal Aviation Administration (FAA) software standard. Its Level A certification requirements demand 100% coverage of all high and low level assurance policies. Some future space exploration projects such as the Mars Science Laboratory (MSL), Project Constellation, and the development of the Crew Launch Vehicle (CLV) and the Crew Exploration Vehicle (CEV) suggest the engineering of some of the most complex man-rated software systems. As stated in the Columbia Accident Investigation Board Report [3], the inability to thoroughly apply the required certification protocols had been determined to be a contributing factor to the loss of STS-107, Space Shuttle Columbia.
Schumann and Visser's discussion in [26] suggests that the current certification methodologies are prohibitively expensive for systems of such complexity. A detailed analysis by Lowry [20] indicates that at the present moment the certification cost of mission-critical space software exceeds its development cost. The challenges of certifying and re-certifying avionics software has led NASA to initiate a number of advanced experimental software development and testing platforms, such as the Mission Data System (MDS) [22], as well as a number of program synthesis, modeling, analysis, and verification techniques and tools, such as The JavaPathFinder [2], the CLARAty project [29], Project Golden Gate [10], The New Millennium Architecture Prototype (NewMAAP) [9]. The high cost and demands of man-rated certification have motivated the experimental development of several accelerated testing platforms [1]. A great number of the experimental faster-than-real-time flight software simulators require the parallelization of previously sequential real-time algorithms. In this work we present the design and implementation of a first *concurrency* and *time* centered framework for *verification* and *semantic parallelization* of real-time C++ within the JPL Mission Data System Framework. Our notion of *semantic parallelization* implies the thread-safe concurrent execution of system algorithms that utilize shared data, based on the application's semantics and invariants. As a practical industrial-scale application, we demonstrate the parallelization and verification of the MDS' Goal Networks, a critical component of the JPL's Mission Data System.

2 Challenges for Mission Critical Autonomous Software

In [21] Perrow studies the risk factors in the modern high technology systems. His work identifies two significant sources of complexity in modern systems: *interactions* and *coupling*. The systems most prone to accidents are those with *complex* interactions and *tight* coupling. With the increase of the size of a system, the number of functions it has to serve, as well as its interdependence with other systems, its interactions become more incomprehensible to human and machine analysis and this can cause unexpected and anomalous behavior. Tight coupling is defined by the presence of time-dependent processes, strict resource constraints, and little or no possible variance in the execution sequence. Perrow classifies space missions in the riskiest category since both hazard factors are present. In this work, we argue that the notions of *concurrency* and *time* are the most critical elements in the design and implementation of an embedded autonomous space system. According to a study on concurrent models of computation for embedded software by Lee and Neuendorffer [18], the major contributing factors to the development and design complexity of such systems are the underlying sequential memory models and the lack of first class representation of the notions of time and concurrency in the applied programming languages.

2.1 Parallelism and Complexity

The most commonly applied technique for controlling the interactions of concurrent processes is the use of mutual exclusion locks. A mutual exclusion lock guarantees thread-safety of a concurrent object by blocking all contending threads trying to access it except the one holding the lock. In scenarios of high contention on the shared data, such an approach can seriously affect the performance of the system and significantly diminish its parallelism. For the majority of applications, the problem with locks is one of difficulty of providing correctness more than one of performance. The application of mutually exclusive locks poses significant safety hazards and incurs high complexity in the testing and validation of mission-critical software. Mutual exclusion locks can be optimized in some scenarios by utilizing fine-grained locks [15] or context-switching. Often due to the resource limitations of flight-qualified hardware, optimized lock mechanisms are not a desirable alternative [20]. Even for efficient locks, the interdependence of processes implied by the use of locks, introduces the dangers of deadlock, livelock, and priority inversion. The incorrect application of locks is hard to determine with the traditional testing procedures and a program can be deployed and used for a long period of time before the flaws can become evident and eventually cause anomalous behavior.

2.1.1 Parallel Programming without Locks

To achieve higher safety and enhance the performance of our implementation, we consider the application of *lock-free synchronization*. As defined by Herlihy [14], a concurrent object is *non-blocking* (lock-free) if it guarantees that *some* process in the system will make progress in a *finite* amount of steps. Non-blocking algorithms do not apply mutually exclusive locks and instead rely on a set of atomic primitives supported by the hardware architecture. The most ubiquitous and versatile data structure in the ISO C++ Standard Template Library [27] is *vector*, offering a combination of dynamic memory management and constant-time random access. In our framework for verification and semantic parallelization of real-time C++ we utilize the design of the first lock-free design and implementation of a dynamically-resizable array in ISO C++ (Section 5). It provides linearizable operations, disjoin-access parallelism for random access reads and writes, lock-free memory allocation and management, and fast execution.

2.2 Motivation and Contributions

As discussed by Lowry [20], in July 1997 The Mars Pathfinder mission experienced a number of anomalous system resets that caused an operational delay and loss of scientific data. The follow-up study identified the presence of a priority inversion problem caused by the low-priority meteorological process blocking the high-priority bus management process. It has been determined that it would have been impossible to detect the problem with the black box testing applied at the time to derive the certification artifacts. A more appropriate priority inversion inheritance algorithm had been ignored due to its frequency of execution, the real-time requirements imposed, and its high cost incurred on the slower flight-qualified computer hardware. The subtle interactions in the concurrent applications of the modern aerospace autonomous software are of critical importance to the system's safety and operation. Despite the challenges in debugging and verification of the system's concurrent components, the existing certification process [25] does not provide guidelines at the level of detail reaching the development, application, and testing of concurrent programs. This is largely due to the process-oriented nature of the current certification protocols and the complexity and high level of specialization of the aerospace autonomous embedded applications. In the near future, NASA plans to deploy a number of diverse vehicles, habitats, and supporting facilities for its imminent missions to the Moon, Mars and beyond. The large array of complex tasks that these systems would have to perform implies their high level of autonomy. In [22] Rasmussen et al. suggest that the challenges for these systems' control is one of the most demanding tasks facing NASA's Exploration Systems Mission Directorate. Some of the most significant challenges that the authors identify are managing a large number of tightly-coupled components, performing operations in uncertain remote environments, enabling the agents to respond and recover from anomalies, guaranteeing the system's correctness and reliability, and ensuring effective communication across the system's components. In the rest of the paper we describe the definition, design, and implementation of a first *concurrency* and *time* centered framework for verification and semantic parallelization of autonomous flight software within the JPL's MDS Framework. We integrate a nonblocking vector in our parallel implementation of the Mission Data System's Temporal Constraint Network Library (TCN) in order to achieve higher thread safety and boost the performance of the MDS Goal Networks component. We demonstrate how to specify, model, and formally verify the TCN algorithms and their semantic invariants. Based on our formal models and the application's semantics, we derive a technique for automatic and semantic parallelization of the TCN library's constraint propagation algorithm.

3 Temporal Constraint Networks

A Temporal Constraint Network (TCN) defines the goal-oriented operation of a control system in the context of a system under control. The Temporal Constraint Networks (TCN) application is at the core of the Jet Propulsion Laboratory's Mission Data System (MDS) [22] state-based and goal-oriented unified architecture for testing and development of mission software. The framework's state- and model-based methodology and its associated systems engineering processes and development tools have been successfully applied on a number of test applications including the physical rovers Rocky 7 and Rocky 8 and a simulated Entry, Descent, and Landing (EDL) component for the Mars Science Laboratory mission. A TCN consists of a set of temporal constraints (TCs) and a set of time points (TPs). In this model of goal-driven operation, a time point is defined as an interval of time when the configuration of the system is expected to satisfy a property predicate. The width of the interval corresponds to the temporal uncertainty inherent in the satisfaction of the predicate. Similarly, temporal constraints have an associated interval of time corresponding to the acceptable bounds on the interactions between the control system and the system under control during the performance of a specific activity. A TCN graph topology represents a snapshot at a given time of the known set of activities the control system has performed so far, is currently engaged in, and will be performing in the near future up to the horizon of the elaborated plan initially created as a solution for a set of goals. The topology of a temporal constraint network must satisfy a number of invariants.

(a) A TCN is a directed acyclic graph where the edges represent the set of all time points (S_{tps}) and the vertices the set of all temporal constraints (S_{tcs})
(b) For each time point $TP_i \in S_{tps}$, there is a set of temporal constraints that are immediate successors (S_{succ_i}) of TP_i and a set, S_{pred_i}, consisting of all of TP_i's immediate predecessors

(c) Each temporal constraint $TC_j \in S_{tcs}$ has exactly one successor TP_{succ_j} and one predecessor TP_{pred_j}
(d) For each pair $\{TP_i, TC_j\}$, where $TP_i \equiv TC_{succ_j}$, $TC_j \in S_{pred_i}$ must hold. The reciprocal invariant must also be valid, namely for each pair of $\{TP_i, TC_j\}$ such that $TP_i \equiv TC_{pred_j}$, $TC_j \in S_{succ_i}$
(e) The firing window of a time point $TP_i \in S_{tps}$ is represented by the pair of time instances $\{TP_{min_i}, TP_{max_i}\}$. Assuming that the current moment of time is represented by T_{now}, then $TP_{min_i} \leq T_{now} \leq TP_{max_i}$, for every $TP_i \in S_{tps}$.

General-purpose programming languages lack the capabilities to formally specify and check domain-specific design constraints. Direct representation and verification of the TCN invariants in the implementation source code would result in a slow and cumbersome solution. However, any implementation (in C++, Java or another programming language) must operate under the assumptions that the basic TCN invariants are satisfied. Thus, prior to implementing a solution to the TCN constraint propagation problem, it is necessary to guarantee the correctness and consistency of the topology of the goal network.

4 Verification and Automatic Parallelization Framework

In this section we describe the design, implementation, and practical application of our framework for verification and semantic parallelization of real-time C++ within JPL's MDS Framework (Figure 1). The input to the framework is the MDS mission planning and execution module that is based on the definition of temporal constraint networks. At the core of the most recent implementations at JPL of this critical module is an optimized iterative algorithm for the real-time propagation of temporal constraints, developed and described by Lou in [19]. Constraint propagation poses performance challenges and speed bottlenecks due to the algorithm's frequent execution and the necessary real-time update of the goal network's topology. The end goal of our work is, given the implementation of the optimized iterative propagation scheme and the topology of a particular goal network, to establish the correctness of the core TCN semantic invariants (see Section 3) and *automatically* derive an implementation that can be executed concurrently on one of the JPL's experimental testbeds for accelerated testing [1]. Our approach for achieving concurrent execution is based on the idea of identifying Time Phases within a goal network, which allow the semantic parallelization of the constraint propagation algorithm. In this work, we define *semantic parallelization* as the thread-safe concurrent execution of an algorithm (whose operation is dependent on shared data), derived from the application's semantics and invariants. In the following sections we describe how we reach our goal of verification and semantic parallelization of the mission planning and control module by constructing and executing a formal verification model in Alloy [16] that represents the implementation's core semantics and functionality. We refine a formal modeling and analysis methodology, initially suggested by Rouquette [24], that helps us analyze the logical properties of the goal network model and automatically derive a meta-model for our parallel solution.

4.1 The Problem of TCN Constraint Propagation

A classic solution to the problem of constraint propagation in TCN is the direct application of Floyd-Warshall's all-pairs-shortest-path algorithm [4], offering a complexity of $O(N^3)$, where N is the number of time points in the TCN topology. Since, by definition, the goal of the TCN propagation algorithm is to compute the real-time values of the network's temporal constraints,

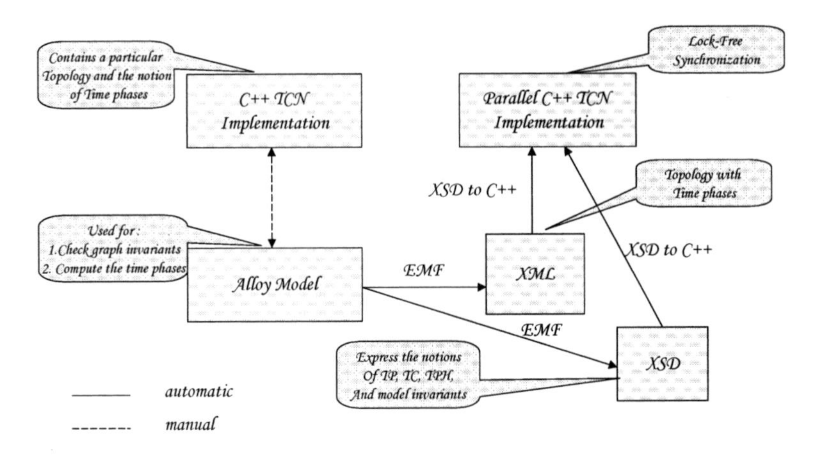

Fig. 1 A Framework for Verification and Semantic Parallelization

the algorithm is frequently executed and, given the massive scale of a real world goal network, can cause significant bottleneck for the overall system's performance. In [19], Lou describes an innovative and effective TCN propagation scheme with a complexity close to linear. Lou's TCN propagation is based on the concept of alternating forward and backward propagation passes. A forward pass updates the time interval at each time point by considering only its incoming temporal constraints (Algorithm 9). Similarly, a backward pass recomputes the time windows at each time point by considering only its outgoing temporal constraints (Algorithm 10). The scheme utilizes a shared container, named a *propagation queue*, to keep track of all time points whose successor time points' windows are about to be updated next (during a forward pass) and all time points whose predecessor time points' windows are about to be updated next (during a backward pass). A forward pass begins by selecting all time points with no predecessors and inserts them into the propagation queue. A backward pass begins by selecting all time points with no successors and inserts them into the propagation queue. Each iteration is carried out until:

(a) An iteration completes without updating any temporal constraints (thus indicating that there are no more updates to be performed during the pass). In this case, the TCN topology is considered to be *temporally consistent*.
(b) The iteration has stumbled upon a time window of negative value and the algorithm terminates with the outcome of having a temporally inconsistent network.

As stated by Lou [19], prior to the execution of the optimized propagation scheme, it is critical to guarantee the validity of the core TCN invariants for the topology of the particular goal network. For example, the propagation scheme operates under the assumption that the goal network graph is cycle free. Should there be cycles, the propagation would enter into an endless loop.

Algorithm 9: Forward Pass. Arguments: a reference to the time point about to be updated (tp) and a reference to the global data structure recording the state updates (vstate)

---

1 $min_{tmp} \leftarrow tp.min$;
2 $max_{tmp} \leftarrow tp.max$;
3 **for** $j = 0$ *to* $tp.preds\_size$ **do**
4 | $min_{tmp} \leftarrow$ std::max(min_{tmp}, $tp.preds[j].pred.min + tp.preds[j].min$);
5 | $max_{tmp} \leftarrow$ std::min(max_{tmp}, $tp.preds[j].pred.max + tp.preds[j].max$);
6 **end**
7 **if** $tp.min! = min_{tmp}$ **then**
8 | ASSERT($tp.min < min_{tmp}$);
9 | $tp.min \leftarrow min_{tmp}$;
10 | vstate.aIncr(vstate.count);
| `/* atomically increment the state vector's counter */`
11 **end**
12 **if** $tp.max! = max_{tmp}$ **then**
13 | ASSERT($tp.max > max_{tmp}$);
14 | $tp.max \leftarrow max_{tmp}$;
15 | vstate.aIncr(vstate.count);
| `/* atomically increment the state vector's counter */`
16 **end**
17 **return** $!(min_{tmp} > max_{tmp})$;

Algorithm 10: Backward Pass. Arguments: a reference to the time point about to be updated (tp) and a reference to the global data structure recording the state updates (vstate)

```
1  min_tmp ← tp.min;
2  max_tmp ← tp.max;
3  for j = 0 to tp.succs_size do
4      min_tmp ← std::max(min_tmp, tp.succs[j].succ.min − tp.succs[j].max);
5      max_tmp ← std::min(max_tmp, tp.succs[j].succ.max − tp.succs[j].min);
6  end
7  if tp.min! = min_tmp then
8      ASSERT( tp.min < min_tmp );
9      tp.min ← min_tmp;
10     vstate.aIncr(vstate.count);
       /* atomically increment the state vector's counter    */
11 end
12 if tp.max! = max_tmp then
13     ASSERT( tp.max > max_tmp );
14     tp.max ← max_tmp;
15     vstate.aIncr(vstate.count);
       /* atomically increment the state vector's counter    */
16 end
17 return !(min_tmp > max_tmp);
```

4.2 Modeling, Formal Verification, and Automatic Parallelization

Alloy [16] is a lightweight formal specification and verification tool for the automated analysis of user-specified invariants on complete or partial models. The Alloy Analyzer is implemented as a front-end, performing the role of a model-finder, to a boolean SAT-solver. Formal verification and modeling of JPL's flight software has been previously demonstrated to be effective and successful by Holzmann [12]. We use the Alloy specification language [16] to formally represent and check the semantics of the temporal constraint networks library (Algorithm 11) and its main invariants (Algorithm 12). In our C++ goal networks implementation we have applied generic programming techniques and concepts [23], so that we can maintain a higher level of expressiveness. As a result we have achieved a significant similarity in the way the main TCN notions and invariants are expressed in our actual implementation and the Alloy verification models. In the future, we intend to utilize a static analysis tool such as The Pivot [28] in order to automate this transition (this is the last non-automated component of the presented framework).
In addition, we utilize the Alloy Analyzer to implement our semantic parallelization approach. Our method for semantic parallelization of the goal network is based on the observation that in a topology we can identify groups of time points that would allow the concurrent execution of the propagation passes. A possible criterion for identifying such groups would be to identify the time points in a topology that allow disjoin-access to the shared data. Given the method used to compute the time window $[TP_{min_i}, TP_{max_i}]$ for each $TP_i \in S_{tps}$, we have observed that

the functionally-independent time points are the time points that are equidistant (with respect to the longest path) from the root of the graph. Thus, in our methodology, we define a *Time Phase* Tph_i as the set of the time points (S_{Tph_i}) in a topology that are equidistant, with respect to the longest path, from the root of the graph. In such a way, by definition, the computations of $[TP_{min_a}, TP_{max_a}]$ and $[TP_{min_b}, TP_{max_b}]$ for every pair of $\{TP_a, TP_b\}$, such that $TP_a \in S_{Tph_i}$ and $TP_b \in S_{Tph_i}$, are mutually independent and allow disjoin-access to the shared data. With the support of Alloy Analyzer we define and identify the time phases in a goal network graph (Algorithm 13 and Algorithm 14). Figure 2 provides an example of a goal network containing 15 time points and 6 time phases.

Algorithm 11: Definition of the notions of Temporal Constraint and Time Point

```
  /* declaration of the Temporal Constraint signature       */
1 sig TC { tc_pred: one TP, tc_succ: one TP} ;
  /* declaration of the Time Point signature                */
2 sig TP { tp_preds: set TC, tp_succs: set TC} ;
```

Algorithm 12: Main TCN invariants expressed in the Alloy Specification Language

```
1 all tc:TC | tc in tc.tc_pred.tp_succs;
2 all tc:TC | tc in tc.tc_succ.tp_preds;
3 all tc:TP | some tp.tp_preds ⇒ tp.tp_preds.tc_succ = tp;
4 all tc:TP | some tp.tp_succs ⇒ tp.tp_succs.tc_pred = tp;
5 no ∧(tc_pred.tp_preds) & iden;
6 no ∧(tc_succ.tp_succs) & iden;
  /* last two lines check for cycles                        */
```

Algorithm 13: Definition of the notions of Time Phase and Temporal Constraint Network (with time phases)

```
  /* declaration of the Time Phase signature                */
1 sig Tph{events: set TP, next: one Tph, tcn: one TCN};
  /* declaration of the TCN signature                       */
2 sig TCN{epoch : TP, tps: set TP, tcs: set TC, init: one Tph};
```

Algorithm 14: Main Time Phase invariants expressed in the Alloy Specification Language

```
1 forall p:Tph do
2 |   p.events.tp_succs.tc_succ in p.∧next.events;
3 |   p.events.tp_preds.tc_pred in p.∧~next.events;
4 |   p in p.tcn.init.*next;
5 |   p.events in p.tcn.tps;
6 |   no p.events & p.∧(next).events;
7 end
```

Having identified the time phases in our temporal constraint network specification in Alloy, the aim of the rest of our tool-chain is to *automatically* derive the C++ implementation of the parallel solution through a number of code transformation techniques. Following Rouquette's methodology [24] for model transformation through the application of the Object Constraint

Language (OCL) and the Eclipse Modeling Framework (EMF), we are able to automatically derive an intermediary XML and XSD representations of the graph's topology and the TCN semantic notions, respectively. We apply an XML parser (XercesC) and a CodeSynthesis XSD transformation tool to deliver the C++ implementation of the goal network and our parallel propagation method.

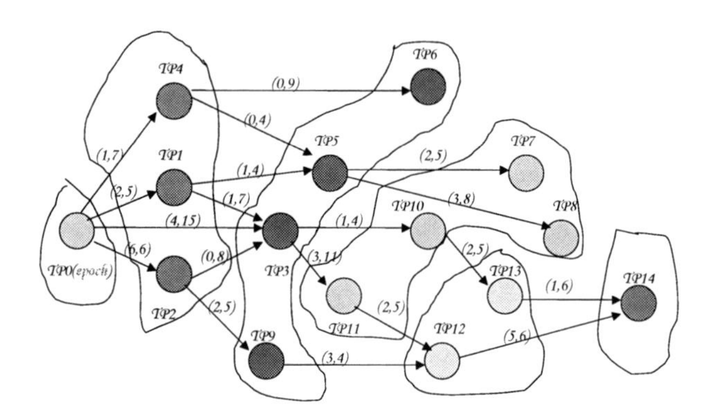

Fig. 2 A Parallel TCN Topology with 15 Time Points and 6 Time Phases

To achieve higher safety and better performance, our parallel propagation scheme employs a number of innovative multi-processor synchronization techniques. In our implementation we have encountered and addressed the following challenges:

(1) Achieving low-overhead parallelization. Our experiments indicated that the wide-spread Pthreads are computationally expensive when applied to the parallel propagation algorithm. Given the frequent real-time changes in the graph topology, employing a thread per iteration for the computations of each time phase comes at a prohibitive cost. To avoid this problem, we have incorporated in our design the application of the Intel tasks from the Threading Building Blocks Library [15]. Our experiments indicate that the Intel tasks provide low-cost overhead when applied in the concurrent execution of the forward and backward passes of the propagation scheme.
(2) Allowing fast and safe access to the shared data. The parallel algorithm requires the safe and efficient concurrent synchronization of its shared data: the propagation queue and the vector containing control data (reflecting the updates during an iteration). By the definition of our algorithm, the propagation queue is synchronized by allowing only disjoint-access writes. While the access to the shared vector is less frequent, its concurrent synchronization is more challenging since we do not have a guarantee that the

concurrent writes would be disjoint. The application of mutual exclusion locks is a possible but likely an ineffective solution due to the risks of deadlock, livelock, and priority inversion. Moreover, the interdependency of processes implied by the use of locks diminishes the parallelism of a concurrent system. A lock-free object guarantees that within a set of contending processes, there is at least one process that will make progress within a finite number of steps. We have employed the implementation of the lock-free vector described in Section 5 in order to meet our goals for thread-safe and effective non-blocking synchronization. The lock-free vector provides the functionality of the popular STL C++ vector as well as linearizable and safe operations with complexity of $O(1)$ and fast execution (outperforming the STL vector protected by a mutex by a factor of 10 or more).

A number of graph properties, in a particular TCN topology, impact the application and performance of the parallel propagation scheme. We expect better performance (with respect to the sequential propagation scheme) when:

(1) The computational load per time point is high. This is the case of a real-world massive-scale goal network. For instance, instructing the Mars Science Laboratory to autonomously find its way in a Martian crater, probe the soil, capture images, and communicate to Mission Control will result in a goal network containing tens or hundreds of thousands of time points. In a small experimental graph topology with a low computational cost per time point (such as a few arithmetic operations), a single processor computation will perform best (when we take into account the parallelization overhead).
(2) Time phases with large number of time points: a topology implying a sequential ordering of the planned events will not benefit from a parallel propagation scheme. The parallel propagation algorithm is beneficial to goal networks representing a large number of highly interactive concurrent system processes.

5 Nonblocking Synchronization

The most common technique for controlling the interactions of concurrent processes is the use of mutual exclusion locks. A mutual exclusion lock guarantees thread-safety of a concurrent object by blocking all contending threads trying to access it except the one holding the lock. In scenarios of high contention on the shared data, such an approach can seriously affect the performance of the system and significantly diminish its parallelism. For the majority of applications, the problem with locks is one of difficulty of providing correctness more than one of performance. The application of mutually exclusive locks poses significant safety hazards and incurs high complexity in the testing and validation of mission-critical software. Mutual exclusion locks can be optimized in some scenarios by utilizing fine-grained locks [15]. Often due to the resource limitations of flight-qualified hardware, optimized lock mechanisms are not a desirable alternative [20]. Even for efficient locks, the interdependence of processes implied by the use of locks, introduces the dangers of deadlock, livelock, and priority inversion.The incorrect application of locks is hard to determine with the traditional testing procedures and a program can be deployed and used for a long period of time before the flaws can become evident and eventually cause anomalous behavior.
To achieve reliability, avoid the dangers of priority inversion, deadlock, and livelock, and at the same time gain performance, we rely on the notion of *lock-free synchronization*. Lock-free systems typically utilize CAS in order to implement a an optimistic speculation on the shared data. A contending process attempts to make progress by applying one or more writes on a local copy of the shared data. Afterwards, the process attempts to swap (CAS) the global data with its updated copy. Such an approach guarantees that from within a set of contending processes, there is at least one that succeeds within a finite number of steps. The system is non-blocking at the expense of some extra work performed by the contending processes.

Linearizability is an important correctness condition for concurrent nonblocking objects: a concurrent operation is linearizable if it appears to execute instantaneously in a given point of time between the time t_1 of its invocation and the time t_2 of its completion. The consistency model implied by the linearizability requirements is stronger than the widely applied Lamport's sequential consistency model [17]. According to Lamport's definition, sequential consistency requires that the results of a concurrent execution are equivalent to the results yielded by *some* sequential execution (given the fact that the operations performed by each individual processor appear in the sequential history in the order as defined by the program). Our vector's nonblocking algorithms are directly derived from the lock-free operations of the first implementation of a lock-free dynamically resizable array presented by Dechev at el. in [5]. The operations of our vector are lock-free and linearizable and in addition they provide disjoin-access parallelism for random access reads and writes and fast execution (outperforming the STL vector protected by a mutex by a factor of 10 or more [5]).

5.1 Practical Lock-Free Programming Techniques

The practical implementation of a hand-crafted lock-free container is notoriously difficult. A nonblocking container's design suggests the update (in a linearizable fashion) of several memory locations. The use of a double-compare-and-swap primitive (DCAS) has been suggest by Detlefs et al. in [7], however such complex atomic operations are rarely supported by the hardware architecture. Harris et al. propose in [13] a software implementation of a multiple-compare-and-swap (MCAS) algorithm based on CAS. This software-based MCAS algorithm has been applied by Fraser in the implementation of a number of lock-free containers such as binary search trees and skip lists [11]. The cost of the MCAS operation is expensive requiring $2M+1$ CAS instructions. Consequently, the direct application of the MCAS scheme is not an optimal approach for the design of lock-free algorithms. The vector's random access, data locality, and dynamic memory management pose serious challenges for its non-blocking implementation. To illustrate the complexity of a CAS-based design of a dynamically resizable array, Table 1 provides an analysis of the number of memory locations that need to be update upon the execution of some of the vector's basic operations.

Table 1 Vector - Operations

	Operations	Memory Locations
push_back	$Vector \times Elem \rightarrow void$	*2: element and size*
pop_back	$Vector \rightarrow Elem$	*1: size*
reserve	$Vector \times size\_t \rightarrow Vector$	*n: all elements*
read	$Vector \times size\_t \rightarrow Elem$	*none*
write	$Vector \times size\_t \times Elem \rightarrow Vector$	*1: element*
size	$Vector \rightarrow size\_t$	*none*

5.2 Overview of the Lock-free Operations

In this section we present a brief overview of the most critical vector's lock-free algorithms (see [5] for the full set of the nonblocking algorithms). To help tail operations update the size and the tail of the vector (in a linearizable manner), the design presented in [5] suggests the application of of a helper object, named "`Write Descriptor (WD)`" that announces a pending tail modifications and allows interrupting threads help the interrupted thread complete its operations. A pointer to the *WD* object is stored in the "`Descriptor`" together with the container's size and a reference counter required by the applied memory management scheme [5]. The approach requires that data types bigger than word size are indirectly stored through pointers and avoids storage relocation and its synchronization hazards by utilizing a two-level array. Whenever `push_back` exceeds the current capacity, a new memory block twice the size of the previous one is added. The remaining part of this section presents the pseudo-code of the tail operations (push_back and pop_back) and the random access operations (read and write at a given location within the vector's bounds). We use the symbols ^, &, and . to indicate pointer dereferencing, obtaining an object's address, and integrated pointer dereferencing and field access respectively.

Algorithm 15: push_back *vector, elem*

1 **repeat**
2 $desc_{current} \leftarrow vector.desc;$
3 $CompleteWrite(vector, desc_{current}.pending);$
4 **if** $vector.memory[bucket] == NULL$ **then**
5 $AllocBucket(vector, bucket);$
6 **end**
7 $writeop \leftarrow new\ WriteDesc(At(desc_{current}.size), elem, desc_{current}.size);$
8 $desc_{next} \leftarrow new\ Descriptor(desc_{current}.size + 1, writeop);$
9 **until** $CAS(\&vector.desc, desc_{current}, desc_{next})$;
10 $CompleteWrite(vector, desc_{next}.pending);$

Algorithm 16: Read *vector, i*

1 **return** $At(vector, i);$

Algorithm 17: Write *vector, i, elem*

1 $At(vector, i)^\wedge \leftarrow elem;$

Algorithm 18: pop_back *vector*

1 **repeat**
2 $desc_{current} \leftarrow vector.desc;$
3 $CompleteWrite(vector, desc_{current}.pending);$
4 $elem \leftarrow At(vector, desc_{current}.size - 1);$
5 $desc_{next} \leftarrow new\ Descriptor(desc_{current}.size - 1, NULL);$
6 **until** $CAS(\&vector.desc, desc_{current}, desc_{next})$;
7 **return** $elem;$

Algorithm 19: CompleteWrite *vector, writeop*

1 **if** $writeop.pending$ **then**
2 $CAS(At(vector, writeop.pos), writeop.value_{old}, writeop.value_{new});$
3 $writeop.pending \leftarrow false;$

Push_back (add one element to end) The first step is to complete a pending operation that the current descriptor might hold. In case that the storage capacity has reached its limit, new mem-

ory is allocated for the next memory bucket. Then, `push_back` defines a new "`Descriptor`" object and announces the current write operation. Finally, `push_back` uses CAS to swap the previous "`Descriptor`" object with the new one. Should CAS fail, the routine is re-executed. After succeeding, `push_back` finishes by writing the element.
Pop_back (remove one element from end) Unlike `push_back`, `pop_back` does not utilize a "`Write Descriptor`". It completes any pending operation of the current descriptor, reads the last element, defines a new descriptor, and attempts a CAS on the descriptor object.
Non-bound checking Read and Write at position i The random access `read` and `write` do not utilize the descriptor and their success is independent of the descriptor's value.

6 Framework Application for Accelerated Testing

The presented design and implementation of our parallel propagation technique enable the incorporation of the optimized propagation approach described by Lou [19] in an experimental framework for accelerated testing currently still under development at NASA. Accelerated testing platforms suggest a paradigm shift in the certification process employed by NASA from system testing with the actual flight hardware and software to accelerated cost-effective certification using hardware simulators and distributed software implementations. Such frameworks aim faster-than-real-time testing and analysis of the complex software interactions in JPL's autonomous flight systems. A number of these platforms require automated refactoring of previously sequential code into modular parallel implementations. Preliminary results reported in academic work [1] as well as experience reports from a number of commercial tools (such as Simics by Virtutech and ADvantage BEACON by Applied Dynamics International) suggest the possible speedup of the flight system testing by a significant factor. We have followed Rouquette's methodology [24] that suggests the application of formal modeling and validation techniques that provide certification evidence for a number of functional dependencies in order to compensate for the added hazards in establishing the fidelity of the simulators. Due to the incomplete status of the accelerated testing framework as well as the lack of the actual flight hardware, it is difficult to measure a priori the effect of our parallel propagation scheme in achieving acceleration (with respect to the execution on the actual flight hardware) in the process of flight software testing. To gain insight of the possible performance gains and the algorithm's behavior we ran performance tests on a conventional Intel IA-32 SMP machine with two 2.0GHz processor cores with 1GB shared memory and 4 MB L2 shared cache running the MAC OS 10.5.1 operating system. In our performance analysis we have measured the execution time in seconds of two versions of our parallel propagation algorithm (one applying mutually exclusive locks and the other relying on nonblocking synchronization) and the original sequential scheme presented by Lou [19]. In the experiments (Figure 3), we have generated a number of TCN graph topologies (each consisting of 4 to 8 Time Phases), in a manner similar to the pseudo-random graph generation methodology described in [8]. In the presented results on Figure 3 the $x-axis$ represents the average measured execution time (in seconds) of each propagation scheme and the $y-axis$ represents the number of time points in the exponentially increasing graph size (starting with a graph of 20000 TPs and reaching a TCN having 160000 TPs). In the experimental setup we observed that the parallel propagation algorithm offers effective execution and a considerable speedup in all scenarios on our dual-core platform. We measured performance acceleration reaching 28% in the case of the nonblocking implementation and 20% for our algorithm relying on mutually exclusive locks. Lock-free algorithms deliver significant speedup in applications utilizing shared data under high contention [5]. In a scenario like our parallel TCN propagation scheme with medium or low contention on the shared data, besides safety and prevention of priority inversion and deadlock, a lock-free implementation can guarantee better scalability. As our experimental results suggest, the gains from the lock-fee implementation gradually progress and we observe

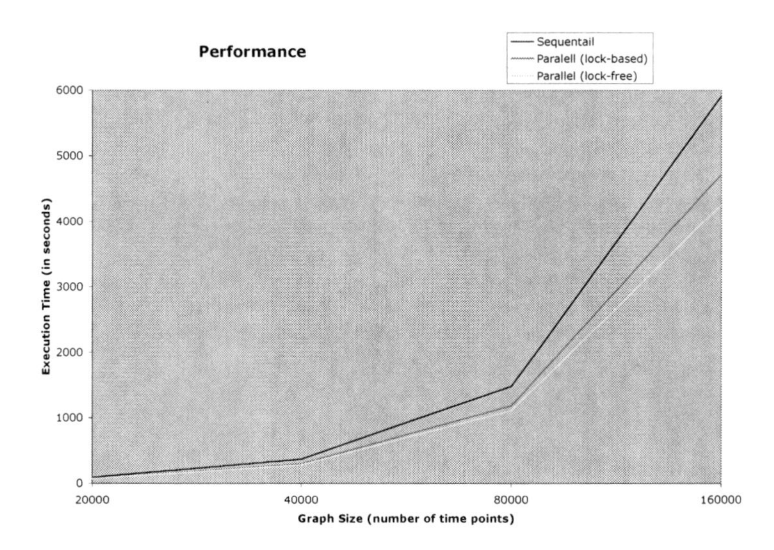

Fig. 3 Performance Analysis. x-axis represents the number of TPs in each experimental TCN topology, y-axis represents the execution time in seconds of each of the three propagation algorithms

better scalability with respect to the blocking propagation scheme. Based on the experimental results, we expect that the integration of our parallel propagation algorithm in the accelerated testing framework (consisting of several dozen processing units) will deliver significant benefits in reaching cost-effective and reliable flight software certification of control modules based on massive real-world goal networks.

7 Conclusion

The notions of time and concurrency are of critical importance for the design and development of autonomous space systems. The current certification methodologies do not reach the level of detail of providing guidelines for the development and validation of concurrent and real-time software. The increasing number of complex interactions and tight coupling of the future autonomous space systems pose significant challenges for their development and man-rated certification. A number of platforms for accelerated testing suggest a paradigm shift by applying a combination of modeling and verification methods, code generation tools, and software parallelization for establishing a cost-effective and reliable certification process. In the light of the challenges posed by the design and development of these highly experimental approaches, we presented in this work a first time- and concurrency-centered framework for validation and semantic parallelization of real-time C++ within JPL's MDS Framework. We demonstrated the application of our framework in the validation of the semantic invariants of the Temporal Constraint Network Library. Temporal constraint networks are at the core of the mission planning and control architecture of the Mission Data System framework. In addition, we presented an approach for automatic semantic parallelization of the propagation scheme establishing the consistency of the temporal constraints in a goal network. Our parallel propagation scheme is based on the identification of time phases within a goal network and is implemented through the application of model transformation and formal analysis techniques to the model specifications of the TCN semantics. We have relied on innovative lock-free synchronization techniques to achieve better performance and higher safety of our parallel implementation. Our preliminary tests indicate that our parallel propagation approach, upon integration in the accelerated testing framework, can support cost-effective and reliable flight software certification of control modules based on massive real-world goal networks. In our future work we plan to focus on developing a component for automatic derivation of the model specification directly from implementation source code. This can be accomplished by utilizing the high-level internal program representation and the analysis tools provided by The Pivot [28], a framework for static analysis and transformations in C++.

References

1. Boehm, B. and Bhuta, J. and Garlan, D. and Gradman, E. and Huang, L. and Lam, A. and Madachy, R. and Medvidovic, N. and Meyer, K. and Meyers, S. and Perez, G. and Reinholtz, K. and Roshandel, R. and Rouquette, N.: Using Empirical Testbeds to Accelerate Technology Maturity and Transition: The SCRover Experience. ISESE '04: Proceedings of the 2004 International Symposium on Empirical Software Engineering (2004)
2. Brat, G. and Drusinsky, D. and Giannakopoulou, D. and Goldberg, A. and Havelund, K. and Lowry, M. and Pasareanu, C. and Venet, A. and Washington, R. and Visser, W.: Experimental Evaluation of Verification and Validation Tools on Martian Rover Software. Formal Methods in Systems Design Journal, September (2005)
3. Columbia Accident Investigation Board: Columbia Accident Investigation Board Report Volume 1, http://caib.nasa.gov/
4. Cormen, T. and Leiserson, C. and Rivest, R. and Stein, C.: Introduction to algorithms. ISBN 0-262-03293-7, MIT Press (2001)
5. Dechev, D. and Pirkelbauer, P. and Stroustrup, B.: Lock-Free Dynamically Resizable Arrays. OPODIS 2006, Lecture Notes in Computer Science, Volume 4305 (2006)
6. Denney, E. and Fischer, B.: Software Certification and Software Certification Management Systems. SoftCement05: In Proceedings of the 2005 ASE Workshop on Software Certificate Management (2005)
7. Detlefs, D. and Flood, C. and Garthwaite, A. and Martin, P. and Shavit, N. and Steele, G.: Even Better DCAS-Based Concurrent Deques. International Symposium on Distributed Computing (2000)
8. Dick, R. and Rhodes, D. and Wolf, W.: TGFF: task graphs for free. CODES/CASHE '98: Proceedings of the 6th international workshop on Hardware/software codesign (1998)
9. Dvorak, D.: Challenging encapsulation in the design of high-risk control systems. Proceedings of the 17th ACM Conference on Object-Oriented Programming, Systems, Languages, and Applications OOPSLA (2002)
10. Dvorak, D. and Bollella, G. and Canham, T. and Carson, V. and Champlin, V. and Giovannoni, B. and Indictor, M. and Meyer, K. and Murray, A. and Reiinholtz, K.: Project Golden Gate: Towards Real-Time Java in Space Missions. In the Proceedings of the 7th IEEE International Symposium on Object-Oriented Real-Time Distributed Computing ISORC (2004)
11. Fraser, K.: Practical lock-freedom. Technical Report UCAM-CL-TR-579, University of Cambridge, Computer Laboratory (2004)

12. Gluck, R. and Holzmann, G.: Using SPIN Model Checker for Flight Software Verification. In Proceedings of the 2002 IEEE Aerospace Conference (2002)
13. Harris, T. and Fraser, K. and Pratt, I.: A practical multi-word compare-and-swap operation. Proceedings of the 16th International Symposium on Distributed Computing (2002)
14. Herlihy, M.: A methodology for implementing highly concurrent data structures. PPOPP '90: Proceedings of the second ACM SIGPLAN symposium on Principles & practice of parallel programming (1990)
15. Intel: Reference for Intel Threading Building Blocks, Version 1.0 (2006)
16. Jackson, D.: Software Abstractions: Logic, Language and Analysis. The MIT Press (2006)
17. Lamport, L.: How to make a multiprocessor computer that correctly executes programs. IEEE Trans. Comput. (1979)
18. Lee, E. and Neuendorffer, S.: Concurrent Models of computation for Embedded Software. IEEE Proceedings on Computers and Digital Techniques (2005)
19. Lou, J.: An Efficient Algorithm for Propagation of Temporal Constraint Networks. NASA Tech Brief Vol. 26 No. 4 from JPL New Technology Report NPO-21098 (2002)
20. Lowry, M.: Software Construction and Analysis Tools for Future Space Missions. TACAS 2002: Lecture Notes in Computer Science, Volume 2280 (2002)
21. Perrow, C.: Normal Accidents. Princeton University Press (1999)
22. Rasmussen, R. and Ingham, M. and Dvorak, D.: Achieving Control and Interoperability Through Unified Model-Based Engineering and Software Engineering. AIAA Infotech at Aerospace Conference (2005)
23. Dos Reis, G. and Stroustrup, B.: Specifying C++ Concepts. ISO WG21 N1886 (2005)
24. Rouquette, N.: Analyzing and verifying UML models with OCL and Alloy. EclipseCon (2008)
25. RTCA: Software Considerations in Airborne Systems and Equipment Certification DO-178B (1992)
26. Schumann, J. and Visser, W.: Autonomy Software: V&V Challenges and Characteristics. In Proceedings of the 2006 IEEE Aerospace Conference (2006)
27. Stroustrup, B.: The C++ Programming Language. Addison-Wesley Longman Publishing (2000)
28. Stroustrup, B. and Dos Reis, G.: Supporting SELL for High-Performance Computing. In Proceedings of the International Workshop on Languages and Compilers for Parallel Computing LCPC (2005).
29. Volpe, R. and Nesnas, I. and Estlin, T. and Mutz, D. and Petras, R. and Das, H.: The CLARATy Architecture for Robotic Autonomy. IEEE Aerospace Conference (2001)

Part III
Applications to Ad-Hoc (Sensor) Networks and Pervasive Systems

Autonomic Communication in Pervasive Multimodal Multimedia Computing System

Manolo Dulva Hina, Chakib Tadj, Amar Ramdane-Cherif, Nicole Lévy

Abstract Autonomic communication in a computing system analyzes the individual system element as it is affected by and affects other elements. In the human-machine communication aspect, the autonomic system performs its services autonomously, adjusting the behavior of its services to suit what the user might request implicitly or explicitly. The pervasive multimodal multimedia computing system aims at realizing anytime, anywhere computing. Its autonomic communication includes the protocols that selects, on behalf of the user, the modalities and media devices that are appropriate to the given interaction context. The modalities are the modes of interaction (i.e. for data input and output) between the user and the computer while the media devices are the physical devices that are used to support the chosen modalities. The interaction context itself is the combined user, environment, and system contexts. In this paper, we present the autonomic communication protocols involved in the detection of interaction context and the multimodal computing system's corresponding adaptation. The heart of this paradigm's design is the machine learning's knowledge acquisition and the use of the layered virtual machine for definition and detection of interaction context.

Manolo Dulva Hina
Université du Québec, École de technologie supérieure, Montréal, Québec, Canada &
PRISM Laboratory, Université de Versailles-Saint-Quentin-en-Yvelines, Versailles, France
e-mail: `manolo-dulva.hina.1@ens.etsmtl.ca`

Chakib Tadj
Université du Québec, École de technologie supérieure, Montréal, Québec, Canada
e-mail: `ctadj@ele.etsmtl.ca`

Amar Ramdane-Cherif
PRISM & LISV Laboratories, Université de Versailles-St-Quentin-en-Yvelines, Versailles, France
e-mail: `rca@prism.uvsq.fr`

Nicole Lévy
PRISM Laboratory, Université de Versailles-Saint-Quentin-en-Yvelines, Versailles, France
e-mail: `nicole.levy@prism.uvsq.fr`

A.V. Vasilakos et al. (eds.), *Autonomic Communication*, DOI: 10.1007/978-0-387-09753-4_10,

1 Introduction

In autonomic communication, a system element learns, in every moment of its existence, about other elements and the world where it belongs through sensing and perception. In the human-machine aspect of autonomic communication, the system performs services autonomously. At the same time, it adjusts the behaviour of its services based on its learned perception of what the user might request, either implicitly or explicitly. We apply this principle in a specific application domain - the pervasive *multimodal multimedia* (MM) computing system. In such a system, various forms of modality for data input and output exist. Also, various media devices may be selected to support these modalities. *Multimodality* is possible and can be implemented if the mechanism for *data input* and *output* exists. Multimodality is essential because it provides for increased usability and accessibility to users, including those with handicaps. With multimodality, the strength or weakness of every media device is decided based on its suitability to a given context[1]. For example, to a user in a moving car, an electronic pen and speech are more appropriate input media than that of keyboard or a mouse. Multimodality can be further enhanced if more media devices (other than the traditional mouse-keyboard-screen combination) and their supporting software are made available. Offering basic services using multimodality (e.g. a multimodal banking services) is not only socially wise but also contributes to the creation of a more humane, inclusive society because the handicapped are given participation in using the new technology.
Slowly, *pervasive computing*, also known as *ubiquitous computing* [1] [2] [3], which advocates anytime, anywhere computing is no longer a luxury but is becoming a way of life. For instance, healthcare [4] is adopting it. Soon, our personal and computing information would "follow" us and become accessible wherever and whenever we want them. This promotes increased productivity as we can continue working on an interrupted task as we please. This has been made possible because the infrastructures for wired, wireless and mobile computing and communications [5] [6] [7] do already exist.
Multimodality also involves the fusion of two distinct data or modalities. For instance, the fusion of two or more temporal data, such as data from a mouse and speech as in simultaneously clicking the mouse and uttering "Put that there" [8] [9], is full of promise, further advancing multimodality. The fusion process, however, is still static - that is, the media and modality in consideration are pre-defined rather than dynamically selected. Also, the fusion process is not adaptive to the changes occurring in the environment (e.g. as in environment becomes noisy); in this case, over time, the effectiveness of a modality (e.g. vocal input) in the fusion process becomes unreliable. In general, it is unwise to predefine a chosen modality. A modality - whatever it may be - should be chosen only based on its suitability to a given context.
Context changes over time. Hence, context should not be viewed as fixed nor should it be pre-defined. Instead, it should be defined dynamically based on the needs and requirement of a system. Our approach, hence, is to define context by considering one context parameter at a time; such parameter may be added, deleted or modified as needed. This leads us to an *incremental context* where context becomes an attribute that suits the needs and requirements of a system. Context parameters may or may not be based on sensors data. For sensor-based context, we propose the adoption of *virtual machine* (VM). In this approach, the real-time interpretation of a context parameter is based on sampled data from sensor(s). The design of our layered VM for incremental user context is robust that it can be adopted by almost any system that takes in context that is based on sensors.
Machine learning (ML) [10] involves the acquisition of knowledge through training or past experiences; this knowledge, when adopted, is ought to improve the system's performance. ML is the heart of this work. Our system's ML component is given:

[1] Here, the term *context* signifies a generic meaning. Later, context will evolve to become *interaction context*. Unless explicitly specified, context and interaction context may be used interchangeably.

1. Functions that (i) define the relationship between context and multimodality, and (ii) define the relationships between modality and media group, and between media group and media devices,
2. Rules and algorithms that (i) determines the media device(s) that replace the faulty one(s), and (ii) re-adapts its *knowledge database* (KD) when a new media device is introduced into the system. The acquired knowledge is then used to optimize configurations and for the system to exhibit fault-tolerance characteristics,
3. Case-based reasoning and supervised learning to find the appropriate solution/adaptation to a new situation, in consultation with the system's stored knowledge.

The rest of this paper is structured as follows: Section 2 surveys related works and highlights the novelty of this work. Section 3 essays on the technical challenges and our approach to address them. Section 4 is all about context - its definition, representation, storage and dissemination. This is essential since adaptation to context is an important aspect of an autonomic and pervasive system . Section 5 is about modalities, media devices and their context suitability. Section 6 is about our system's knowledge acquisition and the use of such knowledge to adapt to a given *interaction context* (IC). The paper is concluded in Section 7.

2 Related Works

Recent research works on multimodality include the application on interface for wireless user interface [11], the static user interface [12], text-to-speech synthesis [13], and a ubiquitous system for visually-challenged user [14]. Related research works that empasize on multimodality data fusion are the combined speech and pen inputs [15], the combined speech and gestures inputs [12] and the combined speech and lips movements [16]. These are a few proofs that multimodality is possible, doable, and feasible. When compared with them, our work, however, is one step further: it provides the infrastructure in which those above-mentioned works can be invoked in, anytime and anywhere.
Context is a vital consideration in determining what modalities are appropriate for the user. Indeed, *"context is the key"* [17]. The evolution of context definitions, including Rey's definition for context-aware computing in [18] [19] and that of contextor [20], is described in Section 4. The federation of context-aware perceptions [17], and context-awareness in wearable computing [18] are some context-aware systems. Our contribution to the domain, however, is we take user's context and relate it to multimodality. While contextor is an interactive context-sensitive system, it does not, however, provide the mechanism to realize an ever-changing context. Our layered VM approach is more adaptable to an ever-changing environment. It has been proven that a layered VM/object-oriented approach and design is an effective paradigm, as in Hughes Aircraft Company [21].
The user profile constitutes an integral part of user's context. Sample works on user profile analysis include [22] and [23]. Our work, however, differs because we consider user handicap as part of a user's profile. This allows our work to cover a much wider spectrum of users. Finally, our objective is to assemble all these beneficial concepts to form a package for ubiquitous computing consumption. In Project Aura [24], the Prism model shows a user's moving aura (profile and task). In comparison, ours include not only the user's ubiquitous profile and task but also an acquired ML knowledge that goes with a mobile user. Such knowledge is used in the detection of changes in IC and resources, and the system's adaptation to these changes by selecting the appropriate modalities and media devices.
This work is intended to contribute to designing paradigms that explores the challenges in technologies that realize that vision wherein devices and applications seamlessly interconnect, intelligently cooperate and autonomously manage themselves, also known as *autonomic communication*.

3 Contribution and Novel Approaches

Our vision is to enhance the use of multimodality through an infrastructure that realizes pervasive MM computing - intelligent, fault-tolerant, rich in media devices, and adaptive to a given context and acting on behalf of the user. To realize this, a system solution must address the key requirements given below.

Requirement 1: *Determine the suitability of various modalities to a given context.* First, it is necessary to classify modalities, and afterwards determine what types of modalities will allow a user to input data and receive output based on a given instance of IC. So, what are the types of modality? What is the relationship between modalities and context?

Proposed Solution: *Modality can be classified into two groups: the input modality and the output modality.* Within the input modality, there exists the visual input, the vocal input, and the manual input. Similarly, within the output modality, the options are visual output, vocal output and manual output. There must be a suitable input modality and also a suitable output modality for multimodality to be implemented. Given a specific IC, a modality has some degree of suitability to it. Such suitability is not just binary (that is, very suitable or not suitable at all) but also includes something in between - medium and low suitabilities. Indeed, our approach takes in a wider spectrum of modalities' suitability. Numerical value for suitability are assigned as follows: High suitability = 100%, Inappropriate = 0. Medium and low suitabilities should have value in between this range. To relate modality to the overall IC, then each type of modality would get a suitability score for each individual context parameter. The final suitability to the overall IC is the normalized product of suitability scores of the individual IC parameters.

Requirement 2: *Provide a relationship between modality and the media devices that are invoked to implement modality. Given that various media devices do exist, then provide a classification of media where all devices could fit.* What should be a generic media classification so that all media devices - those that we know at present and also those that might come in the future - would fit in? What would be the basis of such classification? In which category should, for example, a mouse belongs? What about the eye gaze, etc.?

Proposed Solution: *Media devices may be grouped in the same way as modalities.* Our approach on media's classification is based on man's *natural language processing*; man transmits and receives information through his five senses (e.g. hearing, tasting, etc.) and voice. Therefore, the categorization of media needs to be based on the body part that uses the media device to generate data input, as well as the body part that uses the data generated by the media device. For example, a mouse is a manual input device, and so is the keyboard. A Braille terminal, for the visually-impaired user, is an example of a touch output device. An eye gaze is a visual input device.

Requirement 3: *Determine the parameters that would constitute a context. Also, given that context changes over time then provide a mechanism that allows user to modify (add, change, delete) parameters on the fly.* A mobile user who changes environment over time does not have a fixed context; hence defining a fixed set of parameter that forms the context is incorrect. How do we declare the parameters of a context? Also, if modification of parameters is necessary, what mechanism should be used to effect such modification without producing a ripple effect into the other components of the system?

Proposed Solution: *An IC is the combined user, environment and system contexts, each of which is composed of one or more parameters. Our layered VM for incremental IC is a robust "machine" that can be adapted to suit application domain and in which parameter modification can be done on the fly with minimum system ripple effect.* Also, the context parameter consideration in our layered VM is gradual or incremental. In effect, IC is defined based on the needs of the user.

Requirement 4: *Provide a self-healing mechanism that provides replacement to a faulty media device, and an orderly re-organization if a new device is introduced into the system for the first time.* If two or more media devices are classified as members of the same media group, which one would be given priority in a specific context? What are the guidelines for such priority ranking? If the chosen media device is faulty (missing or defective), how do we determine its replacement? If a new media device is introduced for the first time, how would it affect the priority ranking of other media devices in the same group?

Proposed Solution: *Through training, our ML system acquires knowledge for context detection, determining the suitable modality, determining the appropriate media group and devices. The same system includes knowledge on which devices could replace the defective ones.* The policy of replacement is based on the media devices availability and priority rankings. For example, the devices that are used in usual configuration are given higher priority than those that are not regularly used. The ML training includes user participation that guides the system to recognize positive examples which form system knowledge.

4 The Interaction Context

This section discusses context - the evolution of its definition, its representation, capture, storage and dissemination.

4.1 Context Definition and Representation

In chronological order, some early definitions of context include that of Schilit's [25] in which context is referred to the answer to the questions *"Where are you?"*, *"With whom are you?"*, and *"Which resources are in proximity with you?"*. Schilit defined context as the changes in the physical, user and computational environments. This idea is taken by Pascoe [26] and later on by Dey [27]. Brown considered context as *"the user's location, the identity of the people surrounding the user, as well as the time, the season, the temperature, etc."* [28]. Ryan defined context as the environment, the identity and location of the user as well as the time [29]. Ward viewed context as the possible environment states of an application [30]. In Pascoe's definition, he added the pertinence to the notion of state, stating: *"Context is a subset of physical and conceptual states having an interest to a particular entity"*. Dey specified the notion of an entity: *"Context is any information that can be used to characterize the situation of an entity in which an entity is a person, a place or an object that is considered relevant to the interaction between a user and an application, including the user and application themselves"* [19]. This definition became the basis for Rey and Coutaz to coin the term interaction context: *"Interaction context is a combination of situations. Given a user U engaged in an activity A, then the interaction context at time t is the composition of situations between time t_0 and t in the conduct of A by U"* [31].
We adopted the notion of *"interaction context"*, but define it in the following manner: An *interaction context*, $IC = \{IC_1, IC_2, \ldots, IC_m\}$, is a set of all possible interaction contexts. At any given time, a user has a specific interaction context *i* denoted as IC_i, where $1 \leq i \leq max$, that is composed of variables that are present during the conduct of the user's activity. Each variable is a function of the application domain which, in this work, is multimodality. Formally, an *IC* is a tuple composed of a specific *user context* (UC), *environment context* (EC) and *system*

context (SC). An instance of IC is given as:

$$IC_i = UC_k \bigotimes EC_l \bigotimes SC_m \tag{1}$$

where $1 \le k \le max_k$, $1 \le l \le max_l$, and $1 \le m \le max_m$, and that max_k, max_l and max_m = maximum number of possible user contexts, environment contexts and system contexts, respectively. The Cartesian product (symbol: $\bigotimes$) denotes that IC yields a specific combination of UC, EC and SC at any given time.
The user context UC is composed of application domain-related parameters describing the state of the user during his activity. A specific user context k is given by:

$$UC_k = \bigotimes_{x=1}^{max_k} ICParam_{kx} \tag{2}$$

where $ICParam_{kx}$ = parameter of UC_k, k = the number of UC parameters. Similarly, any environment context EC_l and system context SC_m are specified as follows:

$$EC_l = \bigotimes_{y=1}^{max_l} ICParam_{ly} \tag{3}$$

$$SC_m = \bigotimes_{z=1}^{max_m} ICParam_{mz} \tag{4}$$

4.2 The Virtual Machine and the Incremental Interaction Context

As stated, an instance of IC is composed of specific instances of UC, EC, and SC, which themselves are composed of parameters. These parameters are introduced to the system, one at a time. In our work, a virtual machine is designed to add, modify or delete one context parameter, making the IC parameters reflective of the system's dynamic needs.
A *virtual machine* (VM) is software that creates a *virtualized* environment on computer platform so that the end user can operate the software. *Virtualization* is the process of presenting a group or subset of computing resources so that they can be accessed collectively in a more beneficial manner than their original configuration. In effect, a VM is an *abstract computer*; it accepts input, has algorithms and steps to solve the problem related to the input, and yields an output. The steps taken by the VM are its *"instructions set"* which is a collection of functions that the machine is capable of undertaking. A *layered VM* is a group of VM's in which the interaction is only between layers that are adjacent to one another. The layering is a design choice whose purpose is to limit any error from propagating to other components of the system other than the adjacent layer(s). During program/design modification, a layer's effect is limited only within itself and the adjacent layers. Generally, in layered structure, the *top layer* refers to the interface that interacts with end users while the *bottom layer* interacts with the hardware. Hence, *Layer 0* is a layer that represent the collection of sensors (or machines or gadgets) that generate some raw data representing the value needed by the topmost VM layer. Fig. 1 shows the functionality of such "machine". As shown, the transfer of instruction command is top-down (steps 1 to 4). At Layer 0, the raw data corresponding to IC are collected for sampling purposes. The sampled data are then collated and interpreted, and the interpretation is forwarded to different layers bottom-up (steps 5 to 8).

The VM Layer 4 acts as the human-machine interface; its *"instruction set"* are the four functions found in Layer 3 - the *"Add Parameter"*, *"Modify Parameter"*, and *"Delete Parameter"* are basic commands that manipulate the sensor-based context parameters while *"Determine Context"* yields the values of currently-defined parameters. VM Layer 2 is a *"library of functions"* that collectively supports Layer 3 instructions while Layer 1 is another *"library of functions"* that acts as a link between Layer 2 and Layer 0.

4.2.1 Adding a Context Parameter

Consider using the VM to add a specimen context parameter: the *"Noise Level"*. See Fig. 2. Upon invoking the VM user interface (i.e. Layer 4), the user chooses the *"Add Parameter"* menu. A window opens up, transferring the execution control to Layer 3. To realize adding a new context parameter, at least four functions must exist, namely: (i) *getting context type of the parameter*, (ii) *getting parameter name*, (iii) *getting number of parameter units*, and (iv) *getting number of parameter values and conventions*.

As shown, through Layer 3, the user inputs *"Noise Level"* as parameter name, itself an EC parameter, *"1"* as parameter unit, and *"3"* as parameter values and conventions. When done, two new windows open up, one window at a time, that brings up the functionalities of Layer 2. For each parameter's unit, the VM receives inputs for the unit name and the sensor (or hardware) that supplies its raw data. As shown, the unit of noise is specified as *"decibel"* and the *BAPPU noise measuring device* [32] as the sensor supplying the data. When done, another Layer 2 window opens up for data entry of *"Parameter values and conventions"*. In the diagram, the user specifies the value (range of decibels) that is equivalent to *"Quiet"*,

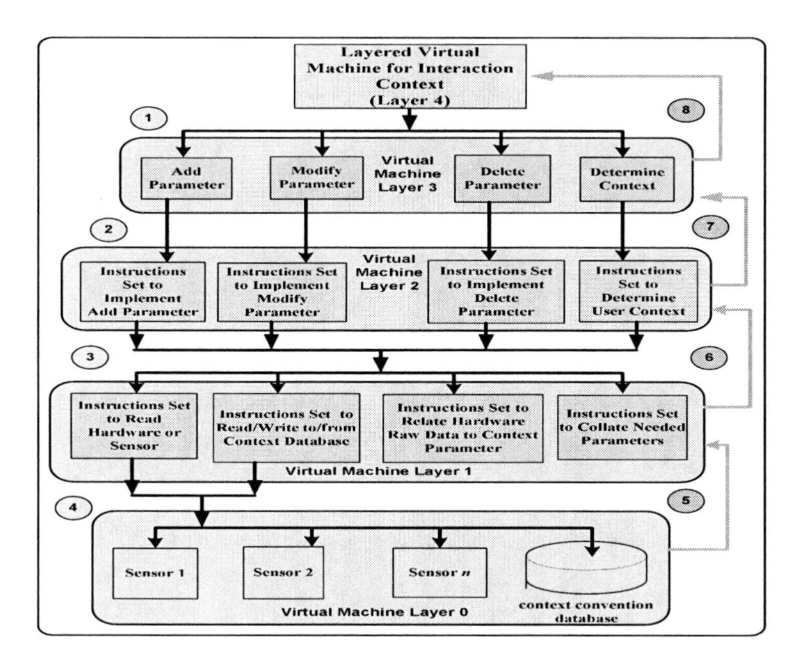

Fig. 1 The design of a layered virtual machine for incremental user context

"Moderate" and *"Noisy"* conventions. When done, a window for Layer 1 opens up to save the newly-added parameter information. This function interacts directly with the hardware (i.e. the context convention database).

4.2.2 Modifying and Deleting a Context Parameter

The VM layers interaction in *"Modify Parameter"* is almost identical to that of *"Delete Parameter"* function. The only thing extra is one that allows the user to select the context parameter that should be modified. Other than that, everything else is the same. The processes involved in *"Delete Parameter"* menu are shown in Fig. 3.

Upon menu selection, the execution control goes to Layer 3, demanding the user to specify the parameter for deletion (i.e. *"Noise level"* is chosen for deletion). When confirmed, the information on the parameter for deletion is extracted and read from database (transfer of control from Layer 2 to Layer 1 to Layer 0). When the information has been obtained, the control goes back to Layer 2 where the information is presented and a re-confirmation of its deletion is required. When parameter deletion is done, the control goes back to Layer 3 which presents the updated list of context parameters. A click to the *"OK"* button transfers the control back to Layer 4.

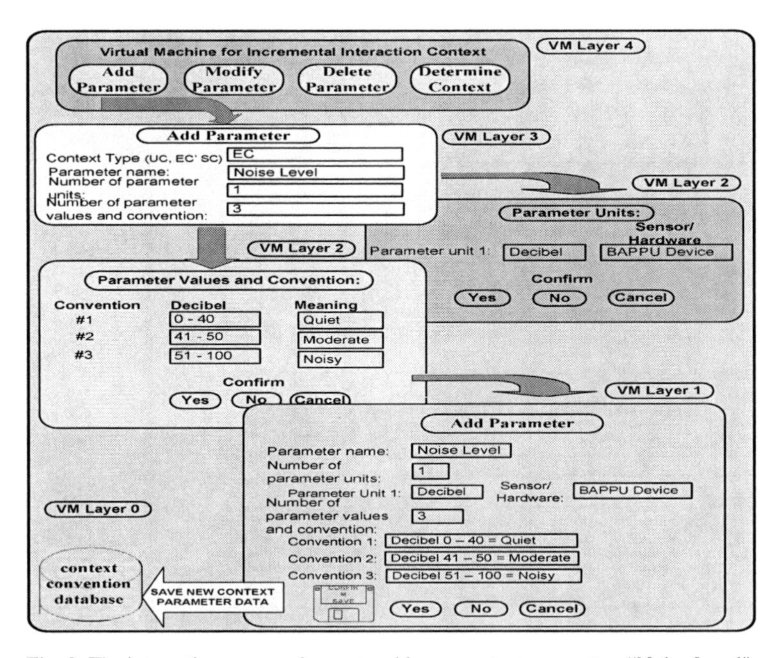

Fig. 2 The interactions among layers to add new context parameter: "Noise Level"

4.2.3 Capturing the User's Context

The interactions of VM layers to *"Determine Context"* are shown in Fig. 4. This is simulated using specimen context parameters, namely (1) *the user location*, (2) *the safety level*, and (3) *the workplace's brightness*.

When the user opts for this menu, the VM execution control goes to Layer 3. The function *"Get User Context"* creates threads equal to the number of parameters currently under consideration. This process produces the thread *"Get Parameter 1"*, assigned to detect user location, the thread *"Get Parameter 2"* assigned to get the user's safety level, and the thread *"Get Parameter 3"* for the user's workplace's brightness (i.e. light intensity). The concepts involved are identical for each thread. Consider the case of *"User Location"*. The thread passes control to Layer 1 wherein the function takes sample data from a sensor (i.e. global positioning system (GPS) [33]) attached to the user computer's USB port. In the VM design, user can specify the number of raw data that need to be sampled and in what frequency (*n samples per m unit of time*). These samples are then collated, normalized and interpreted.
For example, a *specimen GPS data* of 5 samples, taken 1 sample per minute, is shown in Fig. 5. The data are then normalized (averaged). Hence, as shown in the diagram, the user's computer is located at $14°11'$ latitude and $-120°57'$ longitude. Then, this value is interpreted using the convention values for user location parameter. Tables 1, 2 and 3 show the format of the convention values of the specimen parameters. (Recall that the convention value of a parameter is created during the *"Add Parameter"* process.) Using Table 1, the interpretation identifies if the user (who uses the computer equipped with a GPS) is at home, at work or on the go.

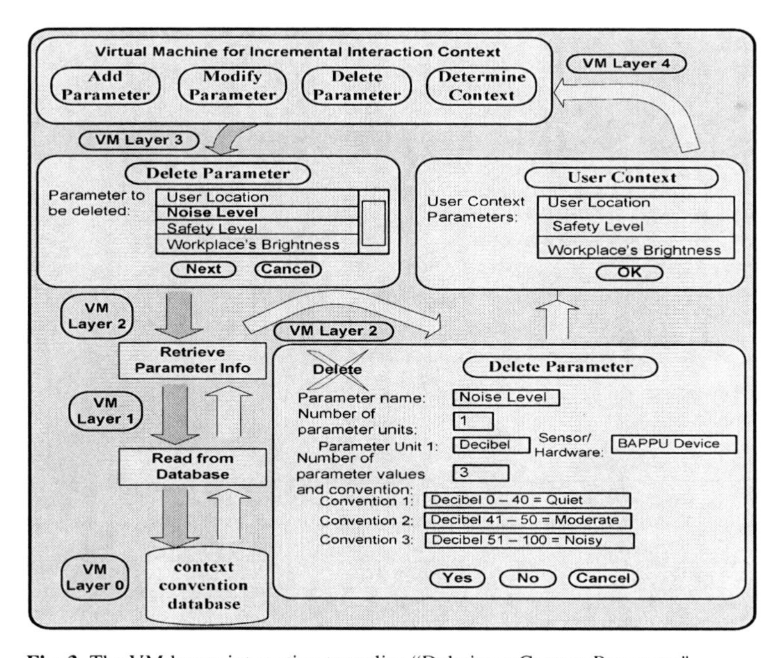

Fig. 3 The VM layers interaction to realize "Deleting a Context Parameter"

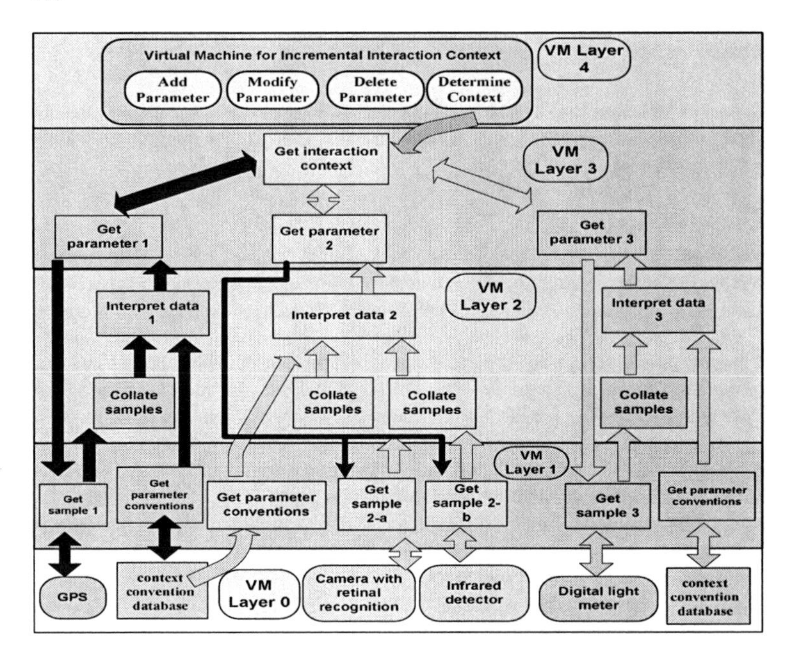

Fig. 4 The interactions among VM layers in detecting the current interaction context

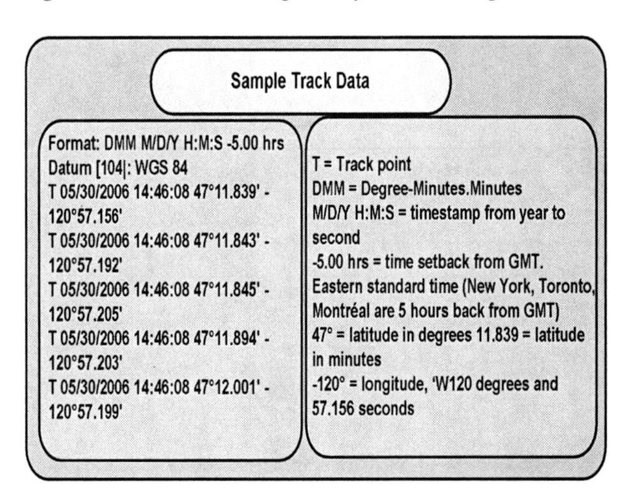

Fig. 5 Sample GPS data gathered from Garmin GPSIII+

Table 1 Convention format for user location

Convention	Longitude	Latitude	Meaning
1	$< Longitude_1 >$	$< Latitude_1 >$	At home
2	$< Longitude_2 >$	$< Latitude_2 >$	At work
3	$!< Longitude_1 >$ $\&\& \, !< Longitude_2 >$	$!< Latitude_1 >$ $\&\& \, !< Latitude_2 >$	On the go

Table 2 Convention format for safety level in user's workplace

Convention	Person in User's Workplace	Other People in User's Workplace	Meaning
2	User	Image	Sensitive
1	User	No Image	Safe
3	Empty	Image	Sensitive
1	Empty	No Image	Safe
3	Other	Image	Risky
3	Other	No Image	Risky

Table 3 Convention format for light intensity in user's workplace

Convention	Foot-Candle	Meaning
1	$< Value - Range_1 >$	Bright
2	$< Value - Range_2 >$	Moderate
3	$< Value - Range_3 >$	Dark

Specimen parameter 2 (the workplace's safety level) is a function of (i) the identity of the person sitting in front of the computer, and (ii) the presence of other people in the user's workplace. For this context parameter, a camera with retinal recognition [34] may be used to identify the person sitting in the user's seat. The identification process would yield three values: (1) *User* - if the legitimate user is detected, (2) *Other* - if another person is detected, and (3) *Empty* - if no one is detected. Also, an infrared detector [35] may be used to identify the presence of other person in front or in either side of the user. The identification process would yield two values: (1) *Image* - if at least one person is detected, and (2) *No Image* - if nobody is detected. (Note that the image and pattern recognition is not the subject of this work; hence, the detection process is not elucidated.). The VM takes $n = 5$ samples, normalizes them and compares the result against the convention values in Table 2. The interpretation yields a result indicating if user's workplace is *safe*, *sensitive* or *risky*. This specimen parameter is useful for people working on sensitive data (e.g. bank manager) but can be irritating to a person working with teammates (e.g. students working on a project). Hence, this specimen parameter can be added or deleted on the user's discretion.

The third specimen parameter in Table 3 (i.e. workplace's brightness) detects the workplace's light intensity. Here, we can assume that a sensor measuring the light's intensity [36] is attached to the computer's USB port. Its measurement unit, the *foot-candle*, is the number of "lumens" falling on a square foot of an inch; lumen is a unit of light used to rate the output of a bulb. For example, we may assume the following conventions in a user's workplace: (i) *0 - 9 foot candles = dark*, (ii) *10 - 20 foot-candles = moderate*, and (iii) *21 - 100 foot-candles = bright*. The processes involved in sampling, collating and interpreting sensor data for parameter 3 is identical with the other 2 parameters just cited earlier. Indeed, given the specimen parameters, when *"determine context"* is performed, the resulting output yield a result that indicates (1) *if the user is at home, at work or on the go*, (2) *if user's workplace is safe, sensitive or risky*, and (3) *if the workplace's light intensity is bright, moderate or dark*.

4.3 Context Storage and Dissemination

In general, if a system must obtain an accurate representation of the user's interaction context, then the system must be introduced to the most number of possible context parameters. As a context parameter is added to the system, the VM's context convention database forms a tree-like IC structure, as shown in generic format in Fig. 6. As shown, every new IC parameter is first classified as if it is a UC or EC or SC parameter and then is accordingly appended as a branch of UC or EC or SC. Then, the conventions of the new parameter are identified.

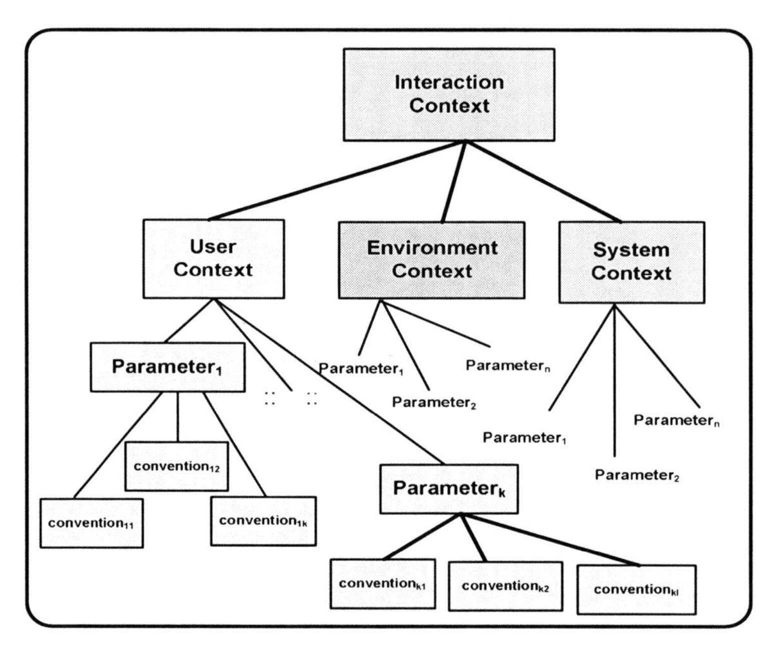

Fig. 6 The structure of stored IC parameters

For the IC information to be propagated in a pervasive system, the data representation used is XML Schema which is based on XML [37]. Fig. 7(Left) illustrates the general XML format of a context parameter (i.e. name, units, source of raw data, and conventions) and Fig. 7(Right) shows the various snapshots of windows involved in adding a parameter in the VM as implemented using Java programming language [38].

Context modeling as well as those of media devices and available services are necessary to inter-relate the context, the services and the context-aware devices. This, however, is a completely broad subject by itself and is not discussed here. [39] is a paper that is related to this area.

5 Modalities, Media Devices and Context Suitability

Here, we formulate the relationships between IC and modalities and between modalities and media group. This includes determining the suitability of a modality to a given IC.

5.1 Classification of Modalities

Here, *modality* refers to the logical interaction structure (i.e. the mode for data input and output between a user and computer). Using natural language processing as basis, we classify modalities into 6 different groups: (1) $VisualInput(VI_{in})$, (2) $VocalInput(VO_{in})$, (3) $ManualInput(M_{in})$, (4) $VisualOutput(VI_{out})$, (5) $VocalOutput(VO_{out})$, and (6) $ManualOutput(M_{out})$. To realize multi-modality, there should be at least one modality for data input and at least one modality for data output, as denoted by the following relationship:

$$Modality = (VI_{in} \vee VO_{in} \vee M_{in}) \wedge (VI_{out} \vee VO_{out} \vee M_{out}) \tag{5}$$

5.2 Classification of Media Devices

In this work, *media* are physical devices that are used to implement a modality. Regardless of size, shape, colour and other attributes, all media - past, present or future - can be classified

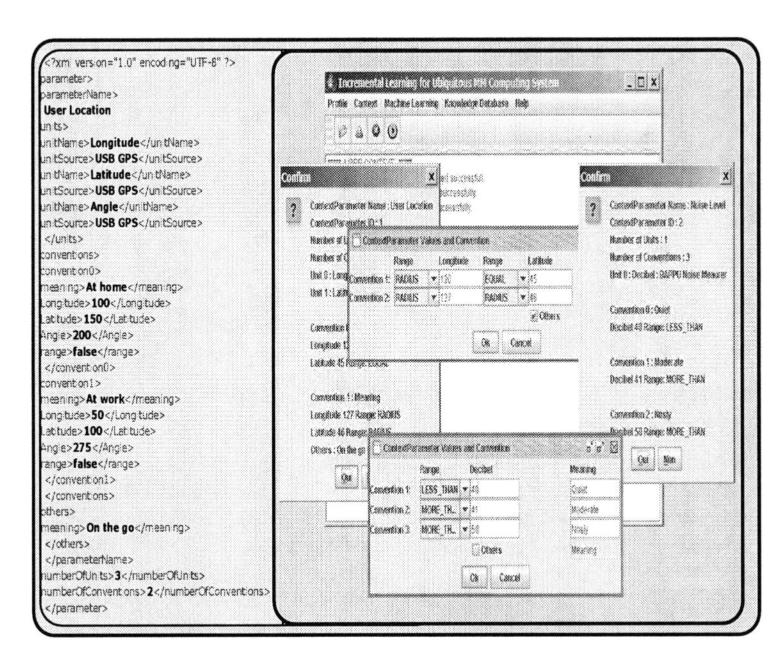

Fig. 7 (Left) Sample context parameter in XML, (Right) snapshots of windows in "Add Parameter" menu

based on the human body part that uses the device to generate data input and the body part that uses the device to consume the output data. Hence, the classifications are as follows:

1. *Visual Input Media (VIM)* - these devices take in inputs from user gaze and sight,
2. *Visual Output Media (VOM)* - these devices generate outputs that are meant to be read,
3. *Oral Input Media (OIM)* - devices that use user's voice to generate input data,
4. *Hearing Output Media (HOM)* - devices that generate outputs that are meant to be heard or listened to,
5. *Touch Input Media (TIM)* - these devices generate inputs through user's touch,
6. *Manual Input Media (MIM)* - these devices generate inputs using hand strokes, and
7. *Touch Output Media (TIM)* - the user touches these devices to receive data output.

5.3 Relationship between Modalities and Media Devices

When a modality is found suitable to a given IC, then the media that support such modality are chosen. Let there be a function g_1 that maps a modality to a media group, given by $g_1 : Modality \rightarrow MediaGroup$. This relationship is shown in Fig. 8. Also, oftentimes, there are many available devices that belong to the same media group. If such is the case then instead of activating them all which is tantamount to redundancy in functionality, devices activation is determined through their priority rankings. To support this scheme, let there be a function g_2 that maps a media group to a media device and its priority rank, and is denoted $g_2 : MediaGroup \rightarrow (MediaDevice, Priority)$. Given below are sample elements that belong to these functions:

- $g_1 = \{(VI_{in}, VIM), (VI_{out}, VOM), (VO_{in}, OIM), (VO_{out}, HOM), (M_{in}, TIM), (M_{in}, MIM), (M_{out}, TOM)\}$
- $g_2 = \{(VIM,$ (eye gaze, 1)), $(VOM,$ (screen, 1)), $(VOM,$ (printer, 1)), $(OIM,$ (speech recognition,1)), $(OIM,$ (microphone, 1)), $(HOM,$(speech synthesis, 1)), $(HOM,$ (speaker, 2)), (HOM, (headphone, 1)), etc.$\}$.

It should be noted, however, that although *media* technically refers to a hardware element, we opted to include a few software elements without which VO_{in} and VO_{out} modalities could not possibly be implemented. These are the speech recognition software and speech synthesis software.

5.4 Measuring the Context Suitability of a Modality

A modality's suitability to an IC is equal to its collective suitability to the IC's individual parameters. Instead of binary suitability (i.e. suitable or not suitable), our measure of suitability is that of *high*, *medium*, *low* or *inappropriate*. *High suitability* means that the modality being considered is the preferred mode for computing; *medium suitability* means the modality is simply an alternative mode, hence, its absence is not considered as an error but its presence means added flexibility on the part of system adaptation to context and events. *Low suitability* means the modality's effectiveness is negligible and is the last recourse when everything else fails. *Inappropriateness* recommends that the modality should not be used at all. If the collective IC is composed of n parameters, then a modality in consideration has n suitability scores, one for each parameter. In this work, the following conventions are adopted:

1. A modality's level of suitability to any context parameter is one of the following: H (high), M (medium), L (low), and I (inappropriate).
2. Mathematically, $H = 100\%$, $M = 75\%$, $L = 50\%$, and $I = 0\%$,
3. Given context parameter $i \in$ interaction context, then a modality's suitability score to the overall context, and its final suitability score are given by:

$$SuitabilityScore_{modality} = \sqrt[n]{\prod_{i=1}^{n}(context\, parameter_i)} \tag{6}$$

$$FinalSuitability_{modality} = \begin{cases} H\ if\ SuitabilityScore_{modality} = 1.00 \\ M\ if\ 0.75 \leq SuitabilityScore_{modality} < 1.00 \\ L\ if\ 0.50 \leq SuitabilityScore_{modality} < 0.75 \\ I\ if\ SuitabilityScore_{modality} = 0.50 \end{cases} \tag{7}$$

5.5 Optimal Modalities and Media Devices' Priority Rankings

Figure 9 shows the algorithm for determining the suitability of modalities to a given IC. Also, in Fig. 12, Algorithm 4 checks if multimodality is possible by checking that not all of input modalities are scored "inappropriate". The same principle is applied for checking for output modalities. The *optimal input modality* is chosen from a group of input modalities, and is one with the highest IC suitability score. The same principle applies to the selection of *optimal output modality*. Subject to the availability of media devices, an optimal modality is ought to be implemented; all other modalities are considered optional. In the absence of supporting me-

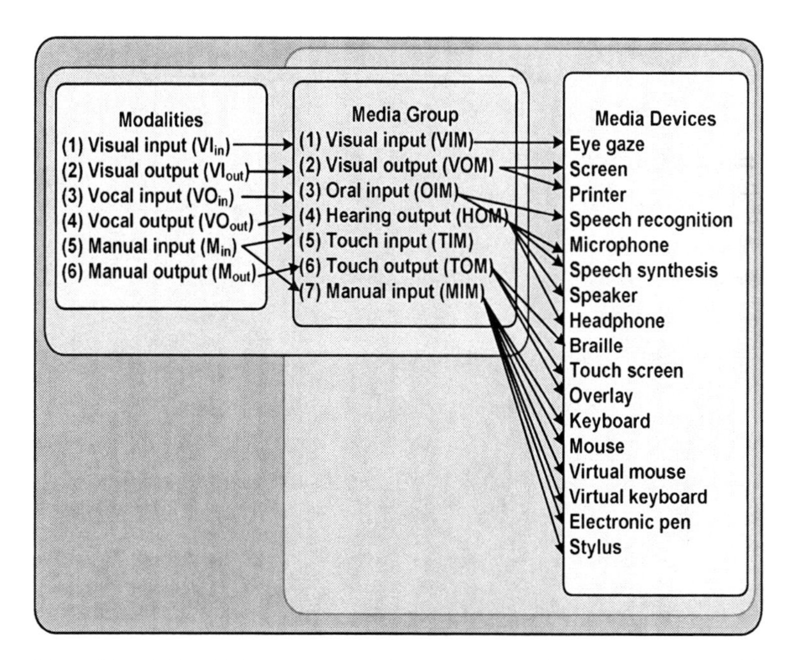

Fig. 8 The relationship between modalities and media, and media group and media devices

dia devices, the chosen optimal modality is a failure and cannot be implemented and hence an *alternative modality* must be chosen and is that one with the next highest score. In its absence, the process of selection is repeated until the system finds a replacement modality that can be supported by currently available media devices. If multimodality is possible and the optimal modalities are selected, then their supporting media devices are checked for availability.

Through function g_1, the media group that support the chosen modality can be identified. Given that $Modality = \{VI_{in}, VO_{in}, M_{in}, VI_{out}, VO_{out}, M_{out}\}$ and $MediaGroup = \{VIM, OIM, MIM, TIM, VOM, HOM, TOM\}$ and that $g_1 : Modality \rightarrow MediaGroup$, then formally, for all media group p, there exists a modality q such that the mapping between p and q is in set g_1, that is $\forall p : Media\,Group, \exists q : Modality | p \rightarrow q \in g_1$.

Using function g_2, the top-ranked media devices that belong to such media group are also identified. Given function g_2, a media device d, priorities p_1 and p_2 where $Priority : N_1$ (positive numbers excluding zero), then the specification for finding the top-ranked device for a media group m is $\exists m : MediaGroup, \forall d : MediaDevice, \exists p_1 : Priority, \forall p_2 : Priority \; | d \bullet m \rightarrow (d, p_1) \in g_2 \wedge (p_1 < p_2)$.

```
//Initialization
Assignment index i ← 1 to 6 to represent modalities
// (VIin, VOin, Min, VIout, VOout, and Mout respectively)
//Evaluate IC suitability of individual modality
Loop i ← 1 to modality_max
   // Calculate modality's IC suitability score
      Loop j ← 1 to parameter_max
      // read suitability level of a modality with respect to parameter i
      Determine suitabilityLevel(j)
          if suitabilityLevel(j) equals
        (1) High then score ← 1.00, (2) Medium then score ← 0.75
            (3) Low then score ← 0.50, (4) Inappropriate then score ← 0.0
      Calculate finalScore = score ↑ (1/parameter_max)
   If finalScore equals
      (1) 1.00 then Suitability ← High
      (2) 0.75 ≤ finalScore < 1.00 then Suitability ← Medium
      (3) 0.50 ≤ finalScore < 0.75 then Suitability ← Low
      (4) < 0.50 then Suitability ← Inappropriate
  Assign modality[i] ← Suitability
```

Fig. 9 Algorithm to determine modality's suitability to IC

Let there be a *media devices priority table* (MDPT) (see Table 4) which tabulates all media groups, and each media group's set of supporting media devices, arranged by priority ranking. Let $T = \{T_1, T_2, \ldots, T_{max_{table}}\}$ be the set of MDPT's. The elements of table $T_n \in T$, where $n = 1$ to max_{table}, are similar to elements of function g_2. It should be noted that every T_n is unique; that no two MDPT's are identical. To create a new table, at least one of its elements is different from all other tables that have already been defined. The priority ranking of a specific media device may be different in each MDPT. By principle, any given IC scenario and its suitable modalities is mapped/assigned to a specific MDPT.

Table 4 A sample media devices priority table (MDPT)

Media Group	Media Devices				
	Priority = 1	Priority = 2	Priority = 3	::	Priority = n
Visual Input	Eye Gaze				
Oral Input	Microphone, Speech Recognition				
Touch Input	Touch Screen	Braille			
Manual Input	Mouse, Keyboard	Virtual Mouse, Virtual KB	Electric Pen	Stylus	Braille
Visual Output	Screen	Printer	Electronic Projector		
Hearing Output	Speaker	Headphone, Speech Synthesis			
Touch Output	Braille	Overlay KB			

5.6 Rules for Priority Ranking of Media Devices

Assuming that an optimal modality has been selected. To implement the modality, the top-ranked media device(s) in the media group that supports the selected modality is/are activated. The rules governing device activation are as follows:

1. If the optimal modality's final suitability = 'H' then the activation of its supporting media group is important. If there are no media devices belonging to such media group are found, then the implementation of the optimal modality is not possible. In such a case, the system searches for a replacement to the optimal modality.
2. A replacement modality (see algorithm in Fig. 9) with 'M' or 'L' suitability score means that the activation of its supporting media group is the last recourse to implement multi-modality. The absence of media devices for such media group means that multimodality failed (due to absence of supporting media devices).

For two or more media devices that belong to the same media group, their priority rankings are governed by the following rules:

1. If their functionalities are identical (e.g. a mouse and a virtual mouse), activating both is incorrect because it is plain redundancy. Instead, one should be ranked higher in priority than the other. The most-commonly-used device gets the higher priority.
2. If their functionalities are complementary (e.g. a mouse and a keyboard), activating both is acceptable. However, if one is more commonly used than the other (i.e. they do not always come in pair), then the more-commonly-used one gets the higher priority. If both devices always come together as a pair, then both are ranked equal in priority.

In the early stage of knowledge acquisition, it is the end user that provides this ranking, which depends on the concerned context. For example, in a quiet workplace, a speaker can be the top-ranked hearing output device. In a noisy environment, however, the headphone gets the top priority. This priority is reflected in every *media devices priority table* (MDPT) associated with every scenario. Initially, there is only one MDPT, similar to Table 4. A second MDPT can be created from the first one by re-organizing the priority order of different devices and by inserting/adding devices into it, as deemed necessary in the scenario. So does follow for the 3rd, the 4th, and the nth MDPT. A MDPT is not static; it can be modified by the user when needed. The MDPT shown in Table 4 is a specimen table and does not contain an exhaustive list of devices. It is merely used for demonstration purposes.

6 Context Learning and Adaptation

After establishing the relationships involving IC, modalities and media devices, we put these relationships to use by considering a specimen IC to which the pervasive MM computing system will adapt and learn.

6.1 Specimen Interaction Context

Our specimen interaction context is based on the following parameters: (1) *user location* - identifies if the user is at home, at work, or on the go, (2) *noise level* - identifies if the user's workplace is quiet, moderate or noisy, (3) *the safety/risk factor* - determines the one sitting in user's workplace and detects the presence of other people; the result identifies if the workplace is safe, sensitive or risky, (4) *the user's handicap* - determines if user is a regular user or is a handicapped, and (5) *the computing device* - identifies if user is using a PC, a laptop or a PDA or cell phone. As to be expected, for each parameter's distinct value, the degree of modality's suitability varies accordingly.

6.1.1 The Context of User Location, Noise Level, and Workplace's Safety

As Table 5 shows, the sample conventions, in generic format, are given for a user's location. The GPS is used as an instrument to detect user's whereabout. Here, the GPS' readings of latitude and longitude provide a specific meaning (i.e. convention). Also, the degrees of suitability of various modalities for each value of user location are also listed (see Table 6). In Table 7, meanings are assigned to a specific range of decibels as observed from the user's workplace. Some sensors, such as [40], can be attached to the computer's USB port to capture the environment's noise level. The table also shows how suitable a certain modality is based on the level of noise in the workplace.

Table 5 User location convention table using GPS values

Convention	Longitude	Latitude	Meaning
1	$< Value_{11} >$	$< Value_{12} >$	At home
2	$< Value_{21} >$	$< Value_{22} >$	At work
3	$!< Value_{11} >$ $\&\& !< Value_{21} >$	$!< Value_{12} >$ $\&\& !< Value_{22} >$	On the go

The context of *safety level* is already briefly discussed in section 4.2.3 - *"Capturing the User's Current Context"*. It is based on two factors: (1) *the person sitting in the user's seat as detected by a camera with retinal recognition*, and (2) *the presence of other people present in the user's workplace as detected by an infrared detector*. [34] is one method by which a legitimate user

Table 6 Suitability of modality based on user location

Type of Modality	User Location = At home	User Location = At work	User Location = On the go
Visual Input	H	H	L
Visual Output	H	H	H
Vocal Input	H	H	H
Vocal Output	H	H	H
Manual Input	H	H	H
Manual Output	H	H	H

Table 7 Sample conventions for noise level

Convention No.	Decibel	Meaning
1	Less than 41	Quiet
2	41 to 50	Moderate
3	Greater than 50	Noisy

Table 8 Suitability of modality based on noise level

Type of Modality	Noise Level = Quiet	Noise Level = Moderate	Noise Level = Noisy
Visual Input	H	H	H
Visual Output	H	H	H
Vocal Input	H	M	I
Vocal Output	H	H	M
Manual Input	H	H	H
Manual Output	H	H	H

can be differentiated from an intruder. Also, [35] provides a wide range of infrared detector products. Fig. 10 shows the safety level detection process.

The combination of the results obtained from infrared detector and of camera indicates how sensitive the user's workplace is. Table 9 provides the workplace's risk/safety convention. Table 10, on the other hand, shows our perception of modalities' suitability with respect to safety level. Note that all modalities are rated *inappropriate* if safety level becomes bad (i.e. risky), not because they are really inappropriate to the context but as a mean to protect the user from unauthorized people's intrusion. As per our view, in a risky setting, the system automatically saves user's information and then logs him out from the system.

6.1.2 The Context of User Handicap and Computing Device

Figure 11 shows the generic format of our user profile. For this paper, some information as shown in the user profile diagram (i.e. user QoS and supplier preferences) are not discussed here since they are not related to this chapter's content. A *user profile* contains, among others,

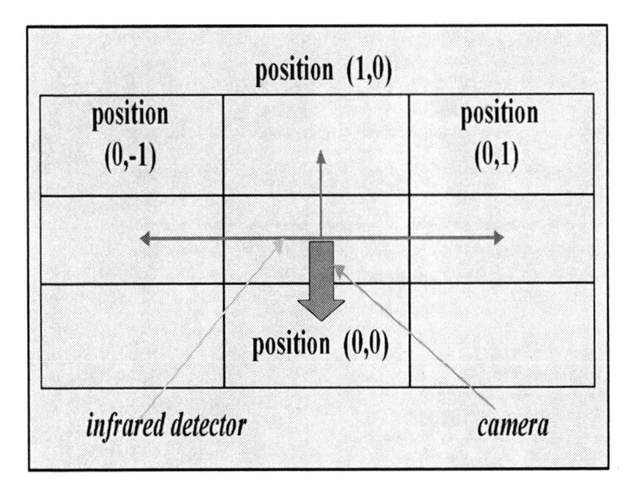

Fig. 10 The safety/risk factor detection using an infrared detector and a camera

Table 9 The safety/risk factor convention table

Convention No.	Detected Person in User's Seat	Detected People in User's Workplace	Meaning
2	User	Image	Sensitive
1	User	No Image	Safe
3	Empty	Image	Sensitive
1	Empty	No Image	Safe
3	Other	Image	Risky
3	Other	No Image	Risky

Table 10 Suitability of modality based on user workplace's safety level

Type of Modality	Safety Level = Safe	Safety Level = Sensitive	Safety Level = Risky
Visual Input	H	M	I
Visual Output	H	M	I
Vocal Input	H	M	I
Vocal Output	H	M	I
Manual Input	H	M	I
Manual Output	H	M	I

the user's username, password and a list of the user's computing devices and their corresponding schedules. Since the user is mobile, his computing device is identified through this component of user profile. In the *special needs* section, the user is identified as either a *regular user* or a handicapped. If the user is indeed handicapped, the disability is specified, indicating if the user is a *mute*, a *deaf*, a *visually impaired*, or a *manually handicapped*. Here, the importance of multimodality is obvious; it provides handicapped users the chance to access informatics through modalities that suit their conditions.

User profile	Special Needs	User QoS preference	Supplier preference
username password computer1 schedule1 computer2 schedule2 :: computer*n* schedule*n*	Disability: No: Yes: - deaf? - mute? - visually impaired? - manually disabled?	Video Player Dimensions: Frame rate: *min=xxx, max=yyy, user_preference=zzz* Frame quality:*min=xxx, max=yyy, user_preference=zzz* Audio quality:*min=xxx, max=yyy, user_preference=zzz* Web Browser Dimensions: Page latency: *min=xxx, max=yyy, user_preference=zzz* Page richness:*min=xxx, max=yyy, user_preference=zzz* Text Editor Dimensions: Etc.	Video: RealOne or QuickTime or MSMedia? *(preference value for each supplier)* Web Browser: Netscape or Internet Explorer or Opera? *(preference value for each supplier)* Text Editor: TextPad or WordPad or NotePad or Microsoft Word *(preference value for each supplier)* Email Browser: Eudora or Outlook or Browser? *(preference value for each supplier)*
(a)	(b)	(c)	(d)

Fig. 11 A sample user profile

Table 11 shows the user profile/handicap convention while Table 12 shows the modalities suitability based on such profile. We also consider the user's computing device as a context parameter because the degree of modality's suitability using a PC, a laptop or a PDA varies. The PDA, for example, has very limited resources (memory, CPU, battery) as compared with a PC or a laptop. Table 13 shows our computing device conventions while Table 14 shows the modalities' suitability based on these computing devices.

6.2 Scenarios and Case-Based Reasoning with Supervised Learning

A *scenario* is an event that needs a corresponding system response. The stimulus that triggers a scenario is called the *pre-condition scenario*. The system response and hence the desired result to the event is called the *post-condition scenario*. In this work, the pre-condition scenario is

Table 11 User profile/disability convention

Convention No.	User Profile
1	Regular User
2	Deaf
3	Mute
4	Visually-Impaired
5	Manually-Impaired

Table 12 Suitability of modality based on user's profile/handicap

Type of Modality	User = Regular User	User = Deaf	User = Mute	User = Visually-Impaired	User = Manually-Impaired
Visual Input	H	H	H	I	H
Visual Output	H	H	H	I	H
Vocal Input	H	M	I	H	H
Vocal Output	H	I	M	H	H
Manual Input	H	H	H	H	I
Manual Output	H	H	H	H	I

Table 13 Computing device convention

Convention No.	Computing Device
1	PC
2	Laptop
3	PDA
3	Cellular Phone

Table 14 Suitability of modality based on user's computing device

Type of Modality	Computing Device = PC	Computing Device = Laptop	Computing Device = PDA/Cellular Phone
Visual Input	H	H	L
Visual Output	H	H	H
Vocal Input	H	H	H
Vocal Output	H	H	H
Manual Input	H	H	H
Manual Output	H	H	L

a specific instance of interaction context $IC_i \in IC$. The desired post-condition is the selection and activation of suitable modalities and their supporting media devices.
Given that $IC_i = UC_k \bigotimes EC_l \bigotimes SC_m$ then the total number of possible scenarios, denoted as $scenTot$, is the product of the number of convention values of each context parameter i, that is,

$$sceneTot = \prod_{i=1}^{param_{max}} card(IC_i) \tag{8}$$

where $card(IC_i)$ =cardinality/total number of conventions for interaction context parameter i. $scenTot$ can also be expressed as $scenTot = card(UC_k) \times card(EC_l) \times card(SC_m)$. The scenario number, $scenNum$, assigned to any specific instance of an interaction context is given by:

$$sceneNum = IC_{param_{max}} + \sum_{i=1}^{param_{max}-1} (IC_i - 1) \cdot \prod_{j=i+1}^{param_{max}-1} card(IC_j) \tag{9}$$

A scenario table is simply a table listing all possible scenarios. An entry in the scenario table is implemented in two ways: (1) through expert intervention (i.e. user) or (2) on the fly as the scenario is encountered. Each entry in the scenario table is composed of pre- and post-condition scenarios. Here's how an entry to the scenario table is performed: (i) the current interaction context parameters and their conventions are listed in the pre-condition scenario, see Fig. 12-Algorithm 2, (ii) the post-condition scenario part lists down the corresponding suitability scores of each modality, calculated using Equations 6 and 7, see Fig. 12-Algorithm 3, (iii) the scenario number is calculated using Equation 9, and (iv) the pointer to MDPT is initially pointed to the very first MDPT, unless it has already been rectified by the expert. A sample snapshot of such table is shown in Table 15.

Table 15 Scenario table containing records of pre and post-conditions scenarios

Scenario Number	Pre-Condition: User Loc.	Noise Level	Safety Level	User Profile	Comp. Device	Post-Condition: Visual Input	Vocal Input	Manual Input	Visual Output	Vocal Output	Manual Output	Media Devices Priority Table
1	1	1	1	1	1	H	H	H	H	H	H	T_1
2	1	1	1	1	2	H	H	H	H	H	H	T_1
3	1	1	1	1	3	M	H	H	H	H	M	T_2
4	1	1	1	2	1	H	M	H	H	I	H	T_a
::	::	::	::	::	::	::	::	::	::	::	::	::
404	3	3	3	5	2	I	I	I	I	I	I	T_{n-1}
405	3	3	3	5	3	I	I	I	I	I	I	T_n

Once a scenario is stored in the scenario table, it becomes an exemplar and is considered as a learned information. An *exemplar* is a stored knowledge. When the ML component encounters a new scenario (i.e. new context), it converts the scenario into a *case*, specifying the problem (i.e. the given context). The ML component searches for a match between the given case and stored exemplars. When a new scenario is converted into a case to be considered, the resulting case would be composed of three elements, namely: (1) *the problem* - the pre-condition scenario in consideration, (2) *the solution* - the final and optimal modality, and (3) *the evaluation* - the rate of relevance of the solution. Using the similarity algorithm, a comparison is made between the problem in the new case against all the available problems in the database. The scenario of the closest match is selected and its solution is returned. The evaluation is the score of how similar it is to the closest match.

Inspired by [41], we modify their similarity scoring scheme to reflect the needs of our system. Hence, given a new case (NC) and an individual case stored in the knowledge database (MC), the similarity of the problem between the two cases, that is NC against MC as designated by corresponding subscripts, is equal to their similarity in the case's UC, EC and SC and is given by:

$$Sim(NC,MC) = \frac{1}{3}Sim(UC_{NC},UC_{MC}) + \frac{1}{3}Sim(EC_{NC},EC_{MC}) + \frac{1}{3}Sim(SC_{NC},SC_{MC}) \quad (10)$$

The similarity between the UC of NC against the UC of MC is given by:

$$Sim(UC_{NC},UC_{MC}) = \frac{\sum_{i=1}^{max_{NC}} Sim(UC_{i_{NC}},UC_{MC})}{max(UC_{NC},UC_{MC})} \quad (11)$$

where UC_i, $i = 1$ to max, is the individual UC parameter, $max(UC_{NC},UC_{MC})$ is the greater between the number of UC parameters between NC and MC, and $Sim(UC_{i_{NC}},UC_{MC}) = max_{j=1}^{max_{MC}} Sim(UC_{i_{NC}},UC_{j_{MC}})$ where $UC_{j_{MC}} \in UC_{MC}$ and $Sim(UC_{i_{NC}},UC_{j_{MC}}) \in [0,1]$ is the similarity between a specific UC parameter i of NC and parameter j of MC.

For the similarity measures of EC of NC against EC of MC, and the SC of NC against SC of MC, the same principle as Equation 11 must be applied, with the formula adjusted accordingly to denote EC and SC, respectively, yielding:

```
Algorithm 1: Knowledge acquisition on context parameter
static int total_param;
while (true) do
begin
  write ('New context parameter entry? ');
  read (answer);
   if (answer = 'Yes') do
        total_param = total_param + 1;
        write ('Number of conventions
              for', total_param);
        read (max);
        for n = 1 to max do
            read (suitability_score, VI_in);
            read (suitability_score, VI_out);
            read (suitability_score, VO_in);
            read (suitability_score, VO_out);
            read (suitability_score, M_in);
            read (suitability_score, M_out);
        end for
   else
      break;
   end if
end while
```

```
Algorithm 2: Pre-condition entry in Scenario Table
for k = 1 to total_param do
  // detect parameter conventions (via LVMIUC)
  read (convention, parameter k)
    // search (convention, parameter k)
    // from a priori knowledge
        write (parameter k, (suitability_score, VI_in));
        write (parameter k, (suitability_score, VI_out));
        write (parameter k, (suitability_score, VO_in));
        write (parameter k, (suitability_score, VO_out));
        write (parameter k, (suitability_score, M_in));
        write (parameter k, (suitability_score, M_out));
end for
```

```
Algorithm 3: Post-condition entry in Scenario Table
// initialization
VI_in = 1; VI_out = 1; VO_in = 1; VO_out = 1, M_in = 1, M_out = 1
for k = 1 to total_param do
   //read (convention, parameter k)
    VI_in = VI_in * (parameter k, (suitability_score, VI_in));
    VI_out = VI_out * (parameter k, (suitability_score, VI_out));
    VO_in = VO_in * (parameter k, (suitability_score, VO_in));
    VO_out = VO_out * (parameter k, (suitability_score, VO_out));
    M_in = M_in * (parameter k, (suitability_score, M_in));
    M_out = M_out * (parameter k, (suitability_score, M_out));
end for
// Application of Equation 2, applicable to all modalities
if (VI_in == 1.0) then VI_in_suitability = 'H' else
  if (0.75 ≤ VI_in < 1.0) then VI_in_suitability = 'M' else
  if (0.50 ≤ VI_in < 0.75) then VI_in_suitability = 'L' else
  VI_in_suitability = 'I'
end if
```

```
Algorithm 4: Checking possibility of multimodality and selection of optimal modalities
//check if multimodality is possible
if ((VI_in != 'I') OR (VO_in != 'I') OR (M_in != 'I')) AND ((VI_out != 'I')
  OR
 (VO_out != 'I') OR (M_out != 'I')) then
  // implement the chosen modalities
  // choose the optimal modality for data input and output
  optimal_InputModality := largest(VI_in, VO_in, M_in);
  optimal_OutputModality := largest(VI_out, VO_out, M_out);
else
  //multimodality is not possible
end if
```

Fig. 12 Algorithms related to knowledge acquisition, entry in scenario table and selection of optimal modality

$$Sim(EC_{NC}, EC_{MC}) = \frac{\sum_{i=1}^{max_{NC}} Sim(EC_{i_{NC}}, EC_{MC})}{max(EC_{NC}, EC_{MC})} \quad (12)$$

$$Sim(SC_{NC}, SC_{MC}) = \frac{\sum_{i=1}^{max_{NC}} Sim(SC_{i_{NC}}, SC_{MC})}{max(SC_{NC}, SC_{MC})} \quad (13)$$

Equation 10 assumes that the weights of UC, EC and SC are all equal (i.e. each is worth 33.3%). This figure is not fixed and can be adjusted to suit the need of the expert (e.g. UC may be worth 40% and each of EC and SC may be worth 30%). Overall, after the comparison has been done, the ideal case match is a perfect match. However, a score of 90% means that a great deal of IC parameters is correctly considered and is therefore 90% accurate. The expert himself, however, decides the acceptable threshold score.
If no match, however, is found (i.e. relevance score is lower than accepted threshold) then the ML component takes the closest scenario as the initial solution of the new case. The user may not accept it. In such a case, a new case with *supervised learning* is produced. If the new case's problem contains more context parameters than those of recorded cases, the expert may decide to include the missing parameter(s) into the a priori knowledge (see Fig. 6). Thereafter, the new case's post-condition scenario is re-evaluated (see Fig. 12-Algorithm 3). The new case in then added to the scenario table, and its scenario number calculated. This whole learning mechanism is called *case-based reasoning with supervised learning*.
As an example, consider the following IC: user location = at home, noise level = quiet, safety factor = safe, user profile = regular user and computing device = PDA. This IC condition $(ic_1, ic_2, ic_3, ic_4, ic_5)$ =(1, 1, 1, 1, 3). It is scenario number 3. The calculated final suitability scores of the modality types are given below and are also stored in scenario table (Table 8).
$VisualInput = (H \times H \times H \times H \times L)^{\frac{1}{5}} = (1 \times 1 \times 1 \times 1 \times 1 \times 0.50)^{\frac{1}{5}} = 0.87$
$VocalInput = (H \times H \times H \times H \times H)^{\frac{1}{5}} = (1 \times 1 \times 1 \times 1 \times 1 \times 1)^{\frac{1}{5}} = 1$
$ManualInput = (H \times H \times H \times H \times H)^{\frac{1}{5}} = (1 \times 1 \times 1 \times 1 \times 1 \times 1)^{\frac{1}{5}} = 1$
$VisualOutput = (H \times H \times H \times H \times H)^{\frac{1}{5}} = (1 \times 1 \times 1 \times 1 \times 1 \times 1)^{\frac{1}{5}} = 1$
$VocalOutput = (H \times H \times H \times H \times H)^{\frac{1}{5}} = (1 \times 1 \times 1 \times 1 \times 1 \times 1)^{\frac{1}{5}} = 1$
$ManualOutput = (H \times H \times H \times H \times L)^{\frac{1}{5}} = (1 \times 1 \times 1 \times 1 \times 1 \times 0.50)^{\frac{1}{5}} = 0.87$
Indeed, given the specified context, the suitability scores of various modalities are: (i) Visual Input = Medium suitability, (ii) Vocal Input = High suitability, (iii) Manual Input = High suitability, (iv) Visual Output = High suitability, (v) Vocal Output = High suitability, and (vi) Manual Output = Medium suitability.
For this specific case, modality is possible. The *optimal input modality* is both Vocal Input and Manual Input. The *optimal output modality* is Visual Output and Vocal Output. All non-optimal modalities are considered *optional*. If this same case reappears again in the future, then using the similarity algorithm (Equation 10), there is an exact match (scenario 3) that can be found in the database, hence, recalculation/decision making is evaded.
Also, Let $M_1, M_2, \ldots, M_6$ be the set of modalities' VI_{in}, VO_{in}, M_{in}, VI_{out}, VO_{out}, and M_{out} respective suitability scores. At any time, the suitability score of M_1 is m_1 = {H, M, L, I} = {1, 0.75, 0.50, 0}. Such suitability scores also apply to $M_2, M_3, \ldots, M_6$. Hence, the modalities selections, Y, as a vectored output is equal to the Cartesian product of the individual modality's suitability score, that is, $Y = M_1 \bigotimes M_2 \bigotimes M_3 \bigotimes M_4 \bigotimes M_5 \bigotimes M_6 = (m_1, m_2, m_3, m_4, m_5, m_6)$ where $m_1 \in M_1$, $m_2 \in M_2$, ..., $m_6 \in M_6$. In the specimen IC, there are $3 \times 3 \times 3 \times 5 \times 3 = 405$ possible context scenario combinations in X and $4^6 = 4096$ possible modality's suitability combinations in Y. Hence, the function $f_1 : X \rightarrow Y$ that maps user context to appropriate modalities is also expressed as $f_1 : (c_1, c_2, c_3, c_4, c_5) \rightarrow (m_1, m_2, m_3, m_4, m_5, m_6)$.

6.3 Assigning a Scenario's MDPT

This process is shown in Fig. 13 using the specimen IC. At the start (step 1), the Context Manager Agent (CMA) gets the current IC. In (step 2), this scenario becomes a case. Using the pre-condition scenario, the case's *scenNum* is calculated and is used as an index to the scenario table. Assuming that a perfect match is found then the post-condition scenario (i.e. the case's solution) is adopted with relevance score = 100%). Since the present case is not new to the system, then steps 3, 4, and 5 are skipped.

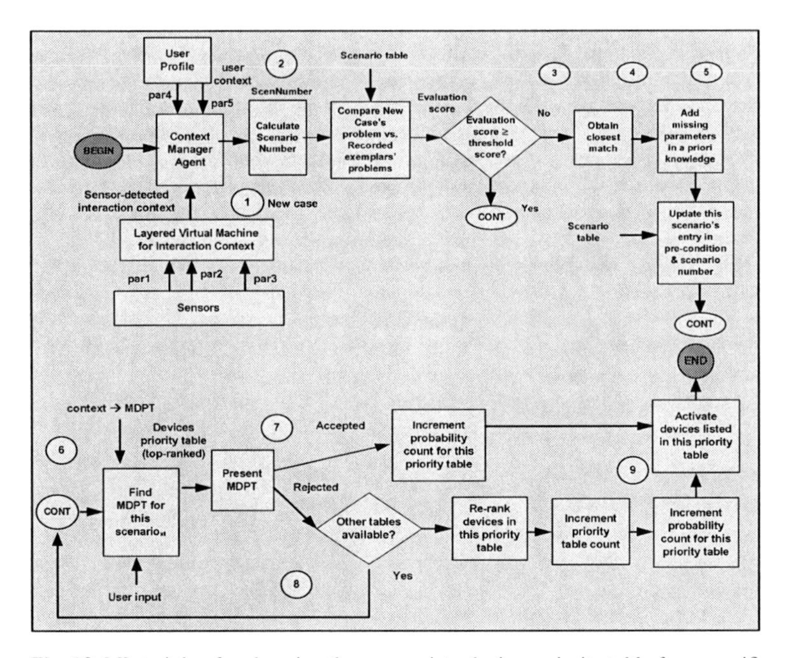

Fig. 13 ML training for choosing the appropriate devices priority table for a specific context

If the similarity/relevance score, however, is low (say, 40%), then no match is found. Hence, the closest match is retrieved and presented to the user. Because the proposed solution is wrong (i.e. 40% accurate vs. 60% erroneous), the new case is treated for adaptation maintenance. The large amount of error is brought by the fact that most of the context parameters in the new case cannot be found in the stored cases of scenario table. Hence, an update of a priori knowledge and scenario table is made; the new context parameters are added and the new case is stored in the scenario table. The new case's corresponding post-condition scenario is recalculated. Due to the newly added context parameter(s) in the scenario table, all scenario numbers of previous entries' are recalculated. In the scenario table, a MDPT for the new case has to be established; hence the available MDPT's are presented to the user, one table at time (step 6). If the user accepts the proposed table (step 7), the table's numerical identification is appended onto the scenario table. The media groups corresponding to the selected modalities are noted and their top-ranked media devices are selected for activation (step 9). If the user rejects such MDPT, then each of the other remaining tables will be presented (step 8). Recall that there is just one MDPT in the beginning. Hence, the user needs to modify the contents of the first table to create a second one. When this is done, the identification number of the newly-created MDPT is appended into the scenario table. And step 9 is executed.

Figure 14 illustrates the format of a completely filled scenario table for specimen user context. Note, however, that as long as new context parameters are being added, the scenario table will keep on growing. This makes our system adaptive to an ever-changing user context.

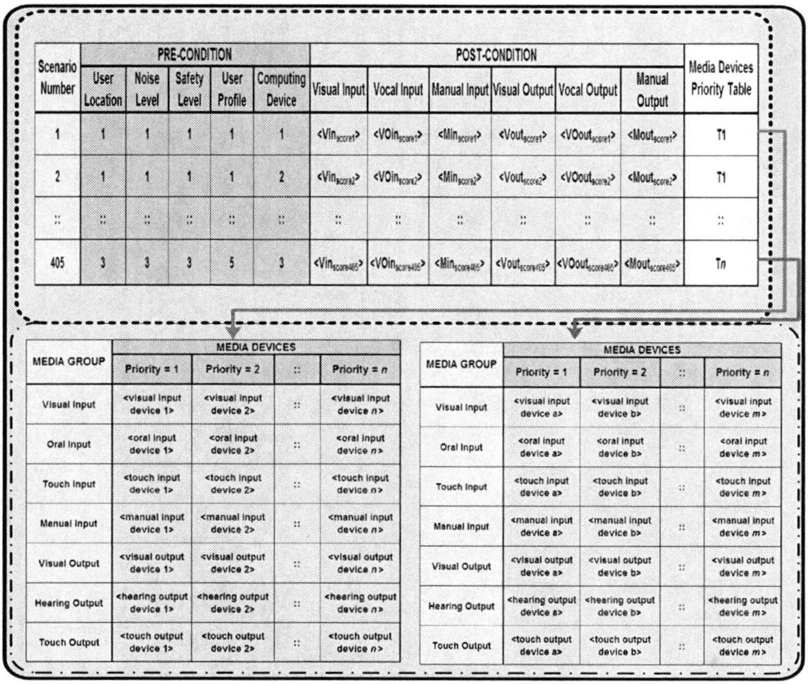

Scenario Number	PRE-CONDITION					POST-CONDITION						Media Devices Priority Table
	User Location	Noise Level	Safety Level	User Profile	Computing Device	Visual Input	Vocal Input	Manual Input	Visual Output	Vocal Output	Manual Output	
1	1	1	1	1	1	$\langle Vin_{score1} \rangle$	$\langle VOin_{score1} \rangle$	$\langle Min_{score1} \rangle$	$\langle Vout_{score1} \rangle$	$\langle VOout_{score1} \rangle$	$\langle Mout_{score1} \rangle$	T1
2	1	1	1	1	2	$\langle Vin_{score2} \rangle$	$\langle VOin_{score2} \rangle$	$\langle Min_{score2} \rangle$	$\langle Vout_{score2} \rangle$	$\langle VOout_{score2} \rangle$	$\langle Mout_{score2} \rangle$	T1
::	::	::	::	::	::	::	::	::	::	::	::	::
405	3	3	3	5	3	$\langle Vin_{score405} \rangle$	$\langle VOin_{score405} \rangle$	$\langle Min_{score405} \rangle$	$\langle Vout_{score405} \rangle$	$\langle VOout_{score405} \rangle$	$\langle Mout_{score405} \rangle$	T*n*

MEDIA GROUP	MEDIA DEVICES			
	Priority = 1	Priority = 2	::	Priority = *n*
Visual Input	<visual input device 1>	<visual input device 2>	::	<visual input device *n*>
Oral Input	<oral input device 1>	<oral input device 2>	::	<oral input device *n*>
Touch Input	<touch input device 1>	<touch input device 2>	::	<touch input device *n*>
Manual Input	<manual input device 1>	<manual input device 2>	::	<manual input device *n*>
Visual Output	<visual output device 1>	<visual output device 2>	::	<visual output device *n*>
Hearing Output	<hearing output device 1>	<hearing output device 2>	::	<hearing output device *n*>
Touch Output	<touch output device 1>	<touch output device 2>	::	<touch output device *n*>

MEDIA GROUP	MEDIA DEVICES			
	Priority = 1	Priority = 2	::	Priority = *n*
Visual Input	<visual input device a>	<visual input device b>	::	<visual input device *m*>
Oral Input	<oral input device a>	<oral input device b>	::	<oral input device *m*>
Touch Input	<touch input device a>	<touch input device b>	::	<touch input device *m*>
Manual Input	<manual input device a>	<manual input device b>	::	<manual input device *m*>
Visual Output	<visual output device a>	<visual output device b>	::	<visual output device *m*>
Hearing Output	<hearing output device a>	<hearing output device b>	::	<hearing output device *m*>
Touch Output	<touch output device a>	<touch output device b>	::	<touch output device *m*>

Fig. 14 A sample snapshot of a completed scenario table, each entry with its assigned MDPT

6.4 Finding Replacement to a Missing or Failed Device

At any time, it is possible that a selected top-ranked media device may be missing or defective. Some techniques for detecting device failures are available in [42]. Hence, a replacement should be found for the system to remain running and operational. The replacement can be found within the same MDPT assigned to the scenario. The algorithm of replacement to a failed device is shown in Fig. 15. In (step 1), using scenario number (*scenNum*), the system determines its assigned MDPT which identifies the media groups' top-ranked devices. In (step 2), the environment profile is consulted to find out the currently available media devices. In (step 3), the system merely activates the top-ranked media device, if available. Otherwise, in (step 4) the second-ranked device is activated, also if available. If it is also missing or defective, then the third-ranked device is searched. In general, the search goes on until a selected device is found. The worse-case event is when no device in a media group in the MDPT is activated due to cascaded failure or collective absence of needed devices (step 5). In such case, the system abandons the selected optimal modality (because it cannot be implemented) and attempts to replace the optimal modality by available non-optimal modality. This process finds again the available media devices, by priority, to support the non-optimal modality. In the worst case that the non-optimal modalities cannot be supported, this simply means that multimodality is impossible in the given computing environment. Given the failed device d of priority p_1, the specification for finding the replacement media device d_{rep} is $\exists m : MediaGroup, \forall d_{rep} : MediaDevice, \exists p_1 : Priority, \forall p_2 : Priority \,|\, (p_1 = p_1 + 1) \wedge (p_1 < p_2) \wedge m \rightarrow (d_{rep}, p_1) \in g_2 \bullet d_{rep}$.

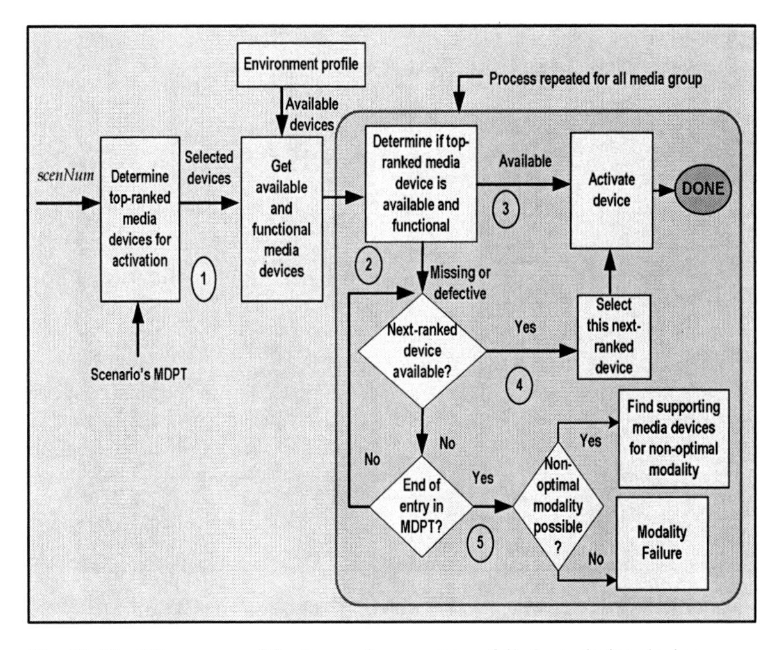

Fig. 15 The ML process of finding replacement to a failed or missing device

6.5 Media Devices' Priority Re-ranking due to a Newly-Installed Device

A newly-installed device affects the priority rankings of media devices in the media group where the new device belongs. Fig. 16 illustrates the update process in a MDPT due to the arrival of this newly-installed device. In (step 1), given that the system has already recognized the new device via environment profile, the user provides the media group where it belongs. In (step 2), the MDPT assigned to scenario number 1 is retrieved and becomes the first MDPT to be updated. This priority table is edited (step 3). The new device's name is inserted into the table (step 4). In (step 5), the rankings of other devices in the same media group are updated by the user. When done, the second MDPT is searched. The update process is repeated on other scenarios until the last of MDPT is also updated. The update process is quite long (i.e. equal to the number of all MDPT's).

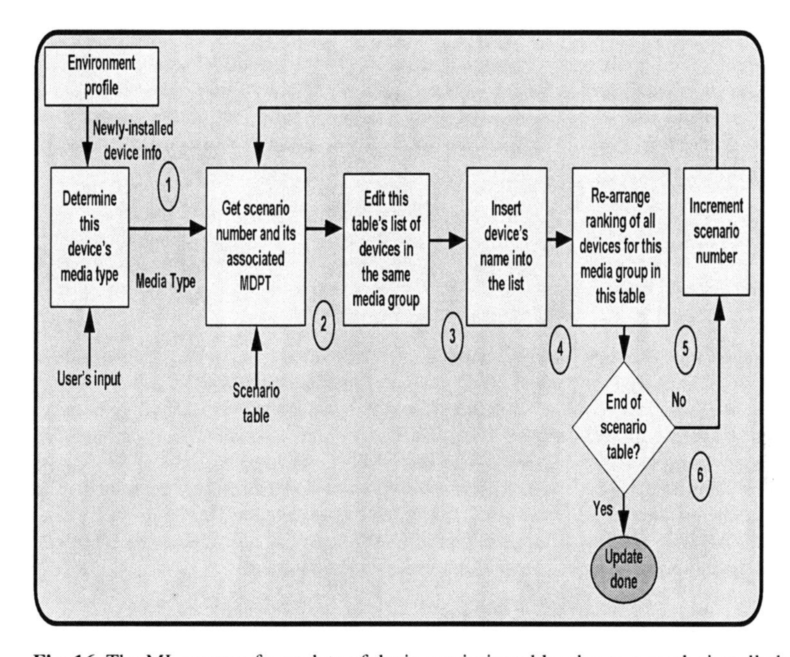

Fig. 16 The ML process for update of devices priority tables due to a newly-installed device

6.6 Our Pervasive Multimodal Multimedia Computing System

Our proposed system is conceived for two purposes: (1) to contribute to MM computing and (2) to further advance autonomic computing system. To achieve the first goal, we develop the model that relates modality with user context, and associate media devices to support the implementation of the chosen modality. For the second goal, we advocate the propagation of knowledge, acquired through training, into the user's computing environment so that such knowledge can be used for system adaptation to user needs, and system restrictions. The major components of our MM computing system are shown in Fig. 17.

The functionality of each component is given below:

1. *The Task Manager Agent (TMA)* - manages user's profile, task and pertinent data and their deployment from a server to the user's computing device, and vice versa.
2. *The Context Manager Agent (CMA)* - detects user context from sensors and user profile, and selects the modality and media apt for the context.

3. *The History and Knowledge-based Agent (HKA)* - responsible for ML training and knowledge acquisition.
4. *The Layered Virtual Machine for Interaction Context (LVMIC)* - detects sensor-based context and allows the incremental definition of context parameters.
5. *The Environmental Manager Agent (EMA)* - detects available and functional media devices in the user's environment

In the diagram, the user (Manolo) can work at home, logs out, and still continue working on the same task at anytime and any place. Due to user's mobility, the variation in interaction context and available resources is compensated by a corresponding variation in modality and media devices selection. Further details of the infrastructure of our pervasive MM computing system is available in [43].

7 Conclusion

In this work, we presented the communication protocols to realize autonomic communication in a pervasive MM computing system. The system detects the current instance of interaction context and accordingly selects the modalities that suit it. We define the relationship between context and modality and between modality and media group. Media devices are identified by their membership to a media group. When two or more media devices of the same group are available, their selection is based on their priority ranking. We assert that defining context through parameters should be incremental and based on the dynamic needs of the system. Hence, we adopted a layered virtual machine to realize incremental interaction context. It allows user to add, modify, and delete one context parameter at a time.

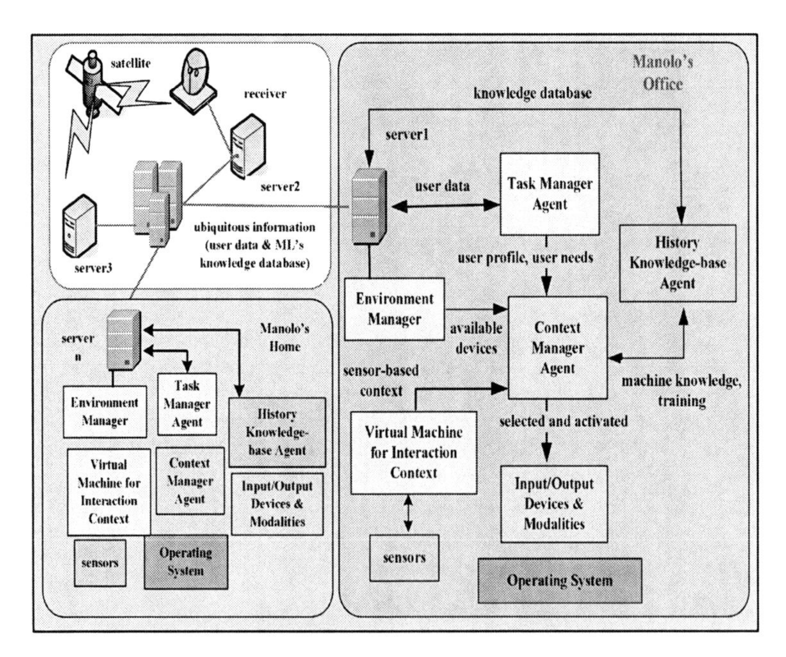

Fig. 17 The architecture of a context-aware ubiquitous multimodal computing system

Using natural language processing as basis, we classify modality as either an input or an output. Then, modalities are further classified based on the body part that uses the modality to input data and the body part that uses the modality to receive output. The same principle is used for media classification, with minor additions. In this work, media are physical devices (and a few software) that support modality. We laid out rules for prioritizing media devices. Device activation and replacement to a failed device depends on this priority ranking.
The system's knowledge acquisition is presented using a specimen interaction context, composed of specimen parameters, namely: user location, noise level, the safety factor, the user profile and the user's computing device. The ML's progressive knowledge acquisition is also applied on context parameters and interaction contexts. When a device failed, a replacement is searched from a list of devices in the same media group within the MDPT. When a new device is introduced onto the system for the first time, all the MDPT's are updated, and the priority rankings of media are updated in each possible scenario.

Acknowledgements We wish to acknowledge the financial support provided by the *Natural Sciences and Engineering Research Council of Canada* (NSERC), the scholarship funds from the *Décanat des études* of Université du Québec, École de technologie supérieure (ÉTS) and the *cotutelle de thèse* funds provided by the Governments of *Qubec* and *France*. We also wish to acknowledge the support provided by *ÉGIDE*, Paris, France and our colleagues in LATIS Laboratory of ÉTS and PRISM Laboratory of UVSQ, France.

References

1. Weiser, M.: The computer for the twenty-first century, 1991. 265(3), pp: 94-104
2. Satyanarayanan, M.: Pervasive Computing: Vision and Challenges. IEEE Personal Communications, 2001. 8(4): pp. 10-17
3. Vasilakos, A.V., Pedrycz, W. (2006) Ambient Intelligence, Wireless Networking, Ubiquitous Computing. ArtecHouse, USA
4. Centre for Pervasive Healthcare, 2007, http://www.pervasivehealthcare.dk/
5. Pahlavan, K., Krishnamurthy, P. (2002) Principles of wireless Networks. Prentice Hall, USA
6. Satyanarayanan, M.: Mobile information access. IEEE Personal Communications, 1996. 3(1): pp. 26-33
7. Satyanarayanan, M.: Scalable, secure and highly available distributed file access. Computer, 1990. 23(5): pp. 9-18
8. Djenidi, H., Tadj, C., Ramdane-Cherif, A., Levy, N.: Generic Multimedia multimodal agents paradigms and their dynamic reconfiguration at the architectural levels. Eurasip Journal of Applied Signal Processing, 2004. 200491): pp. 1688-1707
9. Djenidi, H., Tadj, C., Ramdane-Cherif, A., Levy, N.: Dynamic Multi-agent Architecture for Multimedia Multimodal Dialogs. IEEE Workshop on Knowldege Media Networking, 2002. pp. 107-113
10. Mitchell, T. (1997) Machine Learning. Mc-Graw Hill, USA
11. Ringland, S.P.A., Scahill, F.J.: Multimodality - The future of wireless user interface, 2003. 21(3): pp. 181-191
12. Oviatt, S. L., Cohen, P.R.: Multimodal Interfaces that Process What Comes Naturally. Communications of the ACM, 2000. 43(3): pp. 45-53
13. Schroeter, J. et al: Multimodal Speech Synthesis. IEEE International Conference on Multimedia and Expo, 2000. Vol. 1: pp. 571-574
14. Awde, A., Hina, M.D., Tadj, C. Ramdane-Cherif, C., Bellik, Y.: A Paradigm of a Pervasive Multimodal Multimedia Computing System for the Visually-Impaired Users. GPC 2006, LNCS, Springer-Verlag: pp. 620-633

15. Oviatt, S. L.,: Designing the User Interface for multimodal speech and Gesture Applications: State-of-the-art Systems and Research Directions. Human Computer Interaction, 2000. 15(4): pp. 263-322
16. Rubin, P., Vatikiotis-Bateson, E., Benoit, C.: Audio-Visual Speech Processing. Speech Communications, 1998. 26(1-2)
17. Coutaz, J., et al: Context is key. Communications of the ACM, 2005. 48(3): pp. 49-53.
18. Dey, A.K., Abowd, G.D.: Towards a Better Understanding of Context and Context-Awareness. 1st Intl. Conference on Handheld and Ubiquitous Computing. 1999. Karlsruhe, Germany: Springer-Verlag, LNCS 1707.
19. Dey, A.K.: Understanding and Using Context. Springer Personal and Ubiquitous Computing, 2001. 5(1): pp. 4 - 7.
20. Rey, G., Coutaz, J.: Le contexteur: une abstraction logicielle pour la réalisation de systèmes interactifs sensibles au contexte. 14th French-speaking Conference on Human-Computer Interaction (Conférence Francophone sur l'Interaction Homme-Machine) IHM '02 2002. Poitiers, France: ACM Press
21. Shumate, K.: Layered virtual machine/object-oriented design. 5th Symposium on Ada. 1988: ACM Press, New York, NY, USA.
22. Antoniol, G., Penta, M.D.: A Distributed Architecture for Dynamic Analyses on User-profile Data. 8th European Conference on Software Maintenance and Reengineering. 2004.
23. Bougant, F., Delmond, F., Pageot-Millet, C.: The User Profile for the Virtual Home Environment. IEEE Communications Magazine, 2003. 41(1): pp. 93-98
24. Garlan, D., et al., Project Aura: Towards Distraction-Free Pervasive Computing. Special Issue, Integrated Pervasive Computing Environments, 2002. 21(2): pp. 22-31
25. Schilit, B., Theimer, M.: Disseminating Active Map Information to Mobile Host. IEEE Network, 1994. 8(5): pp. 22-32.
26. Pascoe, J.: Generic Contextual Capabilities to Wearable Computers. 2nd International Symposium on Wearable Computers. 1998.
27. Dey, A.K., et al., An Architecture to Support Context-Aware Applications. GVU Technical Report GIT-GVU-99-23. 1999.
28. Brown, P.J., Bovey, J.D., and Chen, X.: Context-Aware Applications: From the Laboratory to the Marketplace. IEEE Personal Communications. 1997. pp. 58 - 64
29. Ryan, N., Pascoe, J., Morse, D.: Enhanced Reality Fieldwork: the Context-Aware Archeological Assistant. Exxon Computer Applications in Archeology. 1997.
30. Ward, A., Jones, A., Hooper, A.: A New Location Technique for the Active Office. IEEE Personal Communications, 1997: p. 42 - 47.
31. Rey, G.,Coutaz, J.: The Contextor Infrastructure for Context-Aware Computing. 18th European Conference on Object-Oriented Programming (ECOOP 04), Workshop on Component-Oriented Approach to Context-Aware Systems. 2004. Oslo, Norway.
32. Noise Measuring Device, ELK Company, `http://www.bappu.com/`
33. Global Positioning System (GPS), `http://www8.garmin.com/aboutGPS/`
34. Iris/Retinal Recognition, `http://www.biometrics.gov/Documents/IrisRec.pdf`
35. Infrared Detector devices, `http://www.eosystems.com/`
36. Measuring Light Intensity, `http://www.emant.com/316002.page`
37. Extensible Markup Language (XML) Online, `www.w3.org/XML`
38. Developer Resources for Java Technology, `java.sun.com`
39. Miraoui, M., Tadj, C., ben Amar, C.: Context Modeling and Context-Aware Service Adaptation for Pervasive Computing Systems. International Journal of Computer and Information Science and Engineering, 2008. 2(3): pp. 148 - 157
40. Noise Measuring Device.`http://www.bappu.com/`
41. Lajmi, S., Ghedira, C., Ghedira, K.: Une méthode d'apprentissage pour la composition de services web. L'Objet, 2007. 8(2): pp. 1 - 4

42. Hina, M.D., Tadj, C., Ramdane-Cherif, A.: Design of an Incremental Learning Component of a Ubiquitous Multimodal Multimedia Computing System. WiMob 2006, 2nd IEEE International Conference on Wireless and Mobile Computing, Networking and Communications. 2006. Montreal, Canada : pp. 434-441.
43. Hina, M.D., Ramdane-Cherif, A., Tadj, C., Levy, N.: Infrastructure of a Context Adaptive and Pervasive Multimodal Multimedia Computing System. Journal of Information, Intelligence and Knowledge, 2009. 1(3): pp. 281-308

Self-healing for Autonomic Pervasive Computing

Shameem Ahmed[1], Sheikh I. Ahamed[2], Moushumi Sharmin[1], and Chowdhury S. Hasan[2]

Abstract To ensure smooth functioning of numerous handheld devices anywhere anytime, the importance of a self-healing mechanism cannot be overlooked. This is one of the main challenges to growing autonomic pervasive computing. Incorporation of efficient fault detection and recovery in the device itself is the ultimate quest but there is no existing self-healing scheme for devices running in autonomic pervasive computing environments that can be claimed as the ultimate solution. Moreover, the highest degree of transparency, security and privacy should also be maintained. In this book chapter, an approach to develop a self-healing service for autonomic pervasive computing is presented. The self-healing service has been developed and integrated into the middleware named MARKS+ (Middleware Adaptability for Resource discovery, Knowledge usability, and Self-healing). The self-healing approach has been evaluated on a test bed of PDAs. An application has been developed by using the service. The evaluation results are also presented in this book chapter.

Key words: Autonomic computing, Pervasive computing, Autonomic pervasive computing, self healing for pervasive computing

1 Introduction

Ubiquitous computing [1], also known as pervasive computing, has evolved during the last few years due to the rapid developments in portable, low-cost, and lightweight devices. It extends human thought and activity as well as provides a pragmatic world augmented by the behavioral context of its users [2]. Pervasive computing environments focus on integrating computing and communications with the surrounding physical environment for making computing and communication transparent to the users. It aims to design and develop models for next generation computing. Remarkable increases in the number of mobile device users, tremendous developments in wireless and mobile technologies, and low cost availability of handheld devices have contributed to the rapid evolution of this computing platform. In this environment things around us are expected to be able to communicate with each other and at the same time have the capability to collect, process, and transport information [14].

University of Illinois, Urbana-Champaign, IL, USA · Marquette University, 1313 W Wisconsin Avenue, Milwaukee, USA
Email:{sahmed02, iq, msharmin and chasan}@mscs.mu.edu

A.V. Vasilakos et al. (eds.), *Autonomic Communication*, DOI: 10.1007/978-0-387-09753-4_11,

Systems that have the ability to manage themselves and dynamically adapt to change in accordance with policies and objectives are termed as autonomic computing. This system enables computers to identify and correct problems often before they are noticed by the user.
Autonomic systems have the capability to self-configure, self-optimize, self-protect, as well as self-healing. Systems that have these characteristics are termed as self-managing systems. Autonomic pervasive computing [3] maintains characteristics from both autonomic computing and pervasive computing environment. Like pervasive computing, devices running in this area should be context and situation aware and these devices form an ad-hoc ephemeral network. These devices are also expected to have the ability to self-optimize, self-protect, self-configure, and self-heal.

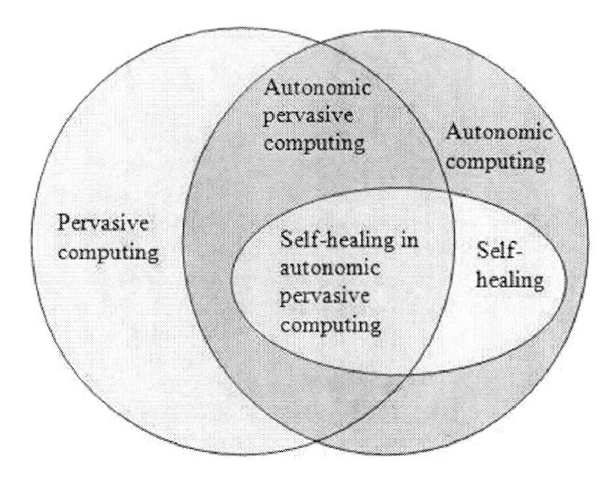

Fig. 1 Scope of self-healing in autonomic pervasive computing

Self-healing describes devices or systems that have the ability to perceive those are not operating correctly and, without human intervention, make the necessary adjustments to restore them to normal operation. A system that continues its operation even in presence of faults is termed as a fault tolerant system [4]. The concept of self-healing goes beyond fault tolerance since it also provides the device with the capability of recovering from fault by itself or with the assistance of other devices present in the network. Fig. 1 depicts the scope of self-healing autonomic pervasive computing.
Considerable research has already been done for fault tolerance in distributed dependable real-time system [5]. Some solutions along with prototype for pervasive computing fault tolerant systems have been proposed [6]. Self-healing autonomous systems are also addressed in [7]. But no solution has been proposed for a self-healing system in autonomic pervasive computing yet, let alone the implementation of such a system. Since the future technology trend lies in pervasive computing, it is of the utmost importance to have an efficient, transparent, and secure self-healing system. Currently, we are developing middleware named MARKS [8, 9, 10, 11] which is suitable for embedded devices running in pervasive computing environments. The Self-healing unit plays a vital role from the above perspectives.
In this chapter, we present an efficient, secure, and transparent self-healing model. Our model is designed for the autonomic pervasive environment, where we assume that the mobile devices would be able to handle necessary computations and communications by themselves without any fixed infrastructure support. We developed its first prototype on a test bed of PDAs, which are connected with short-range ad hoc wireless. Any healing approach will be in vain without proper setup of fault detection and recovery. Efficiency should also not be over-

looked. Our self-healing approach is unique from those perspectives too. A modified secret sharing approach [12], not only to cope with the limited storage capacity of the embedded devices but also to guarantee the security, is being used in our approach. This model provides the third feature (transparency) by performing most of the healing process without users' interference. The contributions of this book chapter are as follows:

1. *Classification of autonomic pervasive computing environment:* We have classified this environment into three categories from infrastructure perspective. Some environments have fixed infrastructure, some environments don't have any infrastructure and some environments follow the hybrid approach. We mainly concentrated on the infrastructure less autonomic pervasive computing.
2. *Classification of fault:* Fault has been defined and classified in different ways in different situation. We have classified from the light of autonomic pervasive computing. We have classified into 4 categories: Hardware Fault, Software fault, Prioritized fault, and Communication Fault.
3. *Attributes of self-healing system for autonomic pervasive computing:* We have proposed some solutions for fault detection, fault notification and faulty device isolation. These solutions have the following attributes:
   a. Less Memory footprint: Our system doesn't require much memory.
   b. Less Time complexity: The overall time complexity of our system is O(n).
   c. Transparent: Our system involves nominal user intervention to do fault detection, notification, and faulty device isolation.
   d. Solution for Infrastructure less system: Our system concentrates only at infrastructure less system.
   e. Scalable system: Even with the increase of the number of the devices in autonomic pervasive computing environment, our solution works fine.

The outline of this book chapter is as follows: Section 2 contains the Motivation. Characteristics of our model with motivations behind designing the system are presented in Section 3. An overview of our proposed approach is illustrated in Section 4. The details of the models are described in section 5. The attributes of our proposed model is presented in section 6. Section 7 contains the current state of the art. The implementation details along with evaluation are provided in section 8. We conclude with some novel research directions of future work in section 9.

2 Motivation

The autonomic pervasive computing area is now strong and powerful as the use of handheld and wearable devices is increasing in a rapid rate. These devices are capable of communicating with other devices and run various applications like powerful devices. People are using these tiny mobile devices all the time and everywhere. If these devices operate incorrectly or prompt users for each little malfunctioning, then the usability of these devices will reduce. Hence, self-healing comes into play as it helps to execute a system uninterruptedly. Self-healing is an integral part of autonomic computing systems. Here, we consider some scenarios to show the importance and applicability of self-healing in autonomic pervasive computing environment.
Scenario 1. A group of high school students appear in a wireless examination. After getting the questions in their PDAs from their teacher Dr. John's PDA (let X, the healing manager), they start their tests. During the exam, all on a sudden, one student's PDA (let Y) starts unusual behavior.
Scenario 2. Returning from a visit of a museum, a group of high school kids want to share their experience (stored in their PDAs) to enrich their knowledge. One device of the network is having a high probability of going down and the device owner wants to store some of its

important information for future use in other devices. The above scenarios present situations where healing is needed. But these scenarios raise the question that do we need to inform the user each time a device is having problem? Can we fix some of these problems transparently and securely? Are we sacrificing security and efficiency issues?

The first scenario can occur in any classroom. This problem can be solved by calculating the rate of changes of all of the status of the faulty device. If the device (Y) finds out that the fault is due to the malfunction of a running application, then without any delay, it can send an SOS message along with the answer files. The teachers' device (X) can isolate Y from the entire network by removing all entries of Y as a service provider. By this time, Y can inform the device owner about the problem. By using the system interrupt, it can kill the problem causing application.

In the second scenario, to avoid the loss of data stored in that device, the healing manager can disseminate the stored information to the remaining devices in a secure manner. By consulting the logbook, necessary measures can be taken to restore the device's prior working state. The disseminated information content can be used later to restore the device to help it to work to its full extent.

A self-healing model can solve the issues of the above situations. To cope with the challenges presented by the pervasive ad-hoc network, we feel that a self-healing model should be lightweight, energy-efficient, and infrastructure less. In this book chapter, we present a self-healing model, which is efficient, secure, and works transparently.

3 Characteristics of Self-healing Model

A self-healing model targeted for autonomic pervasive computing should have the following attributes:

1. Infrastructure less. No infrastructure support (powerful servers, proxies, etc.) should be required. If the focus is on truly pervasive environments then the model should work independently without any external support as in this environment infrastructure support is not always available.

2. Lightweight. The model should be lightweight in terms of executable file size.

3. Non-degradable performance. The model should not put much overhead on the performance of the device.

4. Energy efficient. Self-healing models should be energy efficient. It should not require much battery power for computation or communication purposes.

5. Transparent. The main idea behind designing autonomic systems is transparency. Every self-healing model should have some level of transparency. However, it should also inform the user about critical system information.

6. Secure. Self-healing systems require information distribution and backups to recover from faults. The information distribution and storage should be highly secure to maintain user privacy and security.

4 Design Overview

4.1 Self-healing System of Autonomic Pervasive Computing

Our self-healing service is an integral part of MARKS+, which is the extended version of MARKS [3] (Middleware Adaptability for Resource Discovery, Knowledge Usability, and

Self Healing) which is our developed middleware that incorporates different kind of services. The fault detection, fault notification, and faulty device isolation are taken care of by the healing manager of the self-healing service using system monitoring unit of MARKS+. It uses the system monitoring unit. Fig. 2 shows the MARKS+ architecture along with the self-healing service.

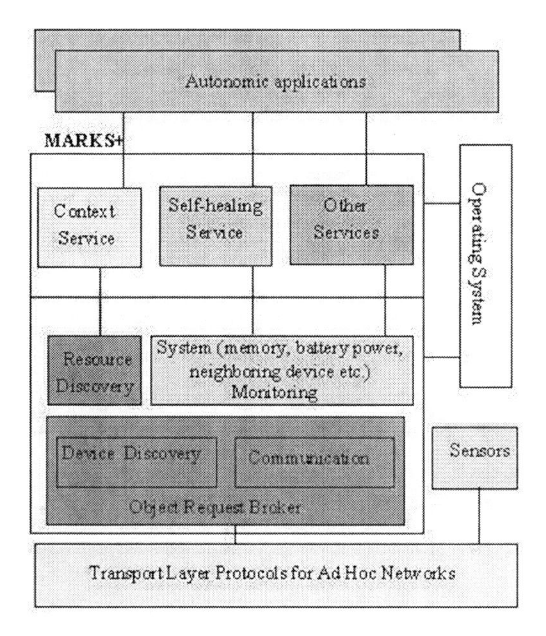

Fig. 2 MARKS+ architecture for autonomic pervasive computing

An effective self-healing service should address the following challenges:

- No regular functionality of the network will be hampered due to any fault of any device.
- All significant information of the faulty device should be preserved in secure fashion.
- The device will be facilitated to heal its fault by itself or at best with the assistance of other devices of that network.
- After reviving, the faulty device should be able to regain its previous states in such a way that it should feel there was no fault.

To address the above challenges in an apposite manner, our proposed self-healing pursues quite a few steps:
1. Fault detection
2. Fault notification
3. Faulty device isolation
4. Fault healing

Fig. 3 shows the architecture of our proposed self-healing service. The fault-detection and fault-notification unit is used by each device running MARKS+. The fault detection unit may be utilized by both the healing manager and the device itself. But the other three units (faulty device isolation, alteration, information distribution) should be handled by the healing manager. In this book chapter, we only concentrate on three steps: fault detection, fault notification, and faulty device isolation.

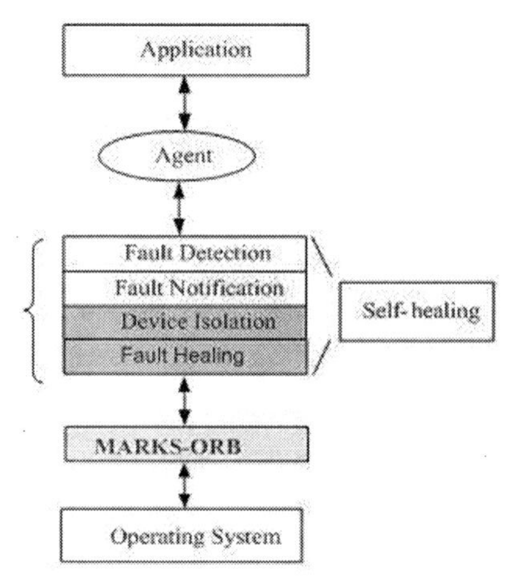

Fig. 3 Architecture of our proposed self-healing service

4.2 Classification of Fault

Fault can be classified from different perspectives:

Hardware Fault

This fault is mainly caused by physical constraints of the devices running in autonomic pervasive computing environment. In our research, we include the following faults as hardware faults:
a. Low battery power: Battery is the main power source for mobile devices. This fault occurs due to insufficient battery power.
b. Limited signal strength: Wireless communication is one of the main communication medium for mobile devices. While the signal strength is low, in most cases, the mobile devices are not able to communicate with each other and the entire network starts malfunctioning.
c. Insufficient disk space: Sometimes mobile devices fail to work properly due to insufficient disk space. We are considering this fault as a hardware fault.
d. Byzantine fault [13]: These faults encompass those faults that are commonly known as "crash failures" and "send and omission failures." A system may respond in an unpredictable way, if there is any Byzantine fault.

Software Fault

It includes the following faults.
a. Application fault: Application fault occurs due to problem in any kind of software application that is running in the mobile devices in autonomic pervasive computing environment.
b. Middleware Fault: Middleware is the software layer between the operating system and the applications. It plays a very important role in autonomic pervasive computing environment. Any fault that occurs due to problem in middleware is called middleware fault.

c. OS fault: There are some faults that occur in OS level. These faults are named as OS fault. Software fault follows the pyramid approach described in Fig. 4.

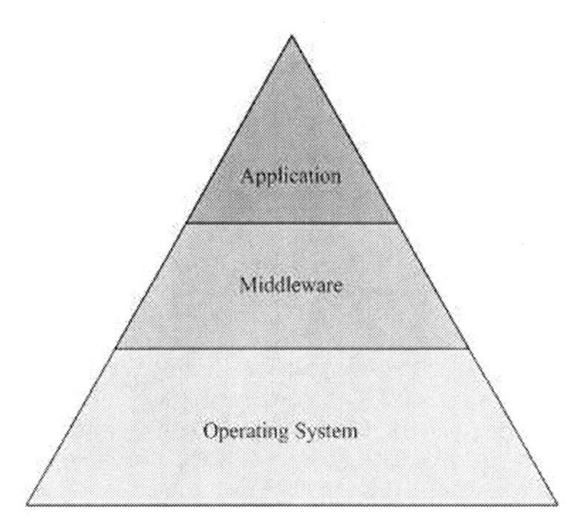

Fig. 4 Scope of software fault for autonomic pervasive computing

Prioritized Fault

According to priority, we have classified faults into the following categories:
a. Low priority fault
b. Medium priority fault
c. High priority fault
The healing manager has a queue for different kind of faults reported by devices. The priority has been assigned to help the healing manager. High priority faults should be handled first, then medium priority faults, and finally low priority faults.

Communication Fault

All types of fault related to wireless communication belong to this category. Network failure is one kind of communication fault.

4.3 Fault Detection

High-quality fault detection, the first step of a self-healing process, not only prevents loss of resources but also lessens healing time. To ensure supreme-quality, our proposed Self-healing approach periodically monitors as well as assembles the status of all of the running applications, memory, power, communication signal etc. Drastic changes in those values will generate faults. By using the rate of change of these over time, it tries to figure out the existence as well as the cause of a fault, if there is any.

4.4 Fault Notification

How can a device inform the healing manager that it is in fault? If the device is in fault then is it really possible to inform of the fault? In a distributed computing system, heart beat message and absence of heartbeat massage are used to notify of the fault. In autonomic pervasive computing, we have applied the similar concept but in a modified way. We have introduced different kinds of message systems like OK message, SOS message etc. to inform the healing manager not only of the presence of a fault but also the cause of the fault.

4.5 Faulty Device Isolation

The isolation of faulty devices from the remaining network, a grand challenge of fault tolerant as well as self-healing system, can be achieved in self-healing in a very simple way. In case of MARKS+, every device is mapped with another one by means of service availability in these devices. It is adequate to remove the entry of the faulty device from that mapping to ensure its isolation from the entire network. The detailed description of our approach is presented in the following section.

5 Self Healing in Autonomic Pervasive Computing

Self-healing is the process of detecting faults, notifying those, and also to recover from the faults without human intervention. This system makes necessary adjustments to restore the devices' normal operating condition. Fig. 5 portrays the high level view of a self-healing system.
According to Fig. 5, all the faults reported from various devices will be compiled first and then will be stored in the fault queue of the healing manager. According to the fault priority, the healing manager decides which fault should be handled first. Then the healing manager will do the analysis of the fault, isolate the faulty device from the existing network, and will also do the service alteration and information distribution. In our approach, we mainly concentrate on first three steps of self-healing: fault detection, fault notification, and faulty device isolation.

5.1 Fault Detection

Fault detection is the first step of any self-healing system. Detecting any kind of fault in a device is normally known as fault detection. Here is the formal definition of fault detection:
Device Status: Let $Z_t(x)$ be the status of a device at time t, where x represents an arbitrary input vector [e.g. rate of change (dy/dt) of power, memory, communicational signal etc. over time]
Test: $T = \{v_1, v_2, \ldots, v_n\}$ where $\{v_1, v_2, \ldots, v_n\}$ are input vectors and $Z_t(v_i)$ represents the status of the device.
Fault Detection: T detects a fault in the device iff $[(Z_t(v_i) \sim Z_{t+1}(v_i)) > predefinedthresholdvalue]$
For example, if the change of signal strength is 10% at time t and change of memory space is 25% at time $(t+1)$ and if the threshold value is 12% then as the rate of change of memory space is greater than the threshold value, according to our approach, there should be a fault.

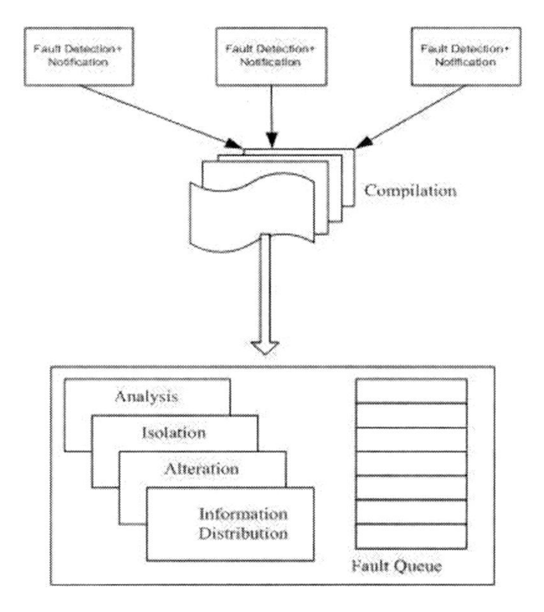

Fig. 5 Scope of self-healing in autonomic pervasive computing

Two Way Message Based Fault Detection. According to Fig. 6, each device (except the healing manager) will send OK/SOS message after a specific time period. This period is called the heartbeat period (H_p). For each device, it takes Δt_i time, where i is the device number. So, the healing manager will get the OK/SOS message after $(H_p + \Delta t_i)$ time. Not only that, the healing manager itself will periodically (ΔT) broadcast another message to inform all the existing devices about its aliveness. If the healing manager doesn't get any message from any device within a specific time period then the healing manager will assume that the device is faulty and it might need some help. On the contrary, if the healing manager doesn't broadcast its message, all the devices along with the service manager will understand that the healing manager itself is in fault and the service manager will take the responsibility of the healing manager.
K. Mills *et al.* [15] described two-way heart beat based failure detection technique for distributed system. Our environment is different than pure distributed system. We concentrate only on the autonomic pervasive computing environment.

Network Bandwidth Consumption. For a small network in autonomic pervasive computing environment, there won't be so many devices and hence the bandwidth consumption might not be prominent for the network itself. However, if we consider a large network, we should have such a system which would be able to utilize the full network bandwidth. If the message size to send to healing manager is M_D and the message size of the healing manager to device is M_H then the system consumes bandwidth, B which is

$$B = (N \times M_D + M_H)/Hp$$

To keep this consumption stable, the healing manager must process N/H_p messages. Since the network is ad-hoc in nature and a device can join or leave at any time, the system should adopt the size of the heartbeat period (H_p). The algorithm to adjust H_p is presented below.

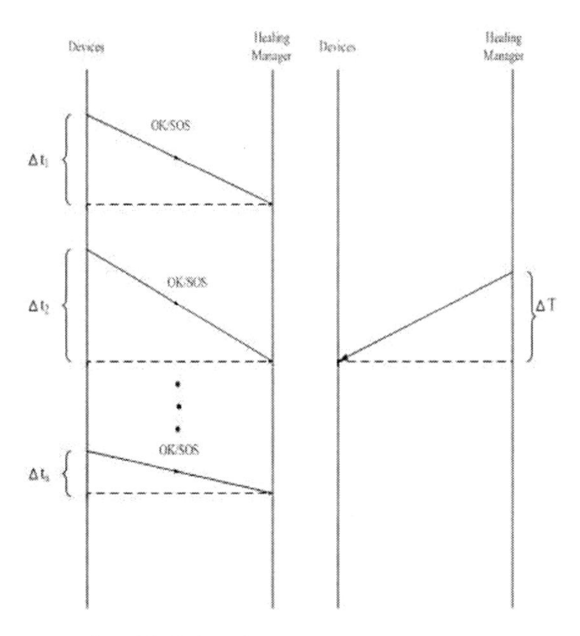

Fig. 6 Two way message based fault detection

5.2 Fault Notification

Not only to push any information but also to notify its aliveness or its fault, the device itself needs to communicate with the healing manager periodically. In Gaia [6], a heart beat message mechanism is used only to indicate the aliveness of any device. The absence of a heart beat message implies the existence of a fault. Thinking ahead a little bit more, we have incorporated a generic message passing scheme not only to facilitate the function of a heart beat message but also the efficacy of an SOS message for helping the healing manager to be informed about the faulty device's current situation. In this scheme, each device will send any one of the following messages to the healing manager:
1. OK message: It simply sends a packet containing "OK" string. It's nothing but a heart beat message.
2. SOS message: After identifying any fault in its own device, the self-healing unit of that device sends SOS message which may include some file names along with that message. An example of such type of message is: "SOS, exam3cosc060, log status". This means that the faulty device is requesting to save files named "exam3cosc060" and "logstatus".
If the healing manager doesn't get such a message for a pre-fixed threshold period of time, right away it will commence the next steps (device isolation, information distribution, alteration) assuming that the device is in fault. If the healing manager gets SOS message along with some file names, then healing manager will initiate to get the files from the device and will store those among other devices in a secured distributed manner. Fig. 7 shows the flow diagram of fault detection and fault notification.

```
procedure AdjustHeartbeat (N, Nmax, C, Hmin, Hmax)
N = Total number of devices (except the healing manager)
Nmax = Maximum allowable number of devices
C = Bandwidth capacity
Hp = Heartbeat period
Hmin = Predefined minimum value of heartbeat message
Hmax = Predefined maximum value of heartbeat message

   L1. if a new device joins then N++;
   L2. if an existing device leaves then N--;
   L3. if (N>Nmax) then
           N--;
           return
   L4. if (N<1) then
           N++;
           return
   L5. Hp = N / C
   L6. Broadcast Hp to all the existing devices
   end procedure
```

5.3 Faulty Device Isolation

We have deployed a very simple approach to isolate a faulty device from the remaining network. In our approach, each device is mapped with another device based on service availability. So if we can remove the entry number from the mapping list, the faulty device can be isolated from the entire network.
Table 1 shows the mapping of the service # and the service provider. We have followed a standard for the service #. For example, service # 1 means "Internet service", service # 2 means "office software", service # 100 means "music software", etc. Self-healing approach incorporates a list named serviceList by which the service can easily be identified. For example, serviceList (100) will return "music software". Table 2 exemplifies the three-dimensional mapping of service provider, service consumer, and service #. Here, D1 means Device 1, D2 means Device 2, etc. These mappings are implemented in our proposed self-healing by using a hash table. By means of a standard "remove" function, our self-healing system can remove the entry of a faulty device from the hash table as well as from the entire network.

5.4 An Illustrative Example

To make our proposed self-healing system more comprehensible, here we are giving an illustrative example. In a wireless exam system, a teacher will conduct an exam. The students will give the answers and will send the answers to the teacher's PDA. The requirements for teacher and students are as follows:

Requirements from Instructor point of view:

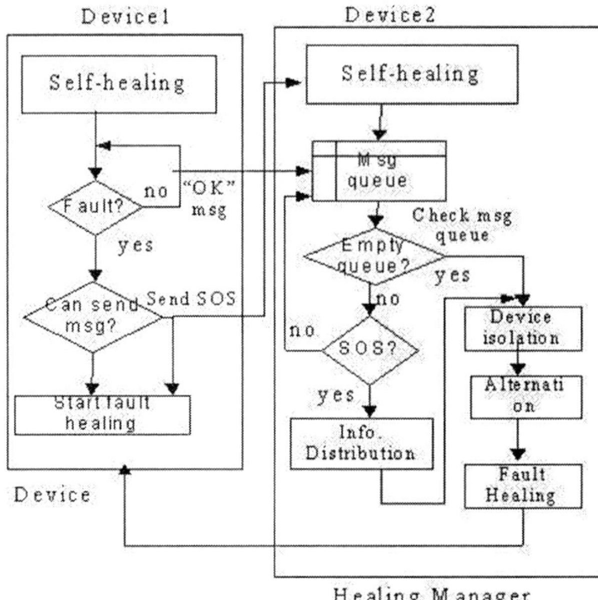

Fig. 7 Flow diagram of self-healing for autonomic pervasive computing

Table 1 Mapping of Service # and service provider

Service #	Service provider			
1	D1	D7	D9	D3
2	D19			
3	D12	D12	D2	
4				
5	D1	D3	D4	D9
.....	D20			
100	D2	D9	D18	

Table 2 Mapping of Service # , service provider and service consumer

Service provider	Service consumer						
	D1	D2	D3	...	D9	..	D20
D1		1			1, 5		
D2			100		100		
D3	5						
D9		1,5					5
..							
D20	15		99				

1. The system should enable the instructors to enter a certain number of questions. We assume that the instructor will be teaching other courses as well and hence may plan to create questions for the other courses concurrently. Hence, one set of questions should not overlap with others. To ensure this, all the questions for a particular course should be grouped together and stored in a data structure or database (whatever is efficient on a PDA).
2. The system should facilitate the creation of exams. It can have any number of multiple choice questions, true/false, short answer (fill-in) and open ended response options.
3. The instructor should be able to create a suitable "answer key" for the questions. Ideally, an instructor will view the question and type the answer so that the answer is visible whenever the question is displayed. Teachers may want the correct answer to be sent to the student after grading is complete.
4. The system should have the capability to do the auto grading (multiple choices, fill in

blank, and true-false) and generate result (text, table and graphical output) of varying format.
5. The distribution of the entire exam should take place when the instructor selects the designated action.
6. The system should also provide the facility to distribute the grades to the students.
7. The instructor would be the only authorized person to specify the time limit for the students to submit their answers. The students can commit their answers to the instructor before the expiration of the given time frame. Otherwise, the system should have necessary mechanisms to deny access to the questions after the pre-defined time is over, and commit all the answers to the instructors PDA identifying the students who answered questions.
8. The instructors PDA must be able to store the data received in an efficient way. The storage constraint imposes the prime challenge in this issue.
9. The instructor will collect feedback from the students during and after any exam. The instructor can exercise the option of filtering stored feedback for repeated use.
10. This system can be easily transformed into a survey instrument. The requirements for designing, administering and collecting survey instruments are easily identifiable by the instructor.

Requirements from Student point of view:
1. The system should facilitate answering of questions. They should be able to submit their answers within a pre-specified time limit.
2. Normally students get tensed up during any exam. It is very reasonable from their perspective to expect the PDAs to be reliable. The devices should not play any tantrums (like denying the students access to network, resetting in the middle of an exam, partial download of file because of unspecified reasons). An occurrence of a single event specified above will undermine the students and instructors confidence in the entire process. All these considerations obviously have to be taken care of.
3. The students should also able to submit their comments during and after the exam. While they would not be able to submit the answer outside the time limit imposed by the instructor, they can comment at any time at any topic related to the course and question.

Fault Detection in Wireless Exam. Each student's device is running our self-healing system. If a student's device becomes faulty (e.g. hardware fault, software fault, communication fault, etc.), it will be detected through our proposed system.

Fault notification in wireless exam. As soon as the student's device finds the problem, it will send OK or SOS message to the healing manager (here teacher's PDA). The healing manager will take necessary steps according to the message it gets from the faulty device.

Faulty Device Isolation in Wireless Exam. After identifying the faulty device, the healing manager will remove that faulty device from the current network temporarily so that the other devices of that network can work perfectly.

6 Attributes of Our Proposed Model

6.1 Efficiency

Our system is efficient from both memory and speed points of view. One replica of the healing manager is preserved in service manager. It does not consume much memory in this approach since it does not need to keep a replica of all the devices, while this is a common scenario

for most of other schemes. Moreover, MARKS-ORB and the fault detection and notification system occupy a very little space in the PDA.
According to the flow diagram of a self-healing service, there is a loop for message queue inside of which the fault detection and notification system lies. The overall time complexity of this system is $O(n)$. On the contrary, Device discovery and Communication, the two functions of MARKS-ORB are independent of each other and that is why the time complexity for MARKS-ORB is $O(n)$. This low time complexity makes our approach efficient from the speed perspective.

6.2 Transparency

Fault detection and notification tries to assure the nominal involvement of the user. In most cases, without any kind of user interruption, the healing manager can detect a fault. It involves the users only when decision largely depends upon the users' preference.

6.3 Infrastructure less

Our system doesn't require any fixed infrastructure to run. All the algorithms are running on different mobile devices (e.g. PDA, cell phone, etc.) using wireless communication technology.

6.4 Non degradable performance

Our system doesn't create any adverse affect on the devices. We have proved those from the perspective of different performance measurement metrics like battery power, memory space, and signal strength. The details have been described in the evaluation section.

7 Related Work

Self-healing is an essential component of every computing system. It is an integral part of devices running in autonomic pervasive environment, as the main focus of this area is to make users free of operating details. It is a widely researched topic in the field of distributed computing, grid computing, and autonomic computing. In each of these areas, many schemes are proposed that attack this problem from various standpoints. Jean-Philippe Martin-Flatin proposed a modified organizational model for e-business information systems which is capable of self-management [16]. Distribution of event correlation in this self-managed system has also mentioned. Researchers are working on several policies like architecture-based system [17], [18], infrastructure based approach [19] for long. Most of these proposed models are suitable for physically connected computers in distributed computing environment. There is no established method that provides a solution for devices running in an autonomic pervasive computing environment where it is assumed that the devices are connected to each other wirelessly. The autonomic computing field demands that devices should be able to self-configure, have self-healing, self-optimization, and self-protection capability. On the other hand, the main attributes of pervasive computing are context-awareness, situation-awareness, and ad-hoc networking. The focus of this work is self-healing for autonomic pervasive computing environment that contains all the above-mentioned characteristics.

Soila and Priya [20] presented proactive recovery in Distributed CORBA applications. They did not concentrate on fault-prediction technique; rather they focused on the exploitation of fault prediction in systems that had real time deadlines. Our system deals with pervasive computing and it can predict a fault by calculating different states of the system. To handle transient software failures, a proactive approach named software rejuvenation, was proposed by Huang *et al.* [21]. According to this approach, if errors are accumulated beyond a threshold, then it will kill and re-launch the application. A lot of works about rejuvenation policies, to increase system availability and to reduce the cost of rejuvenation, were done by [22], [23], [24]. However, to hand-off the existing state of the faulty device just after its re-launching was overlooked here. Our approach preserves this state among other devices in a secured manner so that the healing manager can help the faulty device to get its actual state after healing.

Garlan and Schmerl presented a system [17] that uses an architectural model for monitoring, problem detection, and repair. In their model an executing system is monitored to observe its run time behavior. Monitored values are compared with the properties of an architectural model. The architectural model triggers constraint evaluation if the system is operating outside an envelope of acceptable ranges. These violations of constraints are handled by a repair mechanism according to the architecture and these changes are propagated to the running system. This architectural model is represented as a graph of interacting components where nodes are termed as components. These nodes represent the principal computational elements and data stores of the system: clients, servers, databases, user interfaces, etc. These nodes are interconnected by connectors, which are complex bases of middleware and distributed systems. These powerful device requirements make this model unusable in infrastructure-less pervasive computing environment.

Eric *et al.* [18] describes a system based on software architecture that uses software components and connectors for repair. To dynamically repair a system they concentrate on the current architecture of the system; to express an arbitrary change to that architecture; to analyze the result of the repair; and to execute the repair plan on a running system without restarting the system. To follow this approach they need complete information about the devices and software running in those devices. In a autonomic pervasive computing environment this type of information is not available. Their system is targeted for connected distributed environment where system level information is available. They also use infrastructure support for repair purposes. Gordon *et al.* [25] in their paper presented an analysis of the role of "Reflection" to support self-healing systems. They also suggested that middleware would be the appropriate place for including self-healing unit. They offered their primary analysis based on distributed systems. But they did not implement the middleware and also not the self-adaptive, self-healing unit. AMUN (Autonomic Middleware for Ubiquitous Environment) [7] is a middleware that deals with self-healing for ubiquitous environment. The AMUN self-healing and self-organizing unit consists of four main parts: the transport interface, the event dispatcher, the service interface and service proxy, and the autonomic manager. The autonomic manager is the principal unit for managing communication between other units. But the main concentration of this project is an indoor environment like inside an office building. They use "Smart Doorplates" that use and display situational information of the owner of the office. This idea restricts its use only in a smart environment.

L. Kant [26] proposed a self-healing mechanism for wireless networks. He claimed that this mechanism could provide seamless restoration of affected services due to random/sporadic network facility failures. Also this model considers a fixed telecommunications system only and concentrates on restoration of the faulty system in network layer. This approach is not suitable for pervasive computing since the ad-hoc nature is totally overlooked here. Y. Tohma in [27] described the challenges and proposed solutions to achieve fault tolerance in autonomic computing environment. This solution creates groups of similar devices and the neighboring

devices decide whether a particular device is faulty or not. He also proposed to keep three copies of information to recover fault. This proposed method is expensive and not applicable in an ad-hoc environment as in this environment the group formation is difficult and neighbors are changing very frequently.
Mills *et al.* in [15] presented an algorithm for fault detection for autonomic systems. A two-way heartbeat failure-detection technique is used. A monitor is used to receive periodic messages from a number of monitorables. The time length to receive a message is used as the detection criteria for failure. This system is simple and is designed for connected distributed systems where the devices will not change. Device mobility issues are also ignored here.
Brown and Redlin in [28] proposed a benchmarking system for measuring the effectiveness of self-healing mechanism in autonomic systems. They concentrated on powerful systems that are rarely present in pervasive ad-hoc networks.
In [29], Poladian *et al.* proposed a task-based adaptation technique for an ubiquitous computing environment. This system supports heterogeneity, resource variability, mobility, ubiquity, and task-specific user requirements. This self-adaptation infrastructure has three distinctive features that allow explicit representation of user tasks and provides an environment management capability to translate user tasks and also provides a formal basis for understanding the resource allocation, and support optimal allocation at run time. Though this system is targeted for pervasive environments, it needs infrastructure support for fulfilling all the above activities which makes it unusable for truly mobile pervasive environment.
In [6], Chetan *et al.* have classified faults, pointed out various research challenges, and have also proposed solutions to some of the challenges in a pervasive environment. Their fault tolerant system uses context information to tolerate application and device faults. They considered a fail-stop fault model consisting of device and application faults. In this approach, if an application fails, it is restarted. The fault model only considers application failures caused by device failures, network faults, and failures due to faulty usage. This system is designed for active spaces.
Umesh Bellur and N. C. Narendra [1] presented a programming model named reconfigurable programming for run time binding of system components. A middleware architecture has also been designed. All together it gives self configuration capability for systems in distributed computing.
Keller, A. and Badonnel, R. [2] proposed a provisioning system with BPEL4WS workflow language that automates the deployment, installation and configuration of application services. In systems like e-Commerce where the work load varies rapidly an automated system for handling these issues is highly required. But this issue is not quite similar in pervasive computing environment.
Most of the work in the self-healing area is done either for distributed systems or for active/smart spaces. The area of autonomic pervasive computing is void of powerful device support. The devices running in this environment are resource poor and these are expected to handle the necessary computations and communications with limited battery and processing power. These features make self-healing in autonomic pervasive computing a hard and unique problem to solve. To the best of our knowledge, there is no work till now in the area of self-healing for autonomic pervasive computing.
A comparison table of the existing self-healing and fault tolerance models is presented in Table 3.

8 Evaluation

We have evaluated the performance and usability of our self-healing model by implementing a prototype, designing applications and using simulation tool. An application has been designed that uses our self-healing model. We have also measured the battery power consumption and the signal strength after using our model.

Table 3 Comparison of Self-healing and Fault Tolerant Models

Model	Infrastructure Support Needed	Transparent	Secure	Privacy Aware	Lightweight	Smart Space Needed	Consider Mobility
Proactive recovery [20]	Yes	Yes	Yes	N/A	No	No	No
Architectural Model [17]	Yes	Yes	N/A	N/A	No	No	No
Software Architecture [18]	Yes	Yes	N/A	N/A	No	No	No
AMUN [7]	N/A	Yes	Yes	Yes	No	Yes	Yes
SHWN [26]	Limited	Yes	Yes	No	Yes	No	Yes
FTA [27]	Limited	Yes	N/A	N/A	Yes	No	Yes
FDA [15]	No	Yes	Limited	N/A	Yes	No	No
Our Approach	No	Yes	Limited	Yes	Yes	No	Yes

8.1 Prototype Implementation

A prototype including fault detection, fault notification, and faulty device isolation for autonomic pervasive computing environment has been developed and integrated along with our current developed middleware named MARKS+. WinCE running on a set of Dell Axim X30 pocket PCs (Process type is Intel@PXA270 and speed is 624 MHz), to demonstrate our approach, is used as platform. The .NeT Compact framework along with C# is used as implementation language. Bluetooth, as the underlying wireless protocol, has been used though it is also suitable for IEEE 802.11.
To evaluate the performance of our self-healing model, several applications have been developed. Wireless exam is such an application by which one teacher can send questions to the students (PDA to PDA communication) and also can collect the answers from the students. There are some selected screen shots captured from the implemented prototype below.
Fig. 8 presents a log file stored in an embedded device (a pocket pc exploited in the application which used the first prototype of fault detection, fault notification, and faulty device isolation). Such a log file is really necessary for a self-healing system. Fig. 9 illustrates the nature of rate of change of used memory space over time through which self-healing can determine the possibility of a problem between time stamp 14 and 15 due to the sharp change of rate of used memory space.
Fig. 10 presents the status of the battery power for five devices where the prototype of our proposed system and MARKS+-ORB are running. The sharp change of the status of the battery power for D4 indicates that there is some problem in that device and needs healing immediately. By using the "status changing rate" process, self-healing of the student's device itself tries to find out the fault as well as the reason if there is any. Fig. 11 shows a typical message generated by the devices' healing unit. This message is intended to inform the user about the abrupt change of device status and action taken by the self-healing. By using the "status changing rate" process, self-healing of the student's device itself tries to find out the fault as well as the reason if there is any. Fig. 11 shows a typical message generated by the devices' healing unit. This message is intended to inform the user about the abrupt change of device

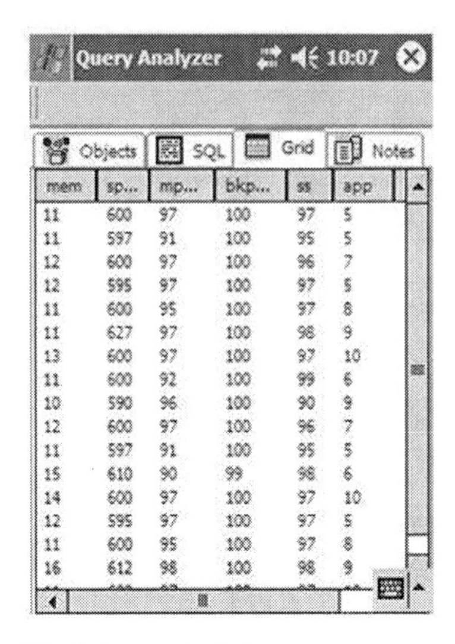

mem	sp...	mp...	bkp...	ss	app
11	600	97	100	97	5
11	597	91	100	95	5
12	600	97	100	96	7
12	595	97	100	97	5
11	600	95	100	97	8
11	627	97	100	98	9
13	600	97	100	97	10
11	600	92	100	99	6
10	590	96	100	90	9
12	600	97	100	96	7
11	597	91	100	95	5
15	610	90	99	98	6
14	600	97	100	97	10
12	595	97	100	97	5
11	600	95	100	97	8
16	612	98	100	98	9

Fig. 8 Status of a device

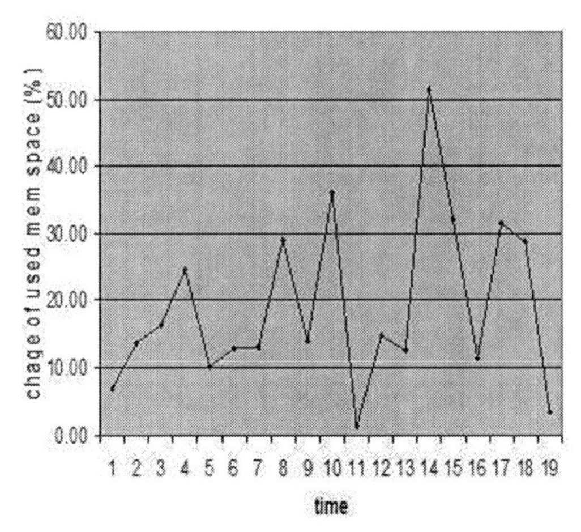

Fig. 9 Rate of change of used memory

status and action taken by the self-healing. There are situations where all the devices operate without any error. Then only the "OK" message is sent to the healing manager periodically. Suppose device 2 has no problem and it periodically sends OK message to teacher, the healing manager of this network. Within a specified period of time, it also sends all the answers to the teacher's PDA. This scenario is shown in Fig. 12.

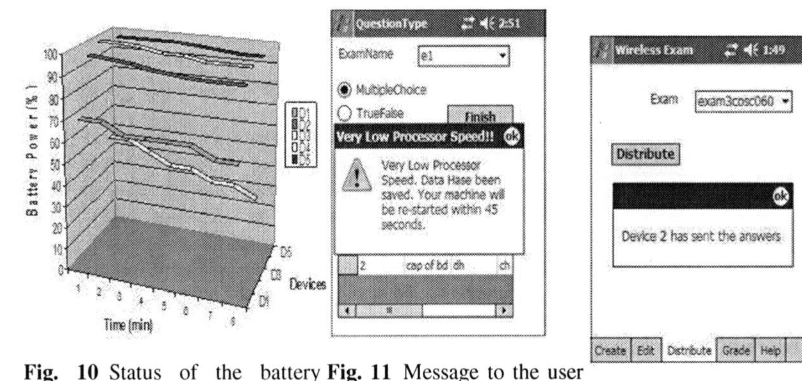

Fig. 10 Status of the battery power of five devices after using proposed system

Fig. 11 Message to the user regarding low processor speed

Fig. 12 Device 2 is running without any problem

Now device 9 finds some problem. It simply sends SOS message including file name exam3cosc060. Without any delay, the healing manager will collect that file from this device. Fig. 13 portrays this event. Another case can occur where the device is unable to send any message due to fault. If the healing manager does not get any message for a long period of time from any device, it will take appropriate action assuming that the device is in fault. Device 5 is unable to send any message for a long time. So, the healing manager takes rapid action regarding device 5. This incident is illustrated in Fig. 14. As both device 5 and 9 are faulty now, the healing manager removes the entry of device 5 and 9 from hash table. It also updates the table to reassign the services. Along with the SOS message, device 9 sends the list of important file names (selected by the user of that device) that need to be saved. This is shown in the above Fig. 15.

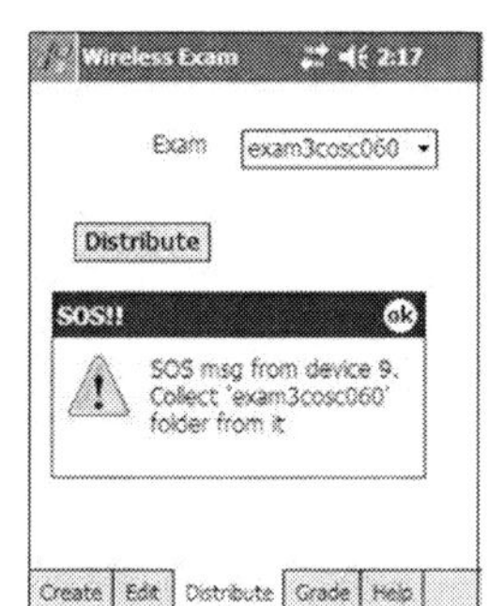

Fig. 13 Device 9 sends SOS message

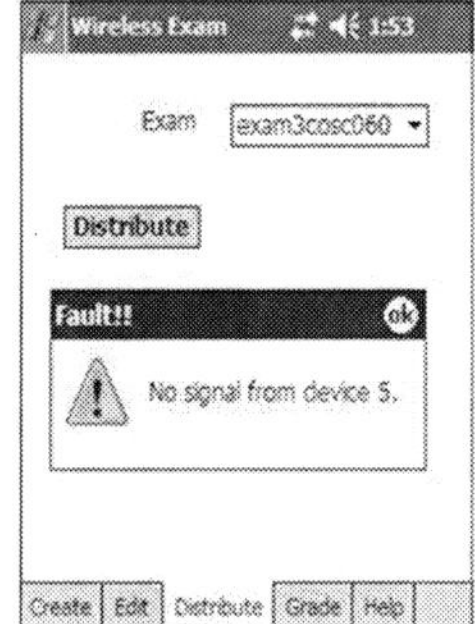

Fig. 14 No signal from device 5

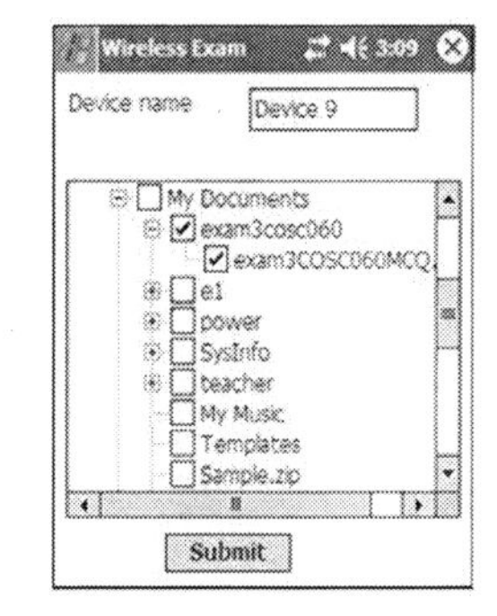

Fig. 15 Important files selected by user

Device 5 and 9 will try to be healed without the help of others. After healing, the healing manager will resend the files that it got before their fault.
A simple yet powerful authentication scheme is provided in our prototype. A random combination of 7 digits is used as the secret code to operate the self-healing unit. To break this authentication system, even if one full secret code is entered within 1 second, it will take approximately 9 years.

8.2 Performance Measurement

We have also developed the MARKS+-ORB, to provide the device discovery and communication functionality of the devices. Fig. 16 shows the battery power consumption with respect to time while MARKS+-ORB is running in that device. Here, S1 indicates the battery power consumption while the Pocket PC is on but the wireless mode is off (no wireless communication via 802.11 or Bluetooth). S2 means that wireless mode is on. S3 indicates that wireless mode is on and MARKS+-ORB is running in the device. It clearly indicates that MARKS+-ORB consumes a very little battery power.
MARKS+-ORB itself transfers data mainly for device discovery. It broadcasts its own IP address and receives the IP addresses of other devices reside in the same ad-hoc network. Fig. 17 shows the data transmission by MARKS+-ORB in every 5 seconds. It clearly shows that it does not need to transmit so much data for device discovery.
Table 4 shows the line of code and the size of the executable file of Self-healing Service and MARKS-ORB+.

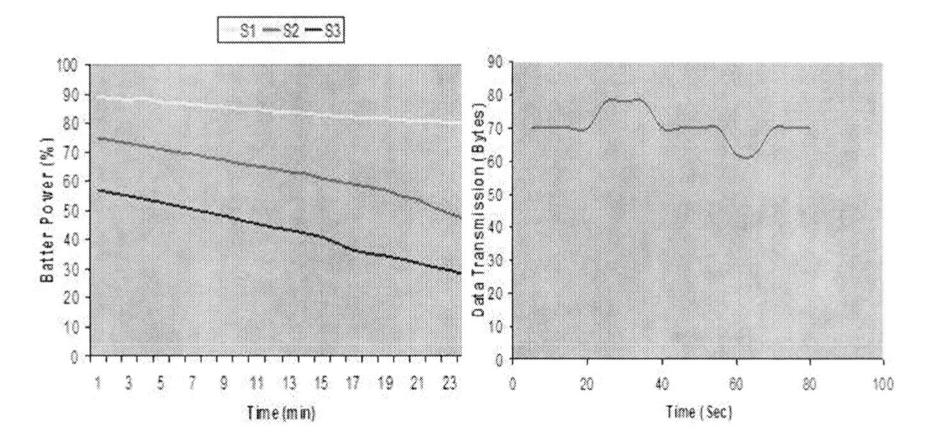

Fig. 16 Time (min) vs. battery power

Fig. 17 Rate of change of used memory

Table 4 Memory Footprint for Our Proposed System and MARKS-ORB

	Line of code (LOC)	Executable file size
Self-Healing Service	1001	36 KB
MARKS-ORB	416	7 KB

8.3 Application that Uses Self-healing Model

We have developed some applications that use a self-healing model as a service. We have used applications to check the practicality of using our proposed self-healing model. The overhead of running this service is also tested by running these applications.

9 Conclusion and Future Work

We have proposed a solution for fault detection and notification and faulty device isolation in autonomic pervasive computing. The fault detection process promises not only the least possible time to detect the fault but also the lowest degree of user intervention and hence makes our solution highly transparent. We have introduced the concept of healing manager in this regard. Though some researchers have already proposed different solutions for fault-tolerance from a distributed computing and autonomic computing perspective, no solution has been proposed yet for autonomic pervasive computing. Researchers have just started addressing self-healing issues in pervasive computing, let alone autonomic pervasive computing. Our solution is a unique one in this regard. We have developed our fault detection, fault notification and faulty device isolation system as a part of MARKS+. MARKS+ is a middleware and the extended version of our previously developed MARKS. MARKS is for pervasive computing and MARKS+ has been developed for autonomic pervasive computing.

At present we are focusing on the last step which is fault healing. The main idea is to store all the crucial information including log status file of the faulty device when the device falls into trouble. After recuperating from a fault, the healing manager re-collects all information and sends it to that device including so that the device can restore easily its previous condition. Information Distribution and Alteration are the two other things under our consideration. Information Distribution is the process of distributing the important information among other existing devices to assist a faulty device for keeping all the important information safe and secured. Secret sharing (N,t) approach [12] can be used in this aspect. Alteration or Responsibility Re-assignment will be responsible for finding an alternate device that is available as well as compatible with the faulty device [6]. This process is needed for smooth functionality. Incorporation of some security features and finding benchmarks for selecting healing manager are also being considered.

References

[1] Weiser, M.: Some Computer Science Problems in Ubiquitous Computing. In: Communications of the ACM, vol. 36, no. 7, pp. 75-84 (July 1993)

[2] Roman, M., Hess, C., Cerqueira, R., Ranganathan, A., Campbell, R.H., Nahrstedt, K.: Gaia: A Middleware Infrastructure for active spaces. In: IEEE Pervasive Computing, pp. 74-83 (2002)

[3] O'Donnell, T., Lewis, D., Wade, V.: Intuitive Human Governance of Autonomic Pervasive Computing Environments. In: Proceedings of the 1st International IEEE WoWMoM Workshop on Autonomic Communications and Computing, pp. 532-536. Giardini Naxos, Italy, (2005)

[4] Garlan, D., Poladian, V., Schmerl, B., Sousa, J.P.: Task-based Self-adaptation. In: Proceedings of the ACM SIGSOFT 2004 Workshop on Self-Managing Systems, CA, USA (2004)

[5] Bracewell, T.D., Narasimhan, P.: A Middleware for Dependable Distributed Real-Time Systems. In: Joint Systems and Software Engineering Symposium, Falls Church, VA (2003)

[6] Chetan, S., Ranganathan, A., Campbell, R.: Towards fault tolerant pervasive computing. In: Workshop on Sustainable Pervasive Computing (Pervasive 2004), Vienna, Austria, (2004)

[7] Trumler, W., Petzold, J., Bagci, F., Ungerer, T.: AMUN - An Autonomic Middleware for the Smart Doorplate Project. In: UbiComp 2004, Nottingham, England, (September 7, 2004)

[8] Ahmed, S., Sharmin, M., Ahamed, S.I.: Knowledge Usability and its Characteristics for Pervasive Computing Environments. In: Proceedings of the 2005 International Conference on Pervasive Systems and Computing (PSC-05) in conjunction with The 2005 International Multi-conference in Computer Science and Engineering, pp. 206-209, NV, USA (2005)

[9] Sharmin, M., Ahmed, S., Ahamed, S.I.: SAFE-RD (Secure, Adaptive, Fault Tolerant, and Efficient Resource Discovery) in Pervasive Computing Environments. In: Proceedings of the IEEE international Conference on Information Technology, pp. 271-276, NV, USA (2005)

[10] Ahamed, S.I., Sharmin, M., Ahmed, S., Havice, M.J., Anamanamuri, S.: An Assessment Tool for Out of Class Learning using Pervasive Computing Technologies. In: Journal of Information, vol. 8, no. 5, pp. 751-768 (2005)

[11] Sharmin, M., Ahmed, S., Ahamed, S.I.: MARKS (Middleware Adaptability for Resource Discovery, Knowledge Usability and Self-healing) for Mobile Devices of Pervasive Computing Environments. In: Third International Conference on Information Technology: New Generations (ITNG 2006), pp. 306-313. Las Vegas, USA (2006)

[12] Secret Sharing
URL: www.cmpe.boun.edu.tr/courses/cmpe471/spring2003/download/cmpe47109-2003.ppt
[13] Schneider, F.B.: Byzantine Generals in Action: Implementing Fail-Stop Processors. In: ACM Transactions on Computer Systems, vol. 2, no. 2, pp. 145-154 (1984)
[14] Basu, J., Callaghan, V.: Towards a Trust Based Approach to Security and User Confidence in Pervasive Computing Systems. In: the IEE International Workshop, Intelligent Environments 2005 (IE05), UK (2005)
[15] Mills, K., Rose, S., Quirolgico, S., Britton, M., Tan, C.: An Autonomic Failure-Detection Algorithm. In: Proceedings of the 4th International Workshop on Software Performance (WoSP 2004), pp. 79-83, CA, USA (2004)
[16] Martin-Flatin, J.P.: Distributed Event Correlation and Self-Managed Systems. In: Proc.of 1st International Workshop on Self- Properties in Complex Information Systems, pp. 61-64, Italy (2004)
[17] Garlan, D., Schmerl, B.: Model-based adaptation for self-healing systems. In: Proceedings of the first workshop on Self-healing systems, pp. 27- 32, South Carolina, USA (2002)
[18] Dashofy, E.M., Hoek, A.V.D., Taylor, R.N.: Towards architecture-based self-healing systems. In: Proceedings of the first workshop on Self-healing systems, pp. 21-26, South Carolina, USA (2002)
[19] Appavoo, K.J., Hui, M.S., Wisniewski, R.W., Silva, D.D., Krieger, O., Soules, C.A.N.: An infrastructure for multiprocessor run-time adaptation. In: Proceedings of the first workshop on Self-healing systems, pp. 3-8, South Carolina, USA (2002)
[20] Pertet, S., Narasimhan, P.: Proactive Recovery in Distributed CORBA Applications. In: IEEE Conference on Dependable Systems and Networks (DSN), Italy (2004).
[21] Huang, Y., Kintala, C., Kolettis, N., Fulton, N.: Software rejuvenation: Analysis, module and applications. In: International Symposium on Fault-Tolerant Computing, pp. 381-390, CA, USA (1995)
[22] Bobbio, A., Sereno, M.: Fine grained software rejuvenation models. In: Computer Performance and Dependability Symposium (IPDS 98), pp. 4-12, (1998)
[23] Garg, S., Moorsel, A.V., Vaidyanathan, K., Trivedi, K.: A methodology for detection and estimation of software aging. In: International Symposium on Software Reliability Engineering, pp. 283-292, (1998)
[24] Yujuan, B., Xiaobai, S., Trivedi, K.S.: Adaptive software rejuvenation: Degradation model and rejuvenation scheme. In: International Conference on Dependable Systems and Networks, pp. 241-248, (2003)
[25] Blair, G.S., Coulson, G., Blair, L., Limon, H.D., Grace, P., Parlavantzas, R.M.N.: Reflection, self-awareness and self-healing in OpenORB. In: Proceedings of the first workshop on Self-healing systems, pp.9-14, South Carolina, USA (2002)
[26] Kant, L.: Design and performance modeling & simulation of self-healing mechanisms for wireless communication networks. In: Proceedings of the 35th Annual Simulation Symposium, pp. 35-42, (2002)
[27] Tohma, Y.: Fault tolerance in autonomic computing environment. In: 2002 Pacific Rim International Symposium on Dependable Computing (PRDC'02), pp. 3-6, (2002)
[28] Brown, A.B., Redlin, C.: Measuring the Effectiveness of Self-Healing Autonomic Systems. In: Proceedings of the Second International Conference on Autonomic Computing (ICAC'05), pp. 328-329, (2005)
[29] Poladian, V., Sousa, J.P., Garlan, D., Schmerl, B., Shaw, M.: Task-based Adaptation for Ubiquitous Computing. In: IEEE Transactions on Systems, Man, and Cybernetics, Part C: Applications and Reviews, Special Issue on Engineering Autonomic Systems, vol. 36, no. 3, (2006).
[30] Umesh, B., Narendra N.: Towards a Programming Model and Middleware Architecture for Self-configuring systems. In: COMSWARE 2006, New Delhi, India (2006)

[31] Keller, A., Badonnel, R.: Automating the Provisioning of Application Services with the BPEL4WS Workflow Language. In: Proceedings of the 15th IFIP/IEEE International Workshop on Distributed Systems, Operations & Management (DSOM 2004), Lecture Notes in Computer Science 3278, Springer Verlag, (2004)

Map-based Design for Autonomic Wireless Sensor Networks

Abdelmajid Khelil, Faisal Karim Shaikh, Piotr Szczytowski, Brahim Ayari and Neeraj Suri

Abstract A prominent functionality of a Wireless Sensor Network (WSN) is environmental monitoring. For this purpose the WSN creates a model for the real world by using abstractions to parse the collected data. Being cross-layer and application-oriented, most of WSN research does not allow for a widely accepted abstraction. A few approaches such as database-oriented and publish/subscribe provide acceptable abstractions by reducing application dependency and hiding communication details. Unfortunately, these approaches ignore the spatial correlation of sensor readings and still address single sensor nodes. In this work we present a novel approach based on a "world model" that exploits the spatial correlation of sensor readings and represents them as a collection of regions called maps. Maps are a natural way for the presentation of the physical world and its physical phenomena over space and time. Our Map-based World Model (MWM) abstracts from low-level communication issues and supports general applications by allowing for efficient event detection, prediction and queries. In addition our MWM unifies the monitoring of physical phenomena with network monitoring which maximizes its generality. From our approach we deduce a general modeling and design methodology for WSNs. Using a case study we highlight the simplicity of the proposed methodology. We provide the necessary tools to use our architecture and to acquire valuable WSN insights in the established OMNeT++ simulator.

1 Introduction and Chapter Structure

Wireless Sensor Networks (WSNs) represent networked autonomous embedded systems. With diversity as a hallmark, WSNs often comprise computing nodes with heterogeneous communication, sensing, processing and storage capabilities. WSNs can be embedded in varied environments with the desired goals of sensing, monitoring, predicting phenomena of interest in

Abdelmajid Khelil, Faisal Karim Shaikh, Piotr Szczytowski, Brahim Ayari, and Neeraj Suri
Technische Universität Darmstadt, Dependable, Embedded Systems and Software Group, Hochschulstr. 10, 64289 Darmstadt, Germany
Tel. +49 6151 16{3414,3711,3414,7066,3513}, Fax. +49 6151 16 4310
e-mail: {khelil, fkarim, piotr, brahim, suri}@cs.tu-darmstadt.de

Research supported in part by DFG GRK 1362 (TUD GKmM), EC CoMiFin, EC INSPIRE and MUET.

A.V. Vasilakos et al. (eds.), *Autonomic Communication*, DOI: 10.1007/978-0-387-09753-4_12,

the physical world. The WSN is considered as a tool for observing states of the real physical world [35] or as a bridge to the physical world [11].

The designer's view on WSN augments this user-centric view with the technical details. From the literature, three main system-level design paradigms arise, namely, considering a WSN as a network, database or an event service. We detail these as:

(a) *WSN as a network*: WSN can be viewed as a self-organized communication platform. The standard WSN communication architecture is layered in a similar way to the OSI layer model. However, a modification of this architecture called a cross-layer model is commonly adopted. The cross-layer design is an envelope of optimizations that benefit from the co-operation of non-adjacent layers [41]. Furthermore, new communication patterns such as data-centric communication have been proposed [17].

(b) *WSN as a database*: Major research issues are in designing efficient query dissemination and in-network selection, projection, join and aggregation. An example being tinyDB [29] that treats sensor data as a single table (sensors) with one column per sensor type. Research focuses here on data-centric communication [5, 8, 28, 50, 53].

(c) *WSN as an event service*: A WSN is usually deployed for missions providing services such as tracking targets or detecting events. The mission determines the overall operation of the WSN including data generation, processing, filtering and transport. To support multiple missions and augment the service flexibility, the publish/subscribe (pub/sub) service architecture has been advocated for WSNs [9, 14, 36, 40].

Most WSN research is application-oriented and relies on cross-layer approaches. Therefore, current research does not allow for a widely accepted abstraction. While the design paradigms mentioned above reduce application dependency and hide low-level communication details, they still address single sensor nodes (although redundancy of nodes and consequently spatial correlation of sensor readings are inherent in WSN). Subsequently, there is a strong need for a *holistic* design methodology. Such a methodology should be flexible/abstract and systemize/simplify the design as well as the deployment phases and involve functional as well as extra-functional attributes. Obviously, the holistic approach should retain the advantages of the existing paradigms. Overall, rather than addressing single nodes, designers should address spatially-correlated and appropriately-grouped nodes. We refer to such groups as *regions*.

A few efforts do exist to address regions instead of single nodes [13, 35, 49]. The *map paradigm* builds on the region principle and therefore, provides excellent modeling primitives for WSNs. Global maps are created for the sake of network monitoring (e.g., residual energy map [54]) or of event detection (e.g., oxygen map [25, 51]). A promising direction consists in using maps to optimize protocols [12, 27], detect [33] or track [55] event boundaries. These papers highlight the map-based methodology as a powerful and promising abstraction. In [21], we propose that maps are the natural step towards a holistic *Map-based World Model (MWM)* and *Map-based WSN design*.

In this chapter we develop the MWM for generalized WSNs. We show how this model retains the advantages of existing design methodologies while augmenting their efficiency and level of abstraction. We also present a step-wise design methodology to build an appropriate MWM and the process driving its usage. We demonstrate the MWM-based design through a case study which highlights the benefits of the MWM approach in predicting a network event, i.e., network partitioning. We further present an implementation of MWM in the OMNeT++ simulator [46] emphasizing its utility for the development of simulation models as well as for network design and validation. In particular, we emphasize the following qualities of our MWM: (1) MWM can appropriately reflect both physical and network worlds while being aware of the strong constraints on network and node resources, and (2) the MWM holistic approach provides a natural way to define, detect, query and predict arbitrarily complex real world situations and events.

The remainder of this chapter is structured as follows. Section 2 provides an overview on models for sensing the real world, the system model that we consider and the main requirements on the MWM. In Section 3, we present our novel MWM-based architecture for WSN and we define essential primitives for MWM management. Subsequently, in Section 4 we highlight the utility of the MWM for enhancing the autonomicity of WSN, and provide a step-wise MWM-based design methodology for WSN. We also validate our design approach using a case study. Section 5 details our MWM implementation in OMNeT++ and sketches some of its uses. In Section 6, we discuss related work. Section 7 concludes our work and gives future research directions.

2 Models and Requirements

First we introduce key models used for sensing the real world through a WSN, while arguing for the need of a holistic world model. We then present the system model and requirements for the proposed world model.

2.1 Models for Sensing the Real World

Typically, the user is interested in observing and controlling a certain physical phenomenon or generally the *physical world* through a WSN representing a dashboard for that physical world. The WSN delivers the required high-level user information with the required Quality of Information (QoI) [4]. Accordingly, a WSN has to create an appropriate model of the physical world of interest, collect the required raw data, synthesize this data and provide the required information to the user. Similarly, an administrator requires a dashboard for the "*network world*", e.g., to show where the energy is suffering more. We refer to both physical and network worlds as the world. We also denote by the user all users of both world models. Therefore, we model a WSN as a (physical/network) *World Model*. The user observes the world through this world model.

Usually, the user (represented by the sink) and the WSN agree on one or more (information) models that should be kept consistent during the deployment (Fig. 1). Common examples of these models include the ambient temperature map and the notification of a certain event. Upon deployment, these models are initialized. The WSN should report changes in the agreed on model (model update). The user can query the implemented model or trigger a model replacement if necessary. An adaptive negotiable model is desirable to allow for evolvable WSN systems. The model update should be incremental in order to minimize the consequent communication overhead. Prediction models allow for predictability while minimizing the communication overhead.

Each sensor node (SN) measures the physical signals of interest at a given sampling rate (with a certain noise level). The sampling rate should be tuned as a function of physical world dynamics (which impact the dynamics of world model data) and the QoI required by the user. Subsequently, a SN produces a time series of the signal (*temporal sampling of the world*). *Spatial sampling of the world* is completed by the set of SNs. Both temporal and spatial sampling play an important role in the accuracy and consistency of the world model, and specifically the achievable QoI. The physical world can be changed through deployed actuators and the network world through network maintenance or reconfiguration.

The design paradigms discussed in the introduction (network, database and event service) are the main existing approaches to realize the world model and its update or query. In order to provide for more comprehensible abstractions, we provide, in Section 3, our Map-based World Model (MWM).

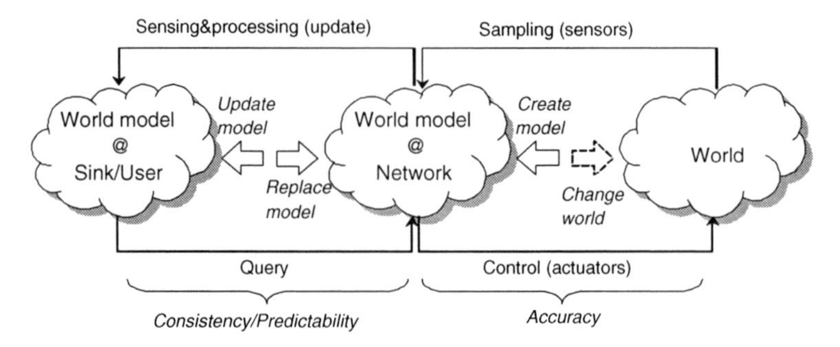

Fig. 1 WSN models - overview

2.2 System Model

Our generalized WSN scenario consists of a network composed of stationary resource-constraint SNs, a static resource-rich sink and mobile resource-moderate assist nodes (ANs). The SNs generate the raw data necessary to create the desired world model. The ANs are usually involved in model management and to a lesser degree in its generation. Commonly, WSNs are built utilizing hundreds to thousands of low-cost SNs. A data sample is characterized by the ID and the location of the SN as well as the sensor reading and its timestamp. We assume that SNs know their own geographic position. The clocks of SNs are synchronized, e.g., via GPS or alternate synchronization protocols [42].

2.3 Requirements on the MWM

In order to develop a unified and adaptable model of the physical and network world, we identify the following requirements on the MWM:
(a) The MWM should incorporate the "network world model" (the WSN) besides the "physical world model" (the science) in a *unified* way. This unification would simplify the design of monitoring techniques for both network and physical world, e.g., by allowing for generic solutions that maximize data piggy-backing. To the best of our knowledge, we are the first to elaborate a unified model for both the physical world and network world. The MWM needs to be generic and independent from modeling of the physical or network world.
(b) The MWM should be *frugal and lightweight*, i.e., its creation, management and use require minimal resources with respect to storage, bandwidth and energy. The management of the MWM is crucial for its usability as it determines its efficiency. The more centralized (e.g., at the sink) is the MWM, the cheaper is its use but the more expensive is its management.
(c) We require the MWM to support different levels of details (i.e., zoom as required), and easy to *customize* to the mission. The modularity and composability of the model are important instruments to reach these goals, and to support diversified, multi-purpose and evolvable applications.
The quality of MWM can be indicated by means of the conventionally accepted metrics of: Accuracy, consistency, efficiency and scalability. There is always a trade-off between accuracy and scalability, and between consistency and efficiency. These tradeoffs need to be investigated by the WSN designers while designing the MWM.

3 The Map-based World Model

We now define the MWM and provide a generalized architecture for WSNs that retains existing architectures while building on MWM. Next, we define core primitives for MWM management and survey the existing map construction techniques.

3.1 MWM Definition

WSNs, on one hand, are inherently embedded in the real world, with their goal being to detect the spatial and temporal world's physical nature, such as temperature, air pressure and oxygen density. On the other hand, maps present a powerful tool to model the spatial and temporal behavior of the physical world being an intuitive aggregated view on it. Therefore, without loss of generality, we model the world as a stack of user maps (uMAP) presenting the spatial and temporal distributions of the sensed attributes of interest in the physical world (Fig. 2). We additionally model the spatial and temporal behavior of system properties as a stack of network maps (nMAP) such as residual energy and connectivity maps. MWM is the superposition of all maps of interest. The unified modeling of both physical/network models maximizes the reusability of concepts and techniques.

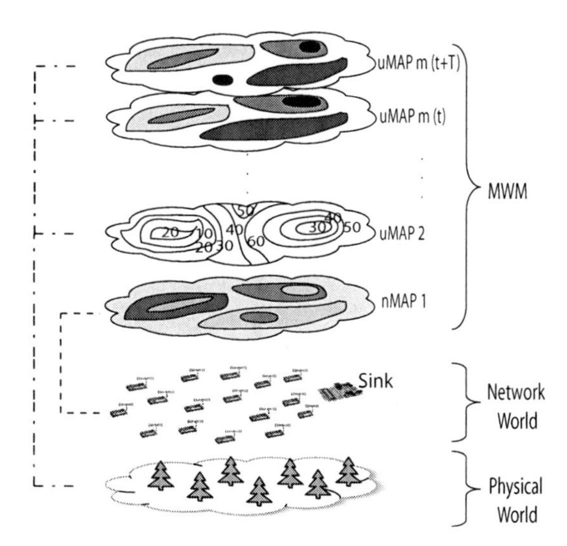

Fig. 2 Map-based World Model (MWM)

A *Map* is an aggregated view on the spatial distribution of a chosen attribute at a specific time. From cartography [34], we identify two main classes of maps, i.e., the choropleths (e.g., nMAP 1 in Fig. 2) and the isomaps (e.g., uMAP 2 in Fig. 2). The map construction groups spatially correlated SNs with similar attribute's values to *regions*. In MWM we define a region by its border (a set of spatial points) and an aggregate (e.g., average) of the attribute's values obtained from all SNs located in the region's area. A map is then the collection of all regions of the WSN. In Section 3.4, we survey existing approaches to create maps for WSNs.

The main benefit from MWM is to abstract from single SNs by addressing regions. While losing sampling details, the map abstraction usually sustains an acceptable accuracy concerning the spatial distribution of sensor readings. Accordingly, the map abstraction presents a natural step to increase the communication efficiency in WSNs. The higher the number of SNs per region, the higher the benefit. The MWM abstraction level simplifies the design and deployment of WSNs as it can be easily accepted by different parties ranging from the user to network designer, programmer and administrator.

3.2 MWM Architecture

Fig. 3 illustrates the MWM architecture which retains established paradigms, builds on top of standard communication protocols and provides valuable support for application design. Network primitives such as unicast, geocast, broadcast and convergecast are essential for query dissemination, event detection and MWM generation and management. The strength of the MWM is to hide the communication and sensing details. This simplifies the design of queries and event detection/prediction on resource-constraint nodes. The required clock synchronization and location information services can easily be realized through existing protocols.
The application interests and requirements define the maps composing the MWM as well as the parameters needed for the construction of these maps. The user may use predicate-based or SQL-like language to specify, delete or modify interests and queries. An example of predicates is "the observed temperature is higher than a certain threshold". The event fire occurs if the predicate value is equal true. Note that the event definition requires fixing the threshold value.
The *query* and *event* services (Fig. 3), provide powerful primitives for the application design and implementation. These services interact as events can be realized through continuous queries, and queries can be triggered by events. The query and event services then act on MWM to answer the user's queries or to notify events of interest. MWM allows for addressing regions instead of single nodes, leading to the optimization of queries, event detection and prediction as detailed in our prior work [21].
From the WSN literature, we identify tinyDB with its extensions as an established query service. TinyDB [29] considers a single table (sensors) and addresses single nodes. TinyDB queries are SQL-based and support selection, projection and aggregation. In MWM we continue using an SQL-like query language. However, we query maps and their regions instead of querying single SNs unlike existing approaches [5, 8, 28, 50, 53]. This may significantly reduce the number of SNs involved in query processing especially if the number of regions constituting the map is very low as compared to the total number of SNs. MWM supports physical world queries and the network world queries in a unified way.
In [21], we defined real-world situations/states as the collection of all relevant maps at the considered time. The extraction of patterns from a given map presents a powerful and generic approach to define simple events [26]. The matching of composed patterns across a stack of maps allows to define composite events. Accordingly to presented unified abstraction, we define simple and composite events in MWM, as situation transitions. Event operations can be matched to comprehensible geometric operations on geometric patterns [21]. Most of existing approaches on event detection [1, 15, 23, 48] are based on network topology and on defining thresholds, and consequently complex and less expressive. MWM further allows for a simple and lightweight prediction of event occurrence. The matching of patterns on a temporal stack of the maps provides for prediction as we will discuss in Section 4.1.
Pub/sub has been proposed as event service for WSN [9, 14, 36, 40]. The main reasoning behind pub/sub is the flexible grouping of nodes as publishers and subscribers by defining interest channels. The grouping of nodes is not fixed and depends on local decisions of nodes. The pub/sub architecture has been defined for WSN to support multiple sinks and to allow for multi-mission systems. Therefore, this architecture is orthogonal to the MWM approach and we propose to deploy it for the management of distributed and/or replicated MWMs.

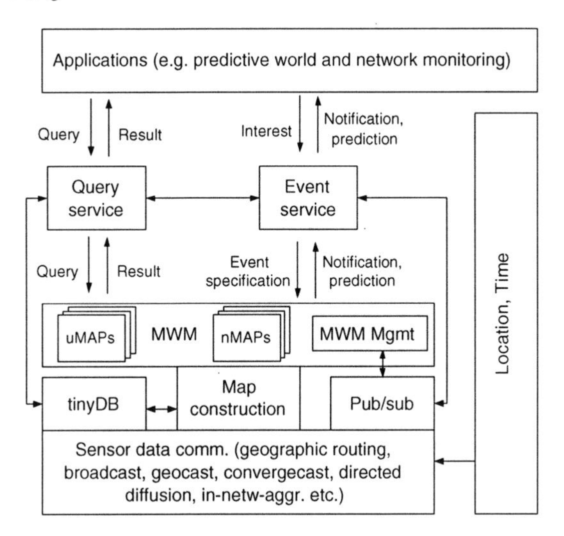

Fig. 3 MWM architecture

Unlike pub/sub and tinyDB the node grouping in MWM depends only on the spatial correlation of the sensor readings, which is desired for most of WSN applications. In the MWM architecture, we allow for the integration of tinyDB and pub/sub (Fig. 3), allowing for an easy add-on of MWM in existing WSN deployments. MWM query service goes into tinyDB if each region is formed by one single SN. TinyDB is mainly a reactive service while pub/sub principally operates pro-actively. Our MWM is hybrid as the map construction can be completed in a pro-active way and the MWM query in a reactive way. There is always a tradeoff between pro-activeness and reactiveness that should be considered depending on the query frequency and MWM dynamics.

3.3 MWM Management

The core question for the MWM management is to address (a) how much pro-activeness is needed in MWM, (b) where to build and optionally replicate it (on sink or elsewhere within the network), and (c) how to maintain the required consistency level through updates. The impetus for investigating these issues is to fix the sources of the MWM and its users.
The MWM layer (Fig. 3) customizes the MWM by specifying the needed maps and the map knowledge needed by each node. This knowledge may range from simple local view to region view (partial view) to the map view (global view). The management of the MWM can easily be transformed into the management of a set of maps. The update of the MWM is completed by the update of the different views on the maps/regions composing it. The MWM approach allows for dynamically inserting and removing maps terming an MWM's re- and pro-activeness besides its reconfigurability and evolvability.
The underlying layer for map construction provides basic primitives for the region and map management at different nodes, i.e., their generation and update (Fig. 3). The management of the individual maps can be considered separately. This allows for a differentiated management for different maps composing the MWM; Some maps can be managed centrally on the sink, others fully distributed, etc. For building MWM, one should investigate opportunistic

construction of maps (e.g., by maximizing message piggy-backing). Update of local views on the map may trigger updates to external view on the map (e.g., at the sink). As mentioned before, it is worth to investigate the pub/sub architecture for the management of the MWM and especially in the presence of mobile ANs and if many nodes manage partial or global MWM views. Here the MWM generators will be publishers, its users the subscribers, and its managers the dispatchers.

3.4 Region and Map Construction Techniques

We now review the existing basic management functions that can be used independently for the maps of interest. These consist in conducting the map construction at the sink. A naive approach for this construction is if each SN reports its value to the sink using multi-hop communication. This is obviously inefficient. Consequently, more efficient approaches have been developed using techniques such as in-network aggregation [16, 52, 54]. Other approaches use suppression mechanisms to reduce the number of SNs reporting their raw readings to the sink [26, 30, 39]. A new approach exploits node mobility to further increase the efficiency of data collection [20].

Aggregation-based Approaches:

eScan [54] and isobar [16] are based on polygon aggregation leading to a choropleth map. First, the sink disseminates the interest/query for a certain map to all SNs (Fig. 3). The query should fix: (a) The map of interest and (b) the update model, i.e., when to send an update and to which node. For eScan [54] the query is on the residual energy of SNs. The query is disseminated using flooding to create a tree having its root at the sink. This tree is used to aggregate the attribute values while being reported. A leaf node sends its raw value to its parent node. An internal node gathers the input of all its children, aggregate it with its value and forward it to its parent node. The aggregation consists in grouping sensor readings that meet a certain criteria (being geographically adjacent and in the same value range). The outcome of the aggregation is a list of (spatial) regions. A region is a polygon that is defined by the line spanning its border nodes. At the sink the aggregation results in a complete map. SNs reply with their current values immediately on query and later only with necessary updates. The update model is aggregation-tree based, i.e., a SN sends its updates to its parent node which aggregates the updates of all its child nodes. *INLR [52]* is an aggregation-based approach similar to eScan but focusing on small scale WSNs. A SN sends its reading or the calculated aggregate not only to its parent node but to all its neighbors that are 1-hop closer to the sink. While using more than one parent increases the accuracy of the map, the efficiency is sacrificed.

Fig. 4 (a) sketches how maps are created and managed in the network by eScan [54], isobar [16] or INLR [52]. The approaches are incremental. Consequently, the closer a SN to the sink, the more global is its view on the map. In Section 5, we show the benefits of the MWM implementation in OMNeT++ to acquire more insights of the eScan approach.

Suppression-based Approaches:

Isoline [39] is an approach based on localized isocluster aggregation. The map building is reduced to the detection of isolines. Neighboring nodes share their readings. A node compares its reading with the readings of all its neighbors and detects an isoline, when the readings lie in different sides of a globally defined isoline. The detection of an isoline needs to be reported to the sink by the closest neighbor to the sink. The isocluster aggregation outperforms polygon aggregation in terms of accuracy with minor energy savings. *Meng et al. [30]* motivate the use of contour (isoline) maps for efficient continuous monitoring in WSN. The authors design a temporal and spatial local suppression mechanism that prohibits some nodes to report their readings. The number of saved reports highly depends on the spatial correlation between sensor readings. SNs report their readings using multi-hop routing without any in-network processing. The map is constructed on the sink using interpolation and smoothing techniques.

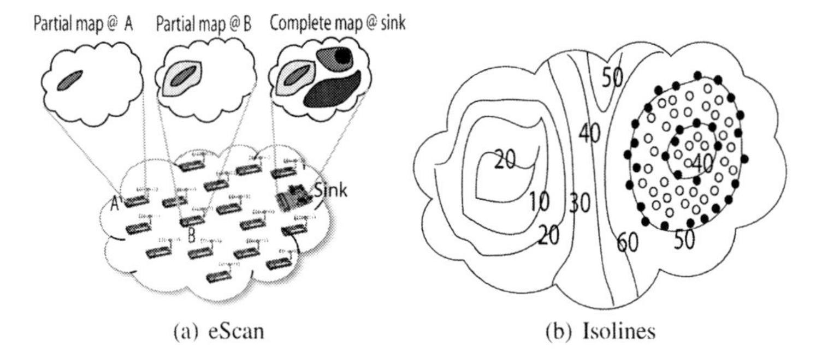

(a) eScan (b) Isolines

Fig. 4 MWM management examples

Iso-Map [26] also does not rely on in-network processing. It uses a suppression mechanism to reduce the number of SNs that report their readings to the sink using multi-hop communication. This approach is very similar to that of Isoline [39]. However SNs need to report the gradient direction of the isolines, which requires excessive processing on SNs.
If one of the suppression-based approaches (Isoline, Iso-Map or [30]) is implemented to create regions, then SNs located on the isolines (filled points) know that they are isoline nodes (see Fig. 4 (b)), and the rest of SNs know only their local views, i.e., own sensor readings (non-filled points).
Mobility-assisted Approach:
In [20], we presented gMAP, an efficient mobility-assisted approach to construct global maps. In gMAP (a) SNs do not need to process readings of other nodes and (b) require to communicate a minimal number of messages compared to the approaches above. This is achieved by opportunistically exploiting the mobility of ANs to collect data of interest, keeping SNs transmit only their own readings on-demand to an AN in their transmission area. Since the data collection lasts long (depending on the mobility of the AN) the gMAP approach is valid only for attributes with values of high time validity.
Abstract Regions Approach:
"Abstract Regions" [49] is a family of spatial operators that allow SNs to form regions in the WSN. Abstract regions define a neighborhood relationship between a particular node and other nodes in the network in terms of radio connectivity, geographic location, or other node properties. Examples include creating regions of k-nearest or n-hop neighbors. Regions can be constructed and maintained depending on the required quality allowing for energy/accuracy trade-offs. The authors demonstrate the effectiveness of abstract regions for varied sensor network applications such as finding spatial contours. Some of the presented operators as well as the contour finding application provide elementary management primitives for MWM.

4 MWM-based WSN Design

In this section, we first emphasize the viability of MWM for enhancing the autonomicity of WSNs. Next, we present a step-wise methodology to benefit from the MWM architecture for design and deployment of WSNs. Consequently, we validate our design methodology through a case study.

4.1 Enhancement of WSN Autonomicity

We present some use scenarios showing how MWM can enhance the WSN's level of autonomicity by supporting pro-active reconfiguration and cooperative WSNs.

4.1.1 Predictive Monitoring for Proactive Reconfiguration

It is challenging to predictively monitor sensor fields over time, i.e., combining data from both the spatial and temporal domains. Efficient long-term data collection from the WSN can enable fine-grained trend analysis and event prediction. There is only limited existing work on event prediction in WSNs [6, 24, 45]. They perform time series forecasting and use autoregressive models for predictability. The main drawback of these techniques is that they act on single SNs, which limits their efficiency in terms of number of messages and processing complexity. However, most of these approaches can be easily integrated into our MWM model acting on regions and maps instead of single nodes. Therefore, protocol designers should manage relevant MWM snapshots in a MWM history for an appropriate time window w. Future snapshots can then be predicted with a certain accuracy. The main benefit of MWM is the appropriate abstraction level for prediction provided by the map without sacrificing much the accuracy of predictability due to the sacrifice of time series accuracy. Especially, when the average number of SNs per region is high, the processing complexity for prediction and the storage overhead for the map history is minimized, allowing for an efficient in-network predictability on resource-limited or resource-moderate devices such as ANs. We present our preliminary work on map-based prediction in [3].
Sensing the physical or the network world constitutes a first step towards a reaction such as actuation back to this world. The maps of the MWM allow for an optimized and goal-oriented spatial intervention (e.g., network maintenance) and navigation (e.g., for ANs and users) in the sensor field. In particular, the pro-active reconfiguration and maintenance are of high interest for future WSN given the growing reconfigurability of entities with respect to hardware and mobility. The MWM allows for an efficient event-driven triggering of reconfiguration and for map-based assessment of reconfiguration options. Examples of MWM supported proactive reconfiguration are: (a) Map-based sensor re-tasking and re-programming, and (b) reactive and pro-active node placement to maximize data collection and to provide for self-healing and graceful degradation (e.g., by delaying network partitioning). Obviously, pro-activeness enhances the self-* capabilities of the WSN and therefore provides for valuable techniques to enhance the autonomicity of WSN systems [19].

4.1.2 Map-based WSN Interoperation/Federation

MWM presents a widely accepted abstraction level as it converts less comprehensive sensor data into understandable information. Furthermore, MWM can be implemented as a middleware which simplifies the integration of varied applications not only intra-WSN but also inter-WSN. The map-based model is potentially a candidate to develop standards for cooperative WSNs. Such standardization will provide a generic interface simplifying the WSN interoperability through the interconnection of heterogeneous and autonomous WSN systems, which can play a major role in future WSN research (e.g., SensorGRID [43] and SensorWEB [7]). This is conform with the trend to integrate sensor data (such as GPS tracking data and webcams) into Geographical Information Systems (GIS) and online services such as earth geographic maps and second life.

4.2 Design Methodology

The main design benefits arise from the MWM being holistic and relying on an abstraction level that is admitted by users as well as application and network designers. WSN designers are usually overwhelmed by a vast number of low-level primitives such as node level communication. Thus, it is invariably left to the designer to implement the details of communication, making it both complex and error-prone. However, the designers would benefit from a higher-level generic view. An understandable higher-level view is given by the maps, which allow for a simplified but holistic design of future WSNs. Maps also provide an intuitive abstraction level for debugging WSNs. Protocol and system design for WSN normally include a heuristic design phase that is followed by validation and optimization phase. Our MWM approach is well-suited for these phases.

We now present top-down approach to benefit from MWM for system design and deployment. We describe a set of five necessary steps for building and configuring a typical MWM in general with the purpose to detect and predict events and situations in the physical and network worlds as well as proactive reconfiguration. We show the application of this step-wise design approach using a case study.

STEP 1: Identify the problem by specifying the situations and events of interest from the physical world (this phase involves domain/sensor experts and users) or the network world (involving network designers and administrators). This is also the appropriate time to fix the user requirements such as on the quality of observation/information (e.g., the event detectability, predictability and accuracy).

STEP 2: Identify the required maps and define events and their operations, queries etc according to the MWM specification.

STEP 3: Sketch a solution for the identified problem assuming an MWM global view given by all maps (e.g., at the sink). The general solution consists of event/situation specifications, and detection/prediction techniques.

STEP 4: Determine the required MWM knowledge of each node in the network while minimizing the needed global knowledge, the overhead for the MWM's creation, distribution and management. The globally sketched solution can now be designed for the fixed MWM specification. The main result is then a set of detection and prediction algorithms.

STEP 5: Select requisite primitives (broadcast, unicast etc.) for intra- and inter-region communication.

4.3 Case Study: Designing a Network Partitioning Prediction Technique

We illustrate some key aspects of the proposed MWM-based design for WSNs by presenting a simple case study, i.e., predicting network partitioning. For simplicity, we consider a WSN composed of static SNs and a single static sink. We assume that the WSN is connected at the time of deployment. We consider that energy depletion leading to a node crash is the primary reason for network partitioning. We deliberately suppress details since our objective is provide a proof of concept.

STEP 1: The problem we consider here is network partitioning, i.e., if either a set of nodes run out of energy, or a set of nodes can not reach the sink. Network partitioning limits the sensing coverage of the WSN and implies the need of counter-measures. For safety and maintainability reasons, we require the prediction of network partitioning.

STEP 2: The map of interest for detecting network partitioning is the connectivity map (cMAP). In order to allow for prediction, the energy map (eMAP) as well as the cMAP are required. For simplicity, we start with a connected WSN and focus only on the prediction of

network partitioning. Subsequently, the MWM is composed of the eMAP. We observe the energy level of the different eMAP regions and as this level approaches a predefined low level (E_{th}), an early warning is reported. A fine-grained strategy for early warning reports is also possible. For instance we can adapt the reporting frequency to the energy level of the region.
STEP 3: We now sketch a solution assuming the knowledge of the eMAP at the sink. Recording the different eMAP regions that run out of energy allows to predict the isolation of some energy-rich regions. The key idea is to monitor the low energy regions of the eMAP and to use a regression algorithm to predict the time the regions run out of energy, and also the time when the "dead" regions isolate other energy-rich regions (Fig. 5).
STEP 4: For prediction we do not need global map knowledge at the sink (spatial suppression). We do not also need the map knowledge during the complete WSN lifetime (temporal suppression). It would be sufficient if only the regions of low energy report this state to the sink that predicts the network partitioning. Furthermore, only a few nodes of each region (e.g., border nodes) have to report their value to the sink . For eMAP, we suggest using the Isolines approach [39] to construct the map and to consider the following global classes of residual energy: 0-E_{th}% and E_{th}-100% (Fig. 5).

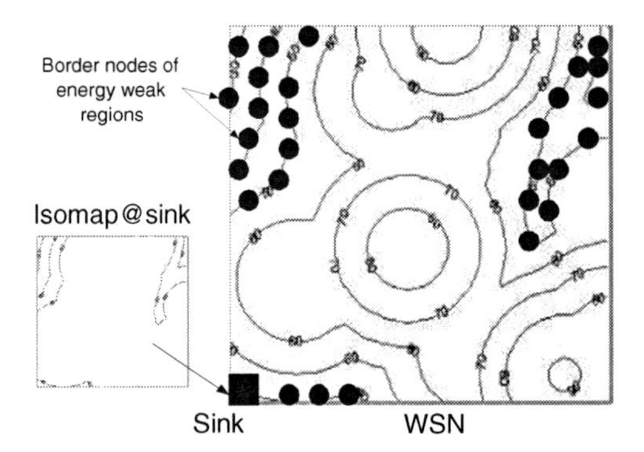

Fig. 5 Prediction of network partitioning

Having the incremental reports (locations and energy values) of border nodes (isoline nodes), the sink can now depict the relevant isolines and predict the coverage drop of the corresponding region. This also allows to predict if some other regions in the network will be isolated as a result of the coverage drops of some regions.
STEP 5: In this step, the designers should fix the communication primitives needed to implement the algorithms sketched in Step 4. They include broadcast for neighbor discovery and border node identification, and convergecast for sending information from border nodes to the sink. Additionally, some existing suppression techniques can be deployed to reduce the number of reporting border nodes.
From the literature, we identify approaches to detect [38] and suspect [37] network partitioning though research for its prediction is lacking. [38] detects linear partitions ,i.e., cuts. Partitioning is detected by monitoring a small subset of SNs, called sentinels. The position of sentinels is calculated at the sink, given the position of all SNs. The sentinels periodically send a heartbeat beacon. When a sentinel is not reachable, the sink concludes that the part of the network where this sentinel is located is partitioned. The main drawbacks of this scheme are (1) the frequent blind reports to the sink, (2) not all shapes of partitions can be detected by this approach (only linear cuts but not holes (dead regions) nor islands (isolated regions)) (3) the poor accuracy of this detection approach as if a sentinel crashes, it is not always a sign of network partitioning. Furthermore, this approach does not provide information about the re-

gion really affected by partitioning, which complicates the maintenance options. The Partition Avoidance Lazy Movement (PALM) protocol for mobile sensor networks [37] is a decentralized approach, where a SN can locally suspect network partitioning and move to avoid it. The PALM scheme assumes that each SN knows its own position and the position of the sink. The SN periodically collects the position of all its neighbors and checks if some neighbors are located in a small angle towards the sink. If no neighbor is located in this "promising zone", the SN suspects network partitioning. The main drawback of this scheme is that every SN has to periodically broadcast its location and to blindly check if network partitioning is suspected. The deficiencies of these approaches express builds the core of the effectiveness of our proposed map-based design.

5 MWM Implementation in OMNeT++

Simulation is a commonly used validation approach for WSNs. Hence, an implementation of the MWM on the established WSN simulators will provide valuable access to maps presenting an understandable, contextual and interactive simulation information representation. OMNeT++ [46] is a discrete event simulator that is getting increased attention in WSN community given its scalability, efficiency and flexibility resulting from its high modularity (OMNeT++ modules are object oriented structures). As we target large scale WSN in MWM, considering OMNeT++ properties and capabilities, we use it to provide the first implementation of MWM.

5.1 MWM Implementation Architecture

The MWM implementation in OMNeT++ consists of a tracing module and a visualization tool. We follow a generic approach, where the tracing module is loosely coupled with OMNeT++ components ensuring its high (re)usability. The tracing module periodically collects the attributes of the network modules and saves the values in XML format. The tracing module can easily be started by including it in topology configuration file (.ned) with a set of parameters such as the specification of attributes to be monitored, the frequency of tracing etc. The generated trace is input to our visualization tool which creates two dimensional representation showing the network topology and renders the desired maps. Next the sensor field is fragmented according to Voronoi polygons, which brings additional benefit of reflecting nodes density. Every polygon is filled by a color corresponding to the value of selected attribute. The visualization tools offers also observation of developing phenomena along the time axis. For every time sample, a separate map is rendered, and user can navigate the visualized trace back and forward in time.

5.2 Uses of Simulator Extension

Our implementation provides a flexible generic aggregation of trace data, and delivers high level information that is acceptable by variety of users involved in the design and simulation of WSN. The major benefit consists of the geographical/contextual visualization of the information hiding the tedious and less comprehensive trace data. In the following, we highlight three main utilizations for our MWM implementation corresponding to three main successive simulation phases: (a) Simulation configuration, (b) design and implementation, and (c) validation.

5.2.1 Simulation Configuration

Like often open simulators OMNeT++ benefits from continuous extensions from the user community. These extensions are usually the results of modeling efforts. In the following we highlight, through the example of energy consumption modeling, how the MWM implementation simplifies the development of OMNeT++ models and the generation of simulation scenarios. Generally the energy consumption within the WSN shows high spatial correlation. This is mainly a result of the inherent node redundancy in WSN. Hence, neighboring nodes participate in the processing of same physical world events as well as network operations. For instance, the closer the SNs to the sink, the more communication traffic they forward, as traffic is mainly directed to the sink. Subsequently, it is crucial to provide appropriate energy consumption models for the simulator such as the hotspot model [54]. Unfortunately, there is no systematic way to validate such models or the simulation scenarios generated according to these models. Using our MWM implementation, it is easy to generate the energy maps for arbitrary scenario settings, energy consumption model parameters and time instant (Fig. 6(a)).

5.2.2 Design and Implementation

We emphasize that the implementation in OMNeT++ simplifies the WSN designing methodology described in Section 4. The MWM implementation aids in understanding the design problem by visualizing the maps of interest for different scenarios. The designer can then interactively test tentative solutions and their preliminary performance. For our case study in Section 4, the MWM implementation can give valuable insights on network partitioning (partition shape, causes, frequency etc.) and the number of border nodes that will report to the sink for different scenarios, e.g., varied energy consumption models. Similarly, the physical world of interest can be modeled and validated using the MWM implementation. These insights allow for an interactive design, for example in the form of manipulation of the map visualization. After discovering regions of interest, the designer can zoom into the visualized map (Fig. 6 (b)) and optionally trace data, only for selected regions.

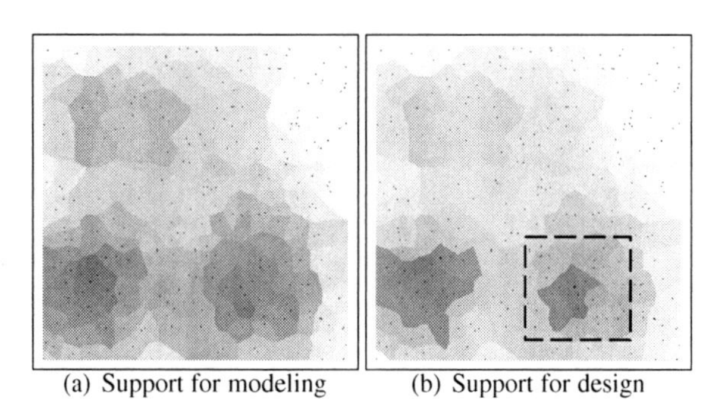

(a) Support for modeling (b) Support for design

Fig. 6 Examples for modeling and design support

5.2.3 Validation

Design validation mainly consists of two activities: Debugging and comparison. Both are strongly enhanced in OMNeT++ through our MWM implementation.
Debugging: Developed concepts and techniques require implementation, which always involves much debugging efforts. The map perspective allows a primary high-level debugging thus simplifying a secondary focused debugging. MWM makes the localization of the spatially correlated problems and the understanding of their nature a straightforward task. For example bugs regarding the implementation of a routing design could manifest in energy-overconsumption in untypical regions. Data aggregation errors in the eScan implementation can be discovered by comparing the map created by eScan to the perfect map, which can be easily generated for arbitrary time points (Fig. 7). Debugging process is also enhanced by the visualization of the maps of the physical world.

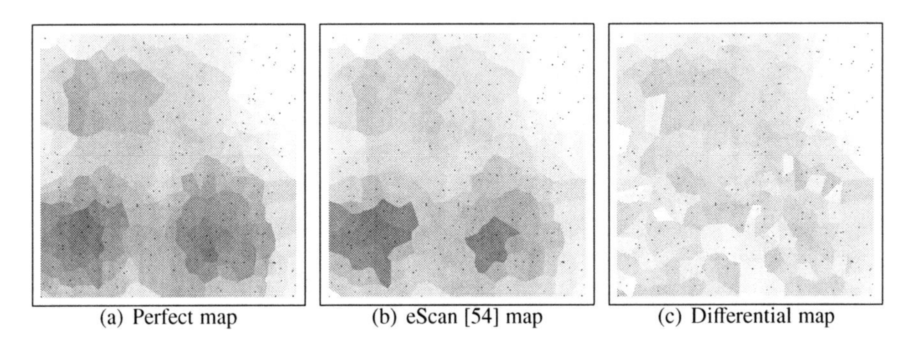

(a) Perfect map (b) eScan [54] map (c) Differential map

Fig. 7 Protocol comparison

Comparison: The MWM implementation actively supports solution comparison. Such a comparison is intuitive given the visualization of the perfect snapshot on one hand and the achieved snapshots by the different solutions under test on the other hand producing a set of differential maps. The map perspective simplifies spatial observation and performance comparison of different solutions, i.e., to identify parts of the sensor field, where solutions perform better.

6 Related Work

Modeling techniques from the GIS literature and from the spatial temporal databases can provide important techniques such as modeling languages and standards for our MWM. The several existing tools such as the Space Time Toolkit [47] can be used for map data analysis. In particular, we cite the existing modeling languages such as SensorML, REACTIVEML and LUSSENSOR, which simplify the specification of our MWM. In this work we do not focus on the MWM specification and formalization and keep it for future work.
We are not the first to define world models. Such a modeling technique is well established for Augmented Reality or Virtual Reality [32]. The most related models to us are those for ubiquitous computing and sentient computing. The ubiquitous computing [31] [10] [18] models are optimized for complex physical worlds and rely on powerful infrastructure that creates and manages the model for mobile clients. Many related concepts can be adopted if our MWM is

centrally stored and managed by the sink. However, in this chapter we argued for a distributed MWM, i.e., only partially centralized at the sink, given the strong resource constraints for the WSN prohibiting a full data collection at the sink. Sentient computing deploys sensors to perceive the environment and react accordingly. One use of the sensors is to construct a world model in the infrastructure, supporting location-aware or context-aware applications [44] [2]. Sentient computing focuses only on indoor scenarios and ambient intelligence applications. All these world models are user-centric and do not fulfil vital requirements for a world model for WSN. In particular, they are not frugal and do not support a unified network/physical world model.

7 Conclusions

Maps provide a widely accepted abstraction. Circumventing the problem of addressing single nodes in a WSN, we have developed a map-based system architecture. The proposed intuitive and lightweight map-based world model (MWM) uniformly models both the physical and the network world using maps. Besides predictive monitoring the MWM provides an important decision base for pro-active WSN reconfiguration to enhance WSN functionality, dependability or security. Our architecture is flexible and presents a powerful tool for both designing and deploying WSNs. This has been elucidated by a case study and an implementation for OMNeT++.
In future work, we are planing to refine MWM by formalizing the definition of the querying language and developing efficient/frugal algorithms for replication and consistency of the MWM on the sensor, assist and sink nodes. In order to simplify the access to MWM for the community, we aim to develop extensions to further simulators such a Tossim and MMulator [22], and platforms such as TinyOS.

References

1. Abadi, D., et al.: REED: Robust, Efficient Filtering and Event Detection in Sensor Networks. In: VLDB (2005)
2. Addlesee, M., et al.: Implementing a Sentient Computing System. IEEE Computer **34**(8) (2001)
3. Ali, A., Khelil, A., Shaikh, F., Suri, N.: MPM: Map-based Predictive Monitoring for Wireless Sensor Networks. In: TR-TUD-DEEDS-05-01-2008 (2008)
4. Bisdikian, C.: On Sensor Sampling and Quality of Information: A Starting Point. In: PerCom Workshops (2007)
5. Bonnet, P., et al.: Towards Sensor Database Systems. In: MDM (2001)
6. Borgne, Y., et al.: Adaptive Model Selection for Time Series Prediction in Wireless Sensor Networks. International Journal for Signal Processing (2007)
7. Chien, S., et al.: Lights Out Autonomous Operation of an Earth Observing SensorWEB. In: RCSGSO (2007)
8. Chu, D., et al.: The Design and Implementation of A Declarative Sensor Network System. In: SenSys (2007)
9. Costa, P., et al.: Publish-subscribe on Sensor Networks: A Semi-probabilistic Approach. In: MASS (2005)
10. Dudkowski, D., et al.: Efficient Algorithms for Probabilistic Spatial Queries in Mobile Ad Hoc Networks. In: COMSWARE (2006)
11. Elson, J., Estrin, D.: Sensor Networks: A Bridge to the Physical World. Chapter in the book Wireless Sensor Networks, Kluwer Academic Publishers (2004)

12. Goussevskaia, O., et al.: Data Dissemination Based on the Energy Map. IEEE Comm. Magazine **43**(7) (2005)
13. Gracanin, D., et al.: On Modeling Wireless Sensor Networks. In: IPDPS (2004)
14. Hauer, J., et al.: A Component Framework for Content-based Publish/Subscribe in Sensor Networks. In: EWSN (2008)
15. Heidemann, J., et al.: Diffusion Filters as a Flexible Architecture for Event Notification in Wireless Sensor Networks. In: USC/Information Sciences Institute (ISI-TR-556) (2002)
16. Hellerstein, J., et al.: Beyond Average: Toward Sophisticated Sensing with Queries. In: IPSN (2003)
17. Intanagonwiwat, C., et al.: Directed Diffusion: A Scalable and Robust Communication Paradigm for Sensor Networks. In: MOBICOM (2000)
18. Kawsar, F.: Prithibi: An Open Platform for Digitizing Real World through Sentient Artefact Model. In: Joint Workshop on Next-Generation Computing Infrastructure (2007)
19. Kephart, J., Chess, D.: The Vision of Autonomic Computing. IEEE Computer Magazine (2003)
20. Khelil, A., Shaikh, F., Ali, A., Suri, N.: gMAP: An Efficient Construction of Global Maps for Mobility- Assisted Wireless Sensor Networks. In: The Sixth Annual Conference on Wireless On demand Network Systems and Services (WONS) (2009)
21. Khelil, A., et al.: MWM: A Map-based World Model for Wireless Sensor Networks. In: ACM AUTONOMICS (2008)
22. Kropff, M., et al.: MM-ulator: Towards a Common Evaluation Platform for Mixed Mode Environments. In: SIMPAR (2008)
23. Kumar, A., et al.: Distributed Collaboration for Event Detection in Wireless Sensor Networks. In: MPAC (2005)
24. Lazaridis, I., Mehrotra, S.: Capturing Sensor-generated Time Series with Quality Guarantee. In: ICDE (2003)
25. Li, M., et al.: Non-Threshold based Event Detection for 3D Environment Monitoring in Sensor Networks. In: ICDCS (2007)
26. Liu, Y., Li, M.: Iso-Map: Energy-Efficient Contour Mapping in Wireless Sensor Networks. In: ICDCS (2007)
27. Machado, M., et al.: Data Dissemination in Autonomic Wireless Sensor Networks. IEEE Journal on Selected Areas in Comm. **23**(12) (2005)
28. Madden, S., et al.: The Design of an Acquisitional Query Processor for Sensor Networks. In: SIGMOD (2003)
29. Madden, S., et al.: TinyDB: an Acquisitional Query Processing System for Sensor Networks. ACM Trans. on Database Systems **30**(1) (2005)
30. Meng, X., et al.: Contour Maps: Monitoring and Diagnosis in Sensor Networks. Computer Networks **50**(15) (2006)
31. Nicklas, D., Mitschang, B.: The NEXUS Augmented World Model: An Extensible Approach for Mobile, Spatially Aware Applications. In: OOIS (2001)
32. Reitmayr, G., Dieter, S.: Semantic World Models for Ubiquitous Augmented Reality. In: SVE (2005)
33. Ren, K., et al.: Secure and Fault-Tolerant Event Boundary Detection in Wireless Sensor Networks. IEEE Trans. on Wireless Comm. **7**(1) (2008)
34. Robinson, A., et al.: Elements of Cartography. John Wiley & Sons, New York (1995)
35. Römer, K., Mattern, F.: Event-Based Systems for Detecting Real-World States with Sensor Networks: A Critical Analysis. In: DEST at ISSNIP (2004)
36. Sharifi, M., et al.: A Publish-Subscribe Middleware for Real-Time Wireless Sensor Networks. In: International Conference on Computational Science (2006)
37. Shih, K., et al.: PALM: A Partition Avoidance Lazy Movement Protocol for Mobile Sensor Networks. In: WCNC (2007)
38. Shrivastava, N., et al.: Detecting Cuts in Sensor Networks. In: IPSN (2005)
39. Solis, I., Obraczka, K.: Isolines: Energy-efficient Mapping in Sensor Networks. In: ISCC (2005)

40. Souto, E., et al.: MIRES: A Publish/subscribe Middleware for Sensor Networks. Personal Ubiquitous Comput. **10**(1) (2005)
41. Srivastava, V., Motani, M.: Cross-layer Design: A Survey and the Road Ahead. IEEE Communications Magazine **43**(12) (2005)
42. Sundararaman, B., et al.: Clock Synchronization for Wireless Sensor Networks: A Survey. Ad-Hoc Networks **3**(3) (2005)
43. Tham, C.: Sensor-Grid Computing and SensorGrid architecture for Event Detection, Classification and Decision-Making. book chapter in Sensor Network and Configuration: Fundamentals, Techniques, Platforms, and Experiments (2006)
44. Town, C., Zhu, Z.: Sensor Fusion and Environmental Modelling for Multimodal Sentient Computing. In: IEEE CVPR (2007)
45. Tulone, D., Madden, S.: PAQ: Time Series Forecasting for Approximate Query Answering in Sensor Networks. In: EWSN (2006)
46. Varga, A.: OMNeT++ Object-Oriented Discrete Event Simulation System User Manual. In: URL http://www.omnetpp.org/doc/manual/usman.html (2006)
47. vast: Space Time Toolkit. http://vast.uah.edu/
48. Vu, C., et al.: Composite Event Detection in Wireless Sensor Networks. In: IPCCC (2007)
49. Welsh, M.: Exposing Resource Tradeoffs in Region-based Communication Abstractions for Sensor Networks. SIGCOMM Comput. Comm. Rev. **34**(1) (2004)
50. Wu, H., et al.: Distributed Cross-Layer Scheduling for In-Network Sensor Query Processing. In: PerCom (2006)
51. Xue, W., et al.: Contour Map Matching For Event Detection in Sensor Networks. In: ACM SIGMOD (2006)
52. Xue, W., et al.: Contour Map Matching For Event Detection in Sensor Networks. In: SIGMOD (2006)
53. Yao, Y., Gehrke, J.: Query Processing for Sensor Networks. In: CIDR (2003)
54. Zhao, Y., et al.: Residual Energy Scan for Monitoring Sensor Networks. In: IEEE WCNC (2002)
55. Zhu, X., et al.: Light-weight Contour Tracking in Wireless Sensor Networks. In: INFOCOM (2008)

An Efficient, Scalable and Robust P2P Overlay for Autonomic Communication

Deng Li and Hui Liu and Athanasios Vasilakos

Abstract The term Autonomic Communication (AC) refers to self-managing systems which are capable of supporting self-configuration, self-healing and self-optimization. However, information reflection and collection, lack of centralized control, non-cooperation and so on are just some of the challenges within AC systems. Since many self-* properties (e.g. self-configuration, self-optimization, self-healing, and self-protecting) are achieved by a group of autonomous entities that coordinate in a peer-to-peer (P2P) fashion, it has opened the door to migrating research techniques from P2P systems. P2P's meaning can be better understood with a set of key characteristics similar to AC: Decentralized organization, Self-organizing nature (i.e. adaptability), Resource sharing and aggregation, and Fault-tolerance. However, not all P2P systems are compatible with AC. Unstructured systems are designed more specifically than structured systems for the heterogeneous Internet environment, where the nodes' persistence and availability are not guaranteed. Motivated by the challenges in AC and based on comprehensive analysis of popular P2P applications, three correlative standards for evaluating the compatibility of a P2P system with AC are presented in this chapter. According to these standards, a novel Efficient, Scalable and Robust (ESR) P2P overlay is proposed. Differing from current structured and unstructured, or meshed and tree-like P2P overlay, the ESR is a whole new three dimensional structure to improve the efficiency of routing, while information exchanges take in immediate neighbors with local information to make the system scalable and fault-tolerant. Furthermore, rather than a complex game theory or incentive mechanism, a simple but effective punish mechanism has been presented based on a new ID structure which can guarantee the continuity of each node's record in order to discourage negative behavior on an autonomous environment as AC.
A detailed measurement study of three popular unstructured P2P overlays and ESR is performed. Our method is to analyze performances of classical searching algorithms in various overlays. Key factors in content locations including scalability, query success rate, query mes-

Deng Li
School of Information Science and Engineering, Central South University, Changsha, Hunan 410083, China, e-mail: d.li@csu.edu.cn

Hui Liu
Department of Computer Science, Missouri State University, Springfield, MO 65897, USA, e-mail: HuiLiu@missouristate.edu

Athanasios Vasilakos
Department of Computer and Telecommunication Engineering, University of Western Macedonia, Greece, e-mail: vasilako@ath.forthnet.gr

A.V. Vasilakos et al. (eds.), *Autonomic Communication*, DOI: 10.1007/978-0-387-09753-4_13,

sages, cost, disturbed times and fault tolerance are considered carefully. The simulation results show some characteristics in unstructured P2P overlay and prove that ESR is a highly efficient, low cost and fault tolerant overlay and a good structure for applications in AC.

1 Introduction

The term Autonomic Communication (AC) [1] addresses such challenging issues as the continuous growth in ubiquitous and mobile network connectivity, together with the increasing number of networked computational devices populating our everyday environments (e.g., PDAs, sensor networks, tags, etc.), by trying to identify novel flexible network architectures, and by conceiving novel conceptual and practical tools for the design, development, and execution of "autonomic"communication services. It is used to enable networks, associated devices and services to work in an unsupervised manner, to self-configure, self-monitor, self-adapt and self-heal – the so-called self-* properties. For such a highly distributed and heterogeneous environment, the classical, centralized client-server communication model seems to be inappropriate because of its inherent limitations in terms of scalability, fault-tolerance and ability to handle highly dynamic environments.
Peer-to-peer (P2P) systems consisting of a dynamically changing set of nodes connected via the Internet have gained tremendous popularity. While initially conceived and popularized for the purpose of file sharing (e.g. Gnutella). P2P has emerged as a general paradigm for the construction of resilient, large-scale, distributed services and applications in the Internet.
The lack of global or central control implies the need for new techniques to design and verify self-* properties. Since many self-* properties (e.g. self-configuration, self-optimization, self-healing, and self-protecting) are achieved by a group of autonomous entities that coordinate in a peer-to-peer (P2P) fashion, thus, it has opened the door to migrating research techniques from P2P systems. P2P's meaning can be better understood with a set of key characteristics similar to AC: Decentralized organization, Self-organizing nature (i.e. adaptability), Resource sharing and aggregation, and Fault-tolerance. The common characteristics shared by P2P overlays and ACs also dictate that both networks are faced with the same fundamental challenges, that is, to provide connectivity in a decentralized, dynamic environment. Thus, there exists a synergy between these two types of networks in terms of the design goals and principles of their routing protocols and applications built on top: both P2P and AC routing protocols and applications have to deal with dynamic network topologies due to membership changes or mobility. The common characteristics and design goals between P2P overlays and ACs point to a new research direction in networking, that is, to exploit the synergy between P2P overlays and ACs to design better protocols and applications.

2 Background on P2P Overlay Networks

The numerous P2P overlay networks for the Internet that have been proposed in the past few years can be broadly classified into two categories:

1. Unstructured. Unstructured P2P overlay networks as exemplified by Gnutella [2] do not have precise control over the overlay topology. The network is typically formed by nodes joining the network following some loose rules, for example, a node joining a Gnutella network starts by connecting to nodes in a host cache file which stores Gnutella nodes learned from the last time the node was part of a Gnutella network, and a Gnutella node typically specifies a default maximal number of neighbors in the Gnutella overlay. In the resulting network topology, the placement of an object or a file is not based on any

knowledge about the topology. Furthermore, the overlay is often not network proximity aware; that is, neighboring nodes in the overlay may be far away from each other in the underlying Internet topology. The typical way of locating an object in an unstructured overlay is to flood the network in which a query is propagated to overlay neighbors within a controlled radius. While the lack of proximity-awareness and flooding-based object location are inefficient, the consequent advantage is that unstructured overlay networks and the companion object location mechanisms that do not rely on any precise structure of the topology are highly resilient to frequent node join and departure.

2. Structured. To overcome the inefficiency with object location in unstructured networks, structured overlay networks have been proposed to combine the inherent self-organization, decentralization, and diversity of unstructured P2P overlays with a scalable and efficient routing algorithm that can reliably locate objects in a bounded number of routing hops, typically logarithmic in the network size, while exploiting proximity in the underlying Internet topology. Numerous structured P2P overlays have been proposed, such as Chord [3], Pastry [4] and CAN [5]. The routing of such structured P2P overlays effectively implements scalable and fault-tolerant distributed hash tables (DHTs): each node in the network has a unique node identifier (nodeID) and each data item stored in the network has a unique key, nodeIDs and keys live in the same namespace, and a message with a key is routed (mapped) to a unique node in the overlay. Thus, DHTs allow data to be inserted without knowing where it will be stored and requests for data to be routed without requiring any knowledge of where the corresponding data items are stored. To maintain efficient routing, nodes in a structured overlay must maintain neighboring nodes that satisfy certain criteria in the namespace. As a result, structured overlays are conceptually less resilient to frequent node join and departure. In the rest of the chapter, we use DHTs and structured P2P overlays interchangeably whenever appropriate for this reason.

3 Challenges and Requirements in Supporting P2P for AC

There are a number of critical challenges and problems which currently prevent the great potential of AC from being revealed: such as information reflection and collection, lack of centralized control, non-cooperation and so on [1]. These key challenges, which are currently the focus of extensive research efforts in the P2P research community are going to be briefly outlined below.

3.1 Information reflection and collection

Search and resource location mechanisms are necessary for information reflection and collection, and are a fundamental and crucial building block of most P2P systems and determine, to a large degree, their efficiency and scalability. The large volume of query traffic generated by the flooding of messages limits the scalability and efficiency of Gnutella-like approaches. There have been a number of attempts like random walk [6, 7] seeking to improve the efficiency of the search for unstructured pure P2P systems.
Currently, many researchers present P2P overlay based on Semantic-similarity [8], content-similarity [9] or share interest [10]. There are, yet, two fundamental problems: (1) in overlay networks the physical underlying infrastructure is not adequately taken into consideration with regard to the construction of an overlay structure. This leads to an inefficient use of the underlying network resources and to a high end-to-end delay for applications; (2) present network

distance estimation services require a certain level of external information for the setup of landmarks. This makes the resulting systems non-self-organizing, and if the landmarks fail, then the entire system is affected. Thus, there is a need for more structure and guidance in this chaos of self-* properties for AC.

3.2 Lack of Centralized Control

Because lack of centralized control, the security requirements of AC differ from the requirements of distributed systems in general, and include traditional security goals such as confidentiality, authentication, integrity and availability.
First of all, the nodes in an AC system are much more autonomous and powerful than in conventionally distributed systems. For example, in numerous P2P fashions, the individual nodes assign their own node identifiers. This makes establishing a level of trust extremely hard for initially unknown entities, since identities can easily be altered. Furthermore, it is possible for a malicious node to create multiple identities in order to gain control of a part of the system. This is referred to as a Sybil Attack [11]. The security of the solution mechanism [12] is based on the secure assignment of node identifiers via a trusted Certification Authority to stop potential Sybil Attacks. Such a trusted entity is typically not available in AC systems. On the contrary, in an open environment with mutually distrusting and potentially malicious participants, the lack of a trusted entity makes it extremely difficult to establish a level of trust mechanisms.

3.3 Non-Cooperation

In AC, open systems architectures allow agents to join networks dynamically and both offer and consume services. The growth of peer-to-peer services and the withering of centralized control make cooperative behavior essential to preventing free riding [13] and other selfish behaviors, which is one of the key challenges of current P2P systems. A free rider is a node that has access to services and consumes resources without participating in any of them. A recent study shows that 85 percent of Gnutella users share no files [14]. We can no longer expect that engineering a network properly can be independent of the economic realities concerning the implementation of that network.
Incentive mechanisms penalizing free riders or rewarding users have been discussed [15–17]. More general reputation mechanisms [18] can be used to obtain a system of wide reputation for each user. Using this information, each user will give priority to the users with a high reputation. In BitTorrent [19], users download pieces of files and at the same time upload the pieces they already have. Similarly, in [20] the users can directly exchange resources between themselves. But BitTorrent can still suffer from free riding [21], because selfishness-proof algorithms are difficult to be designed simply.
Another option is to use monetary incentives to solve the problem of free riding. In this case, users must pay to download files from other peers. The payments may either be in monetary terms where a price is assigned to each file and users exchange files only when they can afford it [22], or in an internal non-monetary currency. In the latter, the budget of a user decreases every time he downloads a file, and increases every time he uploads a file. Recent work [23] studies the system performance as a function of the total amount of internal currency available. The decision to cooperate or not in a P2P context can be modeled with the use of game theory such as [24, 25] in wireless ad hoc networks which are characterized by a distributed, dynamic, self-organizing architecture modeled as a game.

A big problem in proposing an incentive mechanism lies in the ability to track past behaviors in order to determine present service quality. Most of the existing work generally involves a trusted third party for records and query responses, which would suffer from the failure of a single point and in turn result in poor scalability.
That is to say, guaranteeing the history of nodes' behaviors is an extremely challenging problem and an open research issue for AC. Thus, how to make the node ID unique and how to make the records of nodes' behavior consecutive in a highly dynamic and decentralized environment such as AC is a key question.
According to the above, the challenges of AC are because of the high degree of autonomy. This degree of autonomy also makes an incentive mechanism inefficient because there is no trusted global server to secure the assignment and uniqueness of node identifiers. Thus, in AC, it is difficult to deal with Sybil Attacks or the problem called whitewashing [16] – peers leave the system and rejoin with new identities to avoid reputational penalties. In the final analysis, we think that a good P2P system which is compatible with AC should consider and satisfy three correlative requirements: (1) information can be located and routed at a high efficiency; (2) the topology should be scalable and robust; (3) while negative (i.e. the selfish or malicious) behaviors will be prevented. To get a high efficiency in communication, there should be no or few negative users, and too many negative behaviors will lead to bad scalability. Meanwhile, if information can be located and routed at a high efficiency, it will decrease the maintenance cost of the topology so as to improve the robustness. Thus, the three requirements are necessary, important and relative to each other when we evaluate the P2P overlays.
To our knowledge, no related work has addressed all those three requirements for P2P overlays, which are also necessary for AC. This chapter, however, complements many pieces of recent work and firstly aims at satisfying all requirements, presents a novel Efficient, Scalable and Robust (ESR) P2P overlay. The detailed key characteristics of our design are summarized below:

- A 3D overlay with global view combined with limited state, and localized connection and information exchange. That is to say, high efficiency and low cost are combined in our overlay, even when there is a high amount of peer churn. Though there are many researches used overlay construction algorithm in a much more generalized N-dimensional space, rather than a 3-dimensional space described in this paper. They are either used in structured P2P overlay [5, 26] with DHT, which can not suitable for highly dynamic environment with frequent node joining and leaving, or without consideration of heterogeneity of peers [27].To our knowledge, there is no similar research with our overlay.
- Information Center (IC) is presented by considering of physic location and peers' activities to give attention to both efficiency and prevention of the negatives. It differs from most cluster head selection mechanisms only consider nodes' capability such as the end-to-end delay to other nodes [28] in one cluster or outbound bandwidth [29].
- The unique and successive of peers' identities (IDs) without the need of global knowledge or hashing, even though both dynamic and random events like joins, leaves and crashes take place. It is very important for the overlay to be maintained effectively and autonomously so that the corresponding mechanisms (e.g. game theory) could be applied conveniently.

4 The Description of ESR

4.1 The Formation of ESR

In the overlay, nodes are divided into many clusters called Autonomous System (AS). Inspired by the idea of the prime meridian or longitude 0 used to describe longitude and latitude, we propose using a normative node called the Origin Node (ON), in which ASs in the overlay

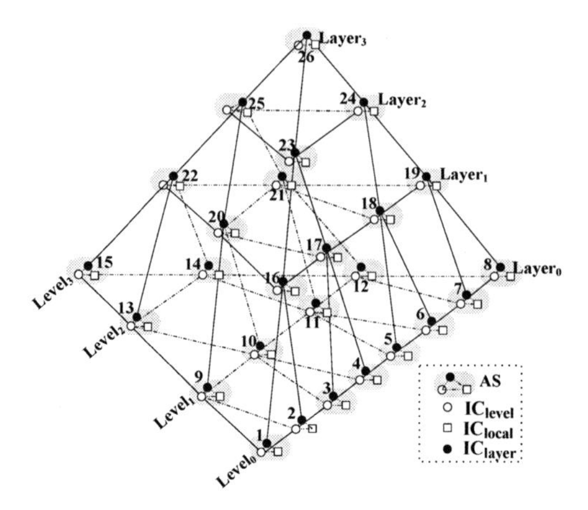

Fig. 1 The perspective of the high efficiency and low cost overlay (H=4 and d=2)

calculate their approximate coordinates in the same virtual space. ON is not a permanent node, since our overlay is a full self-organizing architecture; mainly, it is used to form the overlay through consideration of the physical situation. Moreover, ON disappears when the initial overlay becomes a stable overlay. When ASs can maintain the overlay structure by themselves, the overlay is called stable. Our overlay consists of ASs, levels and layers. The maximum AS size is d+3. That is to say, the maximum peer number in one AS is d+3. The AS with the maximum peer number is called FULL AS (FAS). Before the overlay can become a stable overlay, nodes in one AS are sorted out according to their distances from the ON and the unique node ID is given. Nodes belonging to the same AS have an IP address sharing the longest Common Prefix (CP). The ASs are positioned on a gradient which increases the farther away they are from the ON, and each AS has a unique AS ID. Many grouped ASs are constituent parts of a level, which has a threshold (AN_i), where i is the number of the AS in the level, to the AS number. When one AS is a FAS, a new AS will be created as the next AS of the FAS in the same level. Let L_{max} be the distance between the new node and the farthest node in an AS. If there are many candidate ASs, the new node will choose an AS whose L_{max} is the minimum. Levels are sorted out according to their distance from the ON. A new level (i+1) will be created above the primary level i to set new ASs while the number of ASs on the primary level reaches the predefined threshold. Each level has a unique level ID. The planar space has many levels called layer. Layers are further increasing gradients guided by their distances from the ON. When the number of levels in a layer reaches the predefined threshold (L_i), where i is the number of the layer, the structure grows to a 3D space, i.e. a new $(i+1)^{th}$ layer will be created above the ith layer to set new levels. Thus, new ASs will be initiated in a new layer above the primary layer. At the end of this process, an overlay-structure that matches the underlying network can be constructed based on the clusters of nodes that are close together and in proximity to each other.

Fig. 1 is our logical overlay, where H is the number of levels and the number of layers. Every layer includes many levels and ASs communicating to each other by cluster headers called Information Centers (ICs) appearing in all ASs, levels and layers. The predefined bound d=2. The blank dot, black dot and blank square denote IC_{level}, IC_{layer} and IC_{local}, respectively, which will be introduced in Sect. 4.3.

The edge between neighbor ASs (e.g. the edge between AS 9 and AS 20) is one of the long-range contacts, and each edge in the AS is one of the local contacts. Long-range contacts and local contacts are presented in [30] to build small world networks, while our contacts are not randomly chosen. When the overlay-structure is a stable overlay, the ON ceases to exist. Let

D_{max} be the average distance between the new node and the ICs in an AS. If there are many candidate ASs, the new node will choose an AS whose D_{max} is the minimum. ICs in our work cooperate with each other to provide this geographical partitioning feature.
Each node has the unique 128 bit node identifier in our system. The node ID is used to indicate a node's position in the overlay, which ranges from 0 to 2^{128} -1. The node ID is defined by the layer number, the level number, the AS number and the node number immediately after a node joins the system. The letters *i, j, k* and *t* respectively denote the layer number, the level number, the AS number and the node number, which are shown as follows (i.e. Fig. 2):
$N_{k,t}^{i,j}$ denotes the tth NN of the kth AS at the jth level in the ith layer. It is obvious that the node ID is global unique.

4.2 The Source Ranking

For the selection of IC, we present a mechanism called Source Ranking where a peers' group is formed by the online queuing function *SR*. When the information exchange begins between peer P_i and P_j, it is assumed that query is sent from P_i to P_j. Then $E(P_i, P_j)$ denotes the P_i's evaluation to P_j about the completion of the query.

$$E(P_i,P_j) = \sum_{\forall q \text{ answered by } P_j} \mathrm{Qsim}(q_j,q)^{\alpha} * N(P_i,P_j) \big/ T(P_i,P_j) \qquad (1)$$

Parameter α improves the power of similarity of queries. Since α is bigger, queries that are more satisfied are given a higher evaluation. That is, the more similar queries P_j completes, the larger evaluation P_i gives. In order to find the most likely peers to answer a given query we need a function *Qsim*: $Q^2 \rightarrow [0,1]$ (where Q is the query space), to compute the similarity between different queries. $N(P_i,P_j)$ indicates the number of the communication times between peer P_j and P_j. $T(P_i,P_j)$ denotes the time of information exchange between peer P_j and P_j. When the similarity between the different queries is the same, we can assume that the peer has more communication times in unit information exchange rather than in time joining the AS in order to decrease the percentage of the free riders.
The cosine similarity (formula 2) metric between 2 vectors ($\vec{q}$ and $\vec{q_i}$) has been used extensively in information retrieval, and we use this distance function in our setting. Let L be the set of all requirements for resources or services that have appeared in queries. We define an $|L|$ -dimensional space where each query is a vector. For example, if the set L is the requirements A, B, C, D and we have a query A, B, then the vector that corresponds to this query is (1, 1, 0, 0). Similarly, the vector that corresponds to query B, C is (0, 1, 1, 0). In the cosine similarity model, the similarity *sim* of the two queries is simply the cosine of the angle between the two vectors.

$$sim(q_i,q) = \cos(q_i,q) = \frac{\sum(\vec{q} * \vec{q_i})}{\sqrt{\sum(\vec{q})^2} * \sqrt{\sum(\vec{q_i})^2}} \qquad (2)$$

Thus, the source ranking of P_j is:

$$SR(P_j) = \sum_{\forall i \neq j \in \{\text{thesameAS}\}} (E(P_i,P_j) * D(P_i,P_j)) \qquad (3)$$

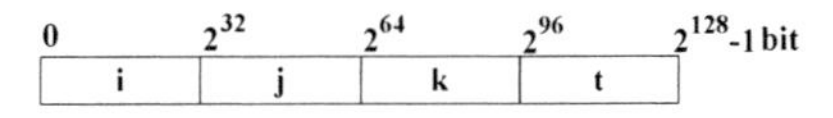

Fig. 2 The structure of node ID

Where $D(P_i,P_j) = 1/dist(P_i,P_j)$, $dist(P_i,P_j)$ denotes the distance between peer P_i and P_j,. That is to say, the closer peers lie in geography, the bigger the function D is. It improves the power of peers adjacent geographically in order to avoid influence of network congestion.
We conducted a 10000-node random physical network topologies generated by using the GT-TTM library [39] by which all simulations in the paper are run based on the different topologies generated. In the topology network, about 10, 50 and 100 near nodes are randomly grouped in one AS respectively (each topology is generated 10 times). Fig. 3 shows the effect of node location on the SR in the AS with different size when the evaluation between any two nodes is similar.
The shadow bar denotes the SR value of each node in the topology, while the solid bar on the right side of the shadow bar denotes the location of the same node. In the simulation, we consider the variance of the delay from the node x to all the other nodes in the same AS rather than the average of the delay in order to more objectively show the relative location of each node in the AS.

4.3 The selection and performance of ICs

There are three ICs called IC_{local}, IC_{level} and IC_{layer} in each AS, and the three nodes are connected to each other. The other nodes in the AS are called normal nodes (NN) and their applications and resources are abstract as the information. In our paper we will define the maximum number of NNs as d. The ICs have different functions but contain the same backup. NN_i^j denotes the jth NN and IC_σ^i denotes the IC_σ ($\sigma \in \{local, level, layer\}$) in AS i ($i \in [1,26]$) in Fig. 1.
IC_{local}: IC_{local} is connected by and receives information such as ID, IP, applications, resources, activity histories (i.e. the values of SRs) and status from most d local NNs. For example, NN_i directly connect to and store their information in IC_{local} (IC_{local}^i). IC_{local} also keeps the backup of information and sends its own information (i.e. information of NNs) to local IC_{level} and IC_{layer}.
IC_{level}: One IC_{level} in the ith level communicates with at most d the $(i-1)^{th}$ $IC_{level}s$. Each IC_{level} is connected with only one IC_{level} in the upper neighbor level. A local IC_{level} submits the application and resource information of its AS (i.e. information from local NNs aggregated in local IC_{local} are backed up to IC_{level}) to the corresponding IC_{level} in the upper neighbor level. E.g. IC_{level}^9 and IC_{level}^{10} are two $IC_{level}s$ connected to the upper neighbor level IC_{level} IC_{level}^{13}. IC_{level}^{13} backs up information in IC_{local}^{13} and IC_{layer}^{13}, and sends information about local NNs (e.g. NN_{13}^j) to its upper neighbor level IC_{level}^{15}. To recover from all ICs' crash, IC_{level} also stores upper neighbor level NNs' IP addresses.
IC_{layer}: One IC_{layer} in the ith layer communicates with at most d $IC_{layer}s$ in the $(i-1)^{th}$ layer while the AS which lies in the highest level has 1 extra subordinate from the highest level AS in the lower neighbor layer. Each IC_{layer} is connected with only one IC_{layer} in the upper neighbor layer. Because of characteristics of our overlay, IC_{layer} in the upper layer only backs up the ID and IP information and activity histories of the nodes of its lower neighbor layer ASs. For example, IC_{layer}^{20} stores the information such as ID, IP and values of SRs receives from IC_{layer}^9, including those information of IC_{local}^9, IC_{level}^9, and NN_9^j etc. Since the distance across layers is longer than the distance between levels in the same layer, the possibility of a fault in transferring large data increases. However, the amount of ID, IP and activity history information is small and it is suitable to be transferred across layers in order to improve the whole protocol's robustness. The IC_{layer} also keeps the backup of information and sends its own information (i.e. ID, IP and value of SR about lower neighbor layer nodes) to local IC_{level} and local IC_{local}. To recover from a tri-IC's crash, the IC_{layer} also stores its upper neighbor layer NNs' IP addresses.

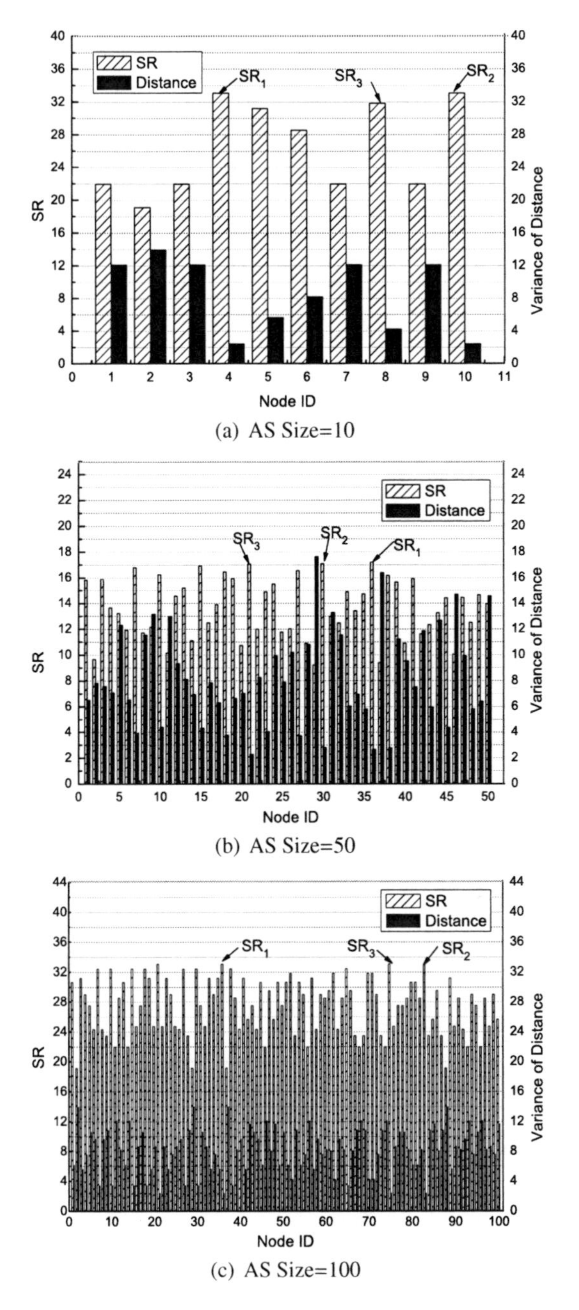

Fig. 3 The effect of node location on the SR

Judging from the condition described above, the load in IC_{level} is the heaviest. So we choose the node with the highest value of SR in an AS as the IC_{level} in this AS. Then the nodes having the ordinal highest (the 2nd highest and the 3rd highest) values of SR are named as IC_{local} and IC_{layer} respectively. The other nodes in this AS are NNs which don't have to keep information tables of each other but only information about the IP and ID of the IC_{local} IC_{level} and IC_{layer}.

5 The maintenance of ESR

In this section, we propose the maintenance of the overlay, including joining, leaving and the mechanism for preventing free riders. Peers in level $j < H-1$ are partitioned into ASs of size in $[\delta, m]$ (The analysis assumed that each AS are connected by at least k lower neighbor ASs. $m \geq \delta \geq 1$ are two constant).

Lemma 1. *it is assumed that k is a constant ($k \geq 2$), to $\forall x > 0$, $k^x \geq kx$.*

Proof. Here we use epagoge.

1. To x=1, k=k is true.
2. To x=2, since $k \geq 2$, $k^2 - 2k = (k-1)^2 - 1 \geq 0$, thus $k^2 \geq 2k$.
3. To $x > 2$, it is assumed that $k^{x-1} \geq k(x-1)$, then

$$k^x = k \cdot k^{x-1} \geq k \cdot k(x-1) \tag{4}$$

4. Since $k \geq 2$, $1 + \frac{1}{k-1} \leq 2 < x$, that is,

$$\frac{k}{x} < k-1 \tag{5}$$

From inequation (5) we conclude

$$k^2(x-1) > kx \tag{6}$$

Combined inequation (4) and (6), $k^x \geq kx$, by the epagoge, lemma 1 is proven.

□

Lemma 2. *it is assumed that k is a constant and ($k \geq 2$), then $k^x > (k-1)x - k$ $(x > 0)$.*

Proof. Since $k \geq 2$, and $x > 0$, then $(k-1)x - k < kx - k < kx$, from lemma 1, lemma 2 is proven. □

Theorem 1. *If one IC_{level} in the ith level communicates with at most d the $(i-1)^{th}$ IC_{level}s, and one IC_{layer} in the ith layer communicates with at most d the $(i-1)^{th}$ IC_{layer}s. The height H of levels or the layers in the overlay is less than $\log_d M + 1$, where M is the number of ASs.*

Proof. To level $H-2$, the group number is at most d, Thus, to layer 0 (the number of levels is $H-1$), the number of all ASs is, $\sum_{n=1}^{H} d^{n-1} = \frac{1-d^H}{1-d}$. Recursively, to layer 1, the number of levels is $H-2$, thus the number of all ASs is, $\sum_{n=1}^{H-1} d^{n-1} = \frac{1-d^{H-1}}{1-d}$. Then, to the overlay, the number of ASs is,

$$\frac{(1-d)+(1-d^2)+(1-d^3)+...+(1-d^H)}{1-d} = \frac{H-d(1+d+...+d^{H-1})}{1-d}$$
$$= \frac{H-d\frac{1-d^H}{1-d}}{1-d} = \frac{H-dH-d+d^{H+1}}{(1-d)^2}$$

Then, $\frac{H-d \cdot H-d+d^{H+1}}{(1-d)^2} = M$. Combined lemma 2, we have

$$M = \frac{H - dH - d + d^{H+1}}{(1-d)^2} > \frac{d^{H+1} - d^H}{(1-d)^2} = \frac{d^H}{d-1} > d^{H-1} \tag{7}$$

From inequation (7), we have $H < \log_d M + 1$. Theorem 1 is proven. □

Theorem 2. *The worst-cast node degree of the multicast overlay is at most $d+4$.*

Proof. Consider a node X in one AS. There are four possibilities:

1. X is the NN: X only links to IC_{local} in the AS. Therefore, the degree of X is 1.
2. X is the IC_{level}: An AS has at most d lower neighbor level ASs and only at most 1 upper neighbor level ASs, thus X has at most d+1 neighbor IC_{level}s (the degree is at most d+3). An exception holds for the highest level where the IC_{level} has no upper neighbor level IC_{level} (the degree is at most d+2).
3. X is the IC_{local}: Each IC_{local} is linked by at most d NNs, thus the degree of X is at most d+2.
4. X is the IC_{layer}: An AS has at most d lower neighbor layer ASs and only at most 1 upper neighbor layer ASs. Thus, X's degree is d+3. The AS which lies in the highest level has 1 extra subordinate from the highest level AS in the lower neighbor layer(the degree is at most d+4). An exception holds for the highest layer where the IC_{layer} has no upper neighbor layer IC_{layer} (the degree is at most d+3). In any case, the degree of a node can not exceed d+4, thus proving the theorem true.

□

Theorem 1 and theorem 2 summarize two properties any P2P structure, such as a tree, should desire. As clients join and leave, we must be able to adjust the 3D overlay without violating the adaptive rules. Overheads incurred by this adjustment should be kept small to keep the system scalable.

5.1 Two rules for maintenance

To prevent the selfish behavior, we present the rule 1 as following:
Rule 1: Each AS has a minimum value $Min(Ef)_k^{i,j}$ of peers' SR, where i, j, k denotes the AS's layer number, level number and its location in the level respectively. $Min(Ef)_k^{i,j}$ is dynamically decided by a number of nodes and the loads in an AS. If one node's value of SR is less than $Min(Ef)_k^{i,j}$, it is considered a selfish peer and will be dropped.
Many researchers have presented the node ID [31]. However, they do not deal with the mechanisms of how to guarantee the uniqueness of IDs in a highly dynamic and distributed environment as AC, and the stability of the system. As analyzed in Sect. 4.1, every node ID is unique. Moreover, from Sect. 4.2, we can see that each AS only knows the global unique IDs of several neighboring ASs in ESR. In ASs communicating with each other by Peer-to-Peer without a global IC, it is important to prevent nodes from registering in several ASs so that one node has several IDs. Thus, we present the rule 2 to prevent malicious nodes.
Rule 2: a threshold distance (T_{dist}) between neighboring ASs in the same level as $T_{dist} = \frac{\sum_{\sigma \in \{IC\}} (p_\sigma^{cri} * Dist_\sigma)}{\sum_{\sigma \in \{IC\}} p_\sigma^{cri}}$, every new joiner must join the nearest AS. If a new node wants to join an AS farther than T_{dist}, the AS will regard it as a malicious peer (e.g. a whitewasher) so that the join request must be refused. In the formula, σ is one type of the three ICs, $Dist_\sigma$ denotes the distance between two neighboring IC_σ at the same level. E.g., $Dist_{local}$ denotes the distance

```
procedure Joining
S ← {r | CP_r ≃ C_x ∧ (∃i ∈ r : D_i < R_x}
 q ← null
 S' ← S
 While ( S' ≠ ∅ ){
   r_j ← ⟨pick a random AS from S'⟩
   Calculate D_avg(x, r_j)
   S' = S' − {r_j}
 }
 While (S ≠ ∅ ∧ q = null){
   r ← {select (min(D_avg(x, r_j)) ∀r_j ∈ S)}
   S = S − {r}
   if num_r + 1 ≤ max_r  /* have room*/
     q ← r
   else
     if ∃i ∈ r : SR_i < Min(Ef)_k^{i,j}
       drop node i
       q ← r
     else
       if ∃i ∈ r : max(D_avg(i, r)) > D_avg(x, r)
         drop node i
         q ← r
 }
```

Fig. 4 The pseudocode of joining algorithm

between IC_{local}s in two neighbor AS at the same level. p_{σ}^{cri} denotes a parameter defined by the position of IC_{σ} in its AS. It should be noted that ICs are changed by the dynamic joining and leaving, Thus, p_{σ}^{cri} is necessary because the position of each IC is dynamic. The discussion of p_{σ}^{cri} could be found in our other article [32]. The rule 2 can limit peers in local situation without the centralize controller as ON.

5.2 Node joining

There are many researches on bootstrapping. In this paper, we do not concentrate on how to bootstrap in our protocol; we just assume that every new node can always satisfy the following condition: For each new joiner, at least one close node existing in the overlay-structure can be found. This assumption allows the new node to join the AS close to itself. It is rational. For example, if the peers in City A are collected in one AS, there is a greater possibility for a new joiner in City A to meet one of the peers in the same city and to become a member of that AS.

- Suppose the node is a new node which has never joined the system before. If the bootstrapping node X is an IC_{local}, the new node submits its information to IC_{local}, and obtains a unique ID from the AS that is also a unique global ID. If the bootstrapping node X is not an IC_{local}, through the IP stored in X, the new node invites X's local IC_{local} to join X's AS and submit all its information to local IC_{local}. Then the new node obtains a unique ID from the AS that must be a unique global ID.
- Suppose that the node had joined the system before. It rejoins the system through its record of the AS. If, upon accepting the new connection, the total number of nodes still is within the preconfigured bound d+3, the connection is automatically accepted. Otherwise, the AS must check if it can find an appropriate existing node to drop and replace the new connection.

Let $D_{avg}(x,r)$ be the average distance between the new node x and the ICs in an AS r. D_i denotes the distance between node x and the node i in the overlay. R_x is the maximum delay that this node x can be tolerable or acceptable. The joining algorithm is shown in Fig. 4.

5.3 Node leaving

Normal leaving. There are two ways that the nodes can normally leave the system: ICs (i.e. IC_{local}, IC_{level} and IC_{layer}) leave; and NNs leave.

- Normal leaving of ICs. In our system, only one IC can normally leave at one time. If all the three ICs want to leave the system, they have to leave one by one following a principle called the first applying first leaving (FAFL). While an IC is preparing to leave the system, it uses the backup information and chooses the NN that has the highest value of SR among NNs as its substitute. (1) If IC_{level} at the jth level is the leaving node, it should broadcast the information of its substitute to (a) local IC_{local} and local IC_{layer}; (b) d the $(j-1)^{th}$ level neighbor IC_{level}s; (c) and to the $(j+1)^{th}$ level neighbor IC_{level}. (2) If IC_{local} is the leaving node, it should broadcast information of its substitute (a) to the local IC_{level} and IC_{layer}; (b) and to d local NNs. (3) If IC_{layer} in the ith layer is the leaving node, it should (a) broadcast its information to local IC_{local} and IC_{level}; (b) and it should broadcast information to the $(i+1)^{th}$ layer IC_{layer} and to d the $(i-1)^{th}$ layer neighbor IC_{layer}s.
- Normal leaving of NNs. It is simple for NNs to leave because the only thing that the node has to do is to submit its leaving request to a local IC_{local} and change its status in the local IC_{local} to be 'offline'

Abnormal leaving. There are also two ways that the nodes abnormally leave the system: ICs leave and NNs leave. In our system, one IC sends life signals to the other corresponding ICs periodically. If, for a long time an IC does not receive life signals from another IC, this means that it has abnormally left the system.

- ICs abnormal leaving. There are two situations as follows: (1) The tri-IC do not leave the system simultaneously. The remaining IC will choose the node which has the highest value of SR as the new IC, and its backup information will be copied to the new IC. Then the new IC broadcasts its information to the corresponding ICs in the system. (2) All tri-ICs leave the system abnormally. The upper level neighbor IC_{level} chooses the node x with the highest value of SR as the new IC_{level}, and copies the backup information to the new IC_{level}. Then the new IC_{level} chooses the new IC_{local} and IC_{layer} following the method described above. The tri-IC broadcast their information to the corresponding ICs in the system as described in Sect. 4.2. For example, as shown in Fig. 1, if IC^{i}_{σ} denotes one of ICs in AS i where $\sigma \in \{local, level, layer\}$, and NN^{i} denotes one of NNs in AS i. When all ICs in AS 20 crashed, (a)IC^{22}_{level} chooses one NN^{20} whose SR value is the highest to be the new IC^{20}_{level}; (b)IC^{20}_{level} receives information about AS 20 backed up in IC^{22}_{level} from the latter; (c)IC^{20}_{level} chooses two NN^{20}s whose SR value is the ordinal highest to be the new IC^{22}_{local} and IC^{22}_{layer} respectively; (d)IC^{20}_{local} receives information about NN^{20}s from IC^{20}_{level} and connect with NN^{20}s; (e) Simultaneously, IC^{16}_{level} and IC^{17}_{level} reconnected to IC^{20}_{level} through inviting any NN^{20} to get the address of the new IC^{20}_{level}; IC^{9}_{layer}, IC^{10}_{layer} and IC^{25}_{layer} reconnected to IC^{20}_{layer} through inviting any NN^{20}.
- NNs abnormal leaving. The necessary information such as ID, IP and activity history are still kept in ICs to distinguish them from the nodes which have never joined the system. In the worst situation, several ASs will break down. Then the new nodes will have to reuse the formation process to recreate ASs based on the existing AS.

6 Evaluation and experimental results

6.1 Modeling and methodology

To evaluate the performance of ESR, we look at three aspects of a P2P system: P2P network topology, query distribution and replication. By network topology, we mean the graph formed by the P2P overlay network; each P2P member has a certain number of "neighbors"and the set of neighbor connections forms the P2P overlay network. By query distribution, we mean the distribution of frequency of queries for individual files. By replication, we mean the number of nodes that have a particular file. During our study of search algorithms we assume static replication distributions. We use four network topologies in our study:

- Random Topology with Power-Law characteristic (RTPL): This is a random graph with different scales. The node degrees follow a power-law distribution: if one ranks all nodes from the most connected to the least connected, then the ith most connected node has ω/i^{α} neighbors, where ω is a constant. Many real-life P2P networks including real Internet network have topologies that are power-law random graphs [33].
- Super-node topology: We ran simulations using a standard super-node topology [34]. It is a two-level hierarchy, consisting of a first level of interconnected peers called super-peers and a second level of so-called leaf nodes or normal peers, which are only connected to a single super-peer. In super-node topology, searches are flooded among super-peers. In the paper, the term "node"is used interchangeably with "peer". Each super-node has two backup nodes. The peers including super-node, its backups and normal peers form a cluster. We set the total number (c_{size}) in any cluster is $c_{size} \in [5, 15]$.
- Square-root topology: Consider a peer-to-peer network with N peers. Each peer k in the network has degree d_k (that is, d_k is the number of neighbors that k has). The total degree in the network is D, where $D = \sum_{k=1}^{N} d_k$. Each peer k maintains two counters: Q^k_{total} , the total number of queries seen by k, and Q^k_{match} , the number of queries that match k's content. g_k denotes the proportion of searches submitted to the system that are satisfied by content at peer k, that is, $g_k = Q^k_{match} \Big/ Q^k_{total}$, then a square-root topology has $d_k \propto \sqrt{g_k}$ for all k. To construct a square-root topology, when peers join the network, they make random connections to some number of other peers. The number of initial connections that peer k makes is denoted d^0_k . Then, as peer k is processing queries, it gathers information about the popularity of its content. From this information, peer k calculates its first estimate of its ideal degree, d^1_k . If the ideal degree d^1_k is more than d^0_k, peer k adds $d^1_k - d^0_k$ connections, and if the ideal degree is less than d^0_k, peer k drops $d^0_k - d^1_k$ connections. Over time, peer k continues to track the popularity of its content, and re-computes its ideal degree (d^2_k, d^3_k,...).Whenever its ideal degree estimate is different from its actual degree, peer k adds or drops connections.
- ESR

We use two types of popular searching algorithms:

- Flooding: When a peer receives a search message, it both processes the message and forwards it to all of its neighbors in the overlay network. Each message is given a time-to-live value *ttl*, and search messages get flooded to every node within *ttl* hops of the source. There are two flooding algorithms: (1)unrepeated flooding, in each query, the query message will not sent to a node which has received that message; (2)repeated flooding, message can be sent to the same node repeatedly.
- Random walk: When a peer receives a search message, it processes the message and then forwards it to one or several randomly chosen neighbors (called walks). Messages continue random walking until either a predefined number of results are found (again, predefined by the user), or a *ttl* is reached. Random walk *ttl* values are high and exist mainly to prevent searches from walking forever. In the paper, walks=4.

Though there are many other unstructured searching algorithms, such as Biased high degree [6], Iterative deepening [35], Most results and Fewest result hops [36] etc., the flooding search method is very robust, flexible and easily supports partial-match and keyword queries. [37] demonstrates analytically that random walks are useful to locate popular content when the topology forms a super-peer network, and have better performance in topologies with power-law characteristic. Moreover, it is proofed in [34] that a square-root topology is optimal for random walk searches. Considering all four topologies being compared in our paper, it is rational that we just use this two search algorithms to evaluate the performance of topologies. Studies have shown that Gnutella, Media and Web queries tend to follow Zipf-like distributions [38]. Thus, in our simulations, the number of each file follows a Zipf-like distribution according to its popular degree. It is assumed that there are m original files. And q_i represents the relative popularity, in terms of the number of queries issued for it, of the ith object. Then, we can get $q_i \propto 1/i^{\alpha}$, where each file f_i is replicated on r_i nodes, and the total number of files stored in the network is R and $\sum_{i=1}^{m} r_i = R$. In our simulations, $\alpha = 0.726$ and $m = 300$, and the replication of a file f_i is proportional to the query probability of the file. If one assumes only nodes requesting a file store the file, then the replication distribution is usually proportional to query distribution (i.e. $r_i \propto q_i$).
We conducted our experiments on three types (i.e. RTPL, super-node topology and ESR) of physical network topologies with different node-number (i.e. N=500, 1000, 1500 and 2000) generated by using the GT-TTM library. For constructing square-root topology, we choose a maximum degree d_{max}, representing the degree we want for a peer whose popularity $g_k = 1$. Of course, it is unlikely that any peer will have content matching all queries, so the actual largest degree will almost certainly be less than d_{max}. Then, we can define D as $D = d_{\max} \cdot \sum_{i=1}^{N} \sqrt{g_i}$.
If the popularity of a peer's content is very low, then d_k will be very small. If peer degrees are too small, the network can become partitioned, which will prevent content at some peers from being found at all. In the worst case, because d_k must be an integer, so the ideal degree might be zero. Therefore, we define a value d_{min}, which is the minimum degree a peer will have. The degree a peer will aim for is:

$$d_k = \begin{cases} round(d_{\max} \cdot \sqrt{Q^k_{match}/Q^k_{total}} & \text{if greater than } \mathrm{d_{min}} \\ \mathrm{d_{min}} \text{ otherwise} \end{cases} \quad (8)$$

The square root constructing progress can be summarized as follows [34]: (1) We choose a maximum degree d_{max} and minimum degree d_{min}, and fix them as part of the peer-to-peer protocol. (2) Peer k joins, and makes some number d_k^0 of initial connections ($d_{\min} \le d_k^0 \le d_{\max}$). (3) Peer k tracks Q^k_{match} and Q^k_{total}, and continually computes d_k according to equation (8). (4) When the computed d_k differs from peer k's actual degree, k adds or drops connections. We ran 10000-time simulations to measure the performance of searches over time as the topology adapted under the square root constructing progress, the parameters for the square-root-construct algorithm are shown in Table. 1.

Table 1 Parameters for constructing square-root topology

Parameter	$N = 500$	$N = 1000$	$N = 1500$	$N = 2000$
d_{max}	40	80	100	160
d_{min}	3	3	3	3
d_k^0	4	4	4	4

We experimented with several parameter settings, and found that these settings worked well in practice.

6.2 Scalability

For simulations in this section, we generate every type of topologies 100 times respectively under various system scales. Node degree information of the four graphs under various node numbers are shown in Fig. 5. The number of layers and levels (are both H) in ESR are both 3.

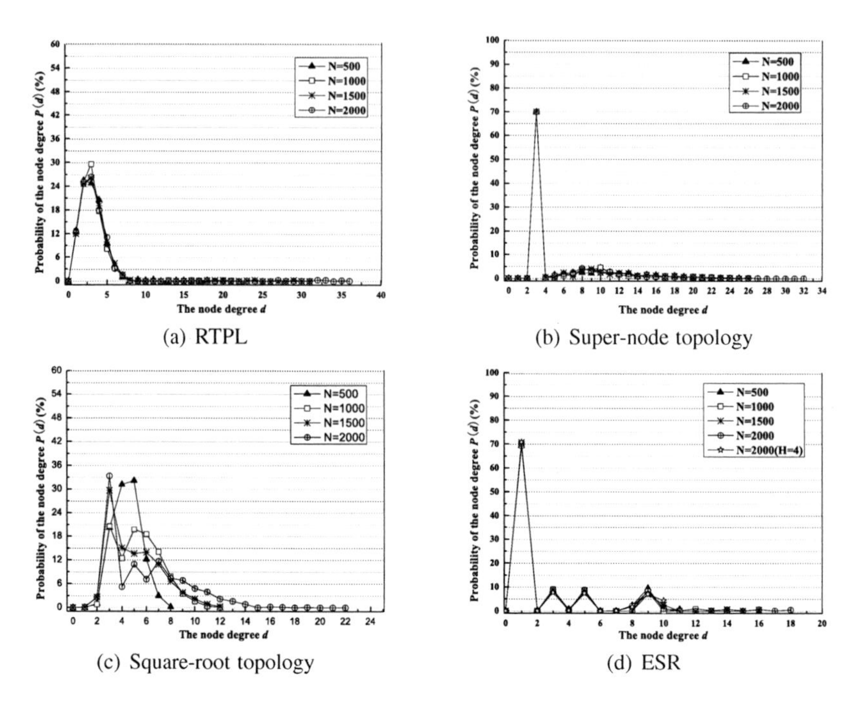

(a) RTPL

(b) Super-node topology

(c) Square-root topology

(d) ESR

Fig. 5 Distribution of node degrees in the four network topology graph

From Fig. 5, we can see that the value of the most degree node's degree respectively in RTPL, super-node topology and square-root topology greatly increases as the system's scale increases. For example, in RTPL, the value of the node with the most degree is 18 when N=500, while that is 36 when N=2000. In super-node topology, the value of the node with the most degree is 25 when N=500, while that is 32 when N=2000. In square-root topology, the former is 8 and the latter is 22. However, in ESR, the max degrees change from 11 to 18 when the N changes from 500 to 2000, that is, the value of the node with the most degree does not change quickly (i.e. the change is limited in unit position). Moreover from Fig. 5, it is clear that the node's degrees in ESR almost do not change as the system scale increases, or the number of layers increases (e.g. H=4). The theorem 1, 2 and Fig. 5 show that we are able to adjust such a 3D structure as ESR without violating the adaptive rules as clients join and leave. Overheads incurred by this adjustment could be kept small to keep the system scalable.

6.3 Query success rate

We evaluate the query success rate of four topologies with different system scales by the popular searching algorithms introduced in Sect. 6.1. In the simulation, there are total 4162 files including both original files and replications. For each experiment, the source is chosen randomly while the requested file is chosen according to Zipf distribution. We report the mean values of results obtained through 120000 runs. From simulation results (not shown in the paper for space limitation), we find that whatever the system scale and topologies are, the query success rate using repeated flooding algorithm is the highest, though the rate is very near to the query success rate using unrepeated flooding algorithm, especially in super-node topology and square-root topology. On the contrary, the rate using random walk algorithm is the lowest. So we just use one middle algorithm (i.e. unrepeated flooding) to evaluate and compare the query success rate in all four topologies when N=500 and N=2000, as shown in Fig. 6.

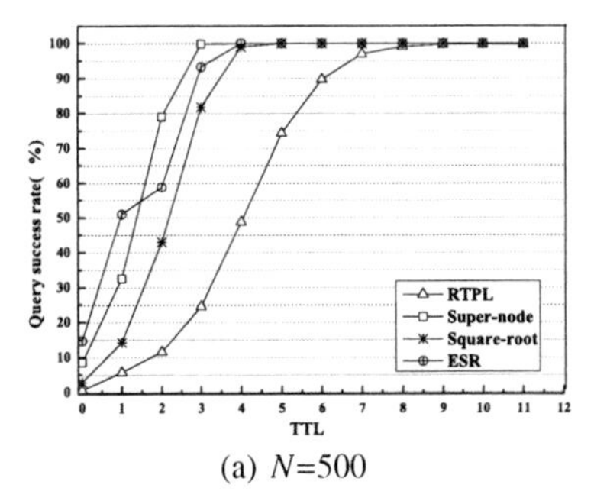

(a) N=500

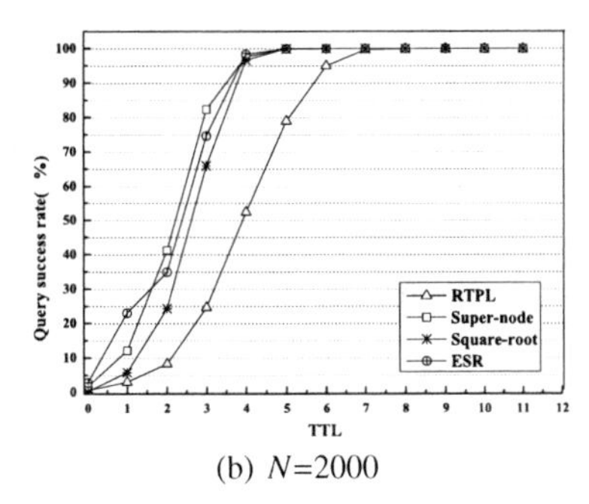

(b) N=2000

Fig. 6 Query success rate for various topologies

From Fig. 6, the query success rate in ESR is the highest when TTL is no more than 1, that is, the satisfied query in one hop in ESR is more than others. We think that it because of the 3D structure so that one AS has more neighbor-ASs.

6.4 Query messages and hops

The number of query messages in one query time is used to be one factor of the evaluation of searching efficiency. The repeated messages across one node are less, the cost should be less. We compute the number of query messages in RTPL, super-node topology and square-root topology under four system scales. We also find a solid trend on the contrary of the trend in query success rate. That is, the number of query messages produced by repeated flooding algorithm is the largest whichever topology and system scale are, while that by random walk algorithm is the lowest.
We just use one middle algorithm (i.e. unrepeated flooding) to evaluate and compare the number of query messages in all four topologies when N=500 and N=2000, as shown in Fig. 7. When one query is satisfied (i.e. the curve is nearly aclinic), the number of query messages in RTPL is much more than those in other three topologies. As shown in Fig. 6 (a) and (b), the query success rata in super-node topology is little higher than that in ESR when TTL=2 at most TTL=3, however, the number of query messages in ESR is much less than that in

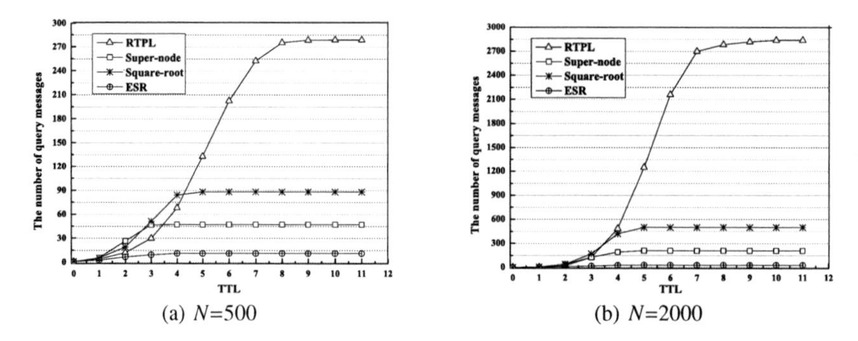

Fig. 7 The number of query messages for various topologies

super-node topology. The searching hop when one query has been satisfied is other important factor for evaluating the searching efficiency. If the hop when the query is satisfied is more, the length of searching path is longer. That is, the cost of searching should be more, which will be proved in Sect. 6.5. Being accordant to simulation results (not shown in the paper), the trend is solid that the hop when one query is satisfied using repeated flooding algorithm is the lowest whichever topology and system scale are, while that by random walk algorithm is the largest. That is understandable considering each algorithm's principle. Thus, we also use unrepeated flooding to evaluate and compare the hop when the query is satisfied in all four topologies when N=500 and N=2000, as shown in Fig. 8.

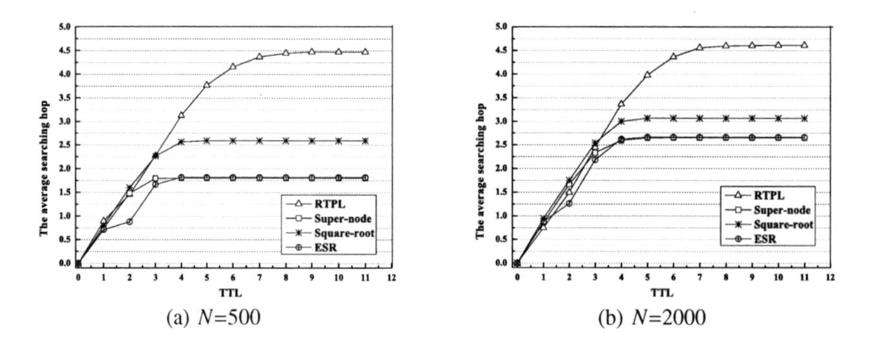

Fig. 8 The average searching hop in various topologies

It is clear that the searching hop in ESR is very near that in super-node topology when the query is satisfied. However, it is obvious that the searching hop in ESR is less than that in super-node topology when the TTL is small.

6.5 Cost and load balancing

In such high dynamic environments as P2P systems, the requirement for system stability is more important than for query success rate. By simulating, we find that there is the same regularity for the cost line distribution of each searching algorithm whichever the topology is. Therefore, we compare the four topology cost using unrepeated flooding algorithm when N=500 and N=2000, the results are shown in Fig. 9. The x-axis shows the upper bound of hops permitted by every searching, the y-axis shows the average system cost. The unit is *kbps*.

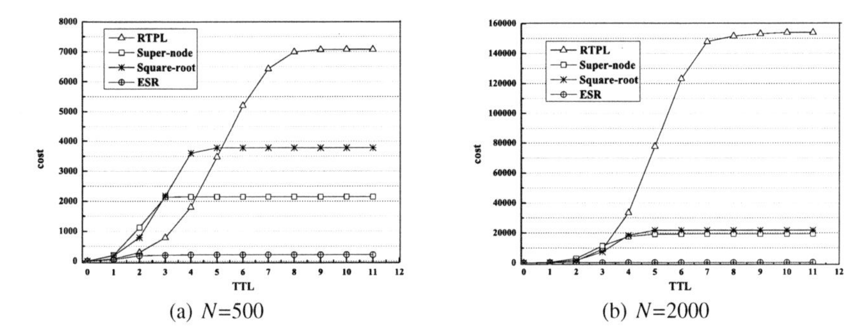

(a) N=500 (b) N=2000

Fig. 9 The searching cost in various topologies

In Fig. 6, it is clear that the query success rates of super-node topology, square-root topology and ESR are all near or equal 100% when TTL=4. However, we can find in Fig. 9 that the searching cost in ESR is much less than that in super-node and square-root topology. And the cost in ESR does not remarkably increase as the searching hops increasing, on the contrary, the costs in other 3 topologies rapidly increase after the first hop. Using different searching algorithms in various topologies, the situation of the disturbing to each node is different. The average disturbing times are more, the load of the system is heavier. Thus, the possibility of system crash is higher. Therefore, a good topology should have small disturbing rate in each search for guaranteeing the stability of the system. Just evaluating the total system cost maybe cover up the load unbalancing that some peers are disturbed too much. Accordingly, we give the average disturbed times of every node in total 120000 querying times in four topologies when N=500 and N=2000, as shown in Fig. 10.

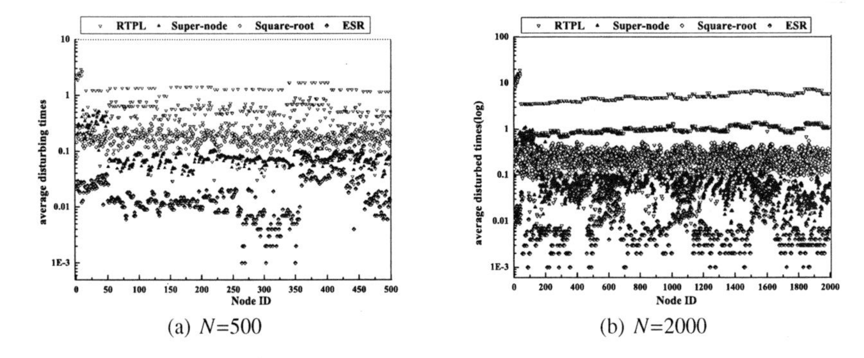

(a) N=500 (b) N=2000

Fig. 10 Disturbed times of every peer

From Fig. 10, it is obvious that to each node in ESR, the average disturbed times are less than other three topologies, some even equal to 0. Being similar to super-node topology, the average disturbed times of some peers are higher than others in ESR. It is understandable that they are ICs which are detailedly described in Sect. 4.3. But it is clear that the average disturbed times of the ICs whose loads are the most are farther less than super peers in super-node topology, and near those of the normal node in the latter. It may be also one of the reasons that the cost of ESR is less.

6.6 Fault-tolerance and robustness

In this section, we concentrate on discussing the performance of every topology using unrepeated flooding algorithm when there are peers to leave system randomly. Fig. 11 shows the performances of topologies, where the x-axis is the ratio of the number of randomly leaving nodes to all nodes' number in the system.

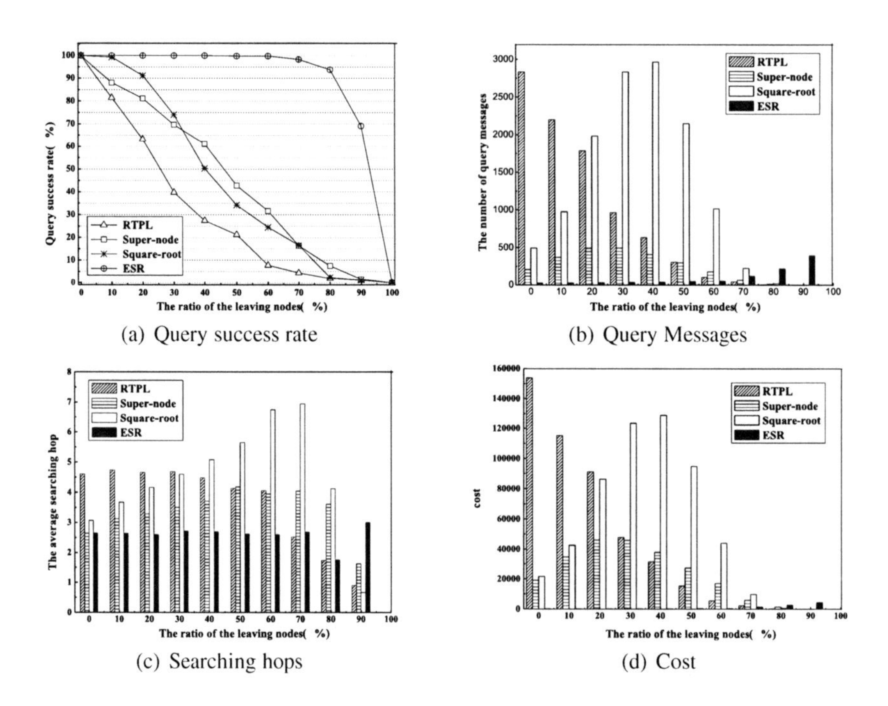

(a) Query success rate (b) Query Messages

(c) Searching hops (d) Cost

Fig. 11 The performance when there is randomly leave

It can be seen from the Fig. 11 that the results are consistent with the conclusion in theory. Fig. 11(a) shows that the query success rate in ESR is very high. Fig. 11(c) shows that the average hops in the RTPL, super-node topology and square-root topology all increase when there are nodes randomly leaving. But the average hops in the ESR are almost unchanged and only increase when the condition is very worse (e.g. 90% nodes leaving). In the worse condition, there are too much nodes leaving the system so that the query hops increase when the query is successful. In addition, it is clear in the Fig. 11(b), (c) and (d) that each performance in the square-root worsens with the number of leaving nodes increasing. We think it is because that the formation of square-root topology is based on the query frequency of the files. In square-root topology, the node whose degree is more is consequentially the node whose own files are more popularity. So each aspect of performance in square-root topology is certainly affected when the number of leaving nodes increases. The number of query message and system cost in the ESR when the query is successful don't greatly wave since the nodes leave. It shows that the ESR topology structure has favorable stability.

It is defined that i is the query initiated by the ith node ($i \in (N - N * rate_{lose})$), $rate_{lose}$ is the failure rate of the node, N is the system scale, n_i is the number of the nodes visited by the ith query. In the system which allows adequate query hops (i.e. the query can visit all nodes connected with the source node directly or not), we define the most coverage rate of nodes is $CoverRate = Max(n_i)/(N - N \cdot rate_{lose})$, while the connectivity rate of the system is $\sum_{i=1}^{N-N \cdot rate_{lose}} n_i \big/ (N - N \cdot rate_{lose})$. Fig. 12 shows the changing of the most coverage rate and the connectivity rate of the system along with the leaving node rate increasing.

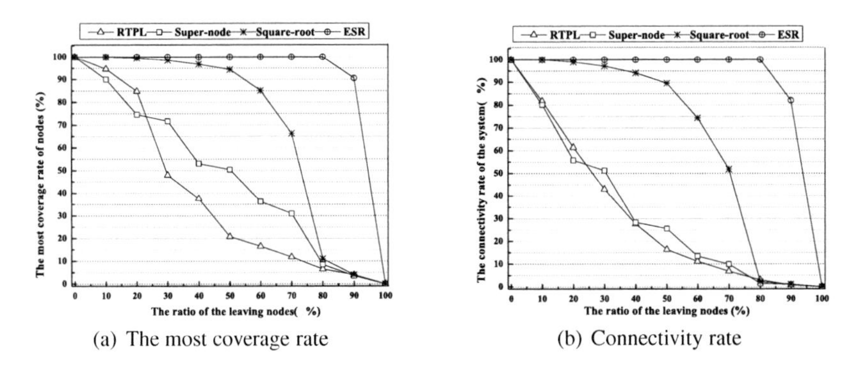

(a) The most coverage rate
(b) Connectivity rate

Fig. 12 The most coverage and connectivity rate when there is randomly leave

It is clear in the Fig. 12 that the robustness of the ESR is far better than other topologies. Its connectivity rate is still 100% even when there are 80% nodes leaving the system. The most coverage rate and the system connectivity have remarkable decrease after 90% nodes leave the system. Combined Fig. 11 and Fig. 12, it is the most different between the intuitive judgment and the real performance of square-root topology. In Fig. 12, it is clear that the node's most coverage rate and the connectivity rate in the square-root topology are far higher than those in the RTPL and super-node topology. To our intuitive judgment, the query success rate of square-root topology should also be far higher than the two latter, while the number of query messages, the query hops and cost should be less than those of the two latter. However, from Fig. 11, we find it is so different from our intuition, e.g. the query success rate of square-root topology is similar to the two latter. This is decided by the characteristic of the square-root topology forming. The physical topology of the square-root is robust enough because of its high average connected degree. But the files on the nodes whose degree are high have higher popularity. When such nodes leave the system more and more, the query success rate will decrease. It reflects that the square-root topology doesn't consider the matching between the physical topology and the logical structure. It is clear through the detailed analyzing that the ESR is better than other three topologies in both the physical topology and logical structure.

7 Conclusion and future directions

This chapter has presented various schemes in structured and unstructured P2P overlay networks that have been proposed by researchers. The P2P overlay network that is best suited depends on the application and its required functionalities and performance metrics for example, scalability, network routing performance, location service, file sharing, content distribution, and so on.
In this chapter, we investigated the challenges facing autonomic communication, showed some researches in P2P overlay to solve it and presented the key problems and requirements in order to build an ideal system for ACs. Then, we presented three requirements for P2P overlay to satisfy those challenges. ESR is the first P2P overlay that can be compatible with AC. And ESR is (1) with global view combined with limited state, and localized connection and information exchange; (2) giving attention to both efficiency and prevention of selfishness; (3) guaranteeing each ID's uniqueness by the particular overlay structure with fault-tolerance and robustness. Comprehensive simulation results show the advantages of the ESR from various aspects. Base on ESR, more challenges for AC will be researched such as management of trust and privacy of users, or information communication algorithms etc. The subject of exploiting peer-to-peer overlays in ACs is relatively new. Many interesting problems require further research, including:

- Given the high dynamics in ACs due to node mobility, which of the unstructured or structured overlay abstractions is more efficient in supporting common distributed applications?
- How can one efficiently integrate an unstructured P2P overlay for the Internet with AC routing protocols other than DHT-like schemes in other applications such as MANETs or Ad Hoc Networks?
- Can incentive techniques developed for P2P overlays in the Internet for encouraging peering nodes to cooperate be applied to ACs?
- Trust and reputation is also important for secured and trustworthy ACs among the peers.

References

1. S. Dobson, F. MASSACCI, F. ZAMBONELLI, *A Survey of Autonomic Communications*, ACM Transactions on Autonomous and Adaptive Systems (TAAS), 1(2), pp.223-259, (2006)
2. Gnutella Protocol Specification, version 0.4. http://www.clip2.com/GnutellaProtocol04.pdf, 2001.
3. I. Stoica, R. Morris, D. Karger, M. F. Kaashoek, and H. Balakrishnan, *Chord: A scalable peer-to-peer lookup protocol for internet applications*, IEEE/ACM Transactions on Networking, 11(1), pp.17-32, (2003)
4. A. Rowstron and P. Druschel, *Pastry: Scalable, distributed object location and routing for large-scale peer-to-peer systems*, In: Proceedings of the Middleware, (2001)
5. S. Ratnasamy, P. Francis, M. Handley, R. Karp, and S. Shenker, *A scalable content addressable network*, In: Processings of the ACM SIGCOMM, pp.161-172, (2001)
6. L.A. Adamic, R.M. Lukose, A.R. Puniyani and B.A. Huberman, *Search in Power-law Networks*, Physical Review, E 64, pp.46135-46143, (2001)
7. N. Sarshar, P.O. Boykin, V.P. Roychowdhury, P*ercolation Search in Power Law Networks: Making Unstructured Peer-To-Peer Networks Scalable*, In: Fourth International Conference on Peer-to-Peer Computing (P2P'04), Zĺźrich, Switzerland, pp.2-9, (2004)
8. A.Q. Al-Namiy, F.S. Majeed, *Improving query answering in peer-to-peer data searching*, Nineteenth International Conference on Advanced Information Networking and Applications, pp.689-694, (2005)

9. L.Rong, and I. Burnett, *Dynamic resource adaptation in a heterogeneous peer-to-peer environment*, Second IEEE Consumer Communications and Networking Conference, pp.416-420, (2005)
10. S. Johnstone, P. Sage, P. Milligan, *iXChange - A Self-Organising Super Peer Network Model*, In: Proceedings of the 10th IEEE Symposium on Computers and Communications (ISCC 2005), pp.164-169, (2005)
11. J.R. Douceur, *The Sybil Attack*, International Workshop on Peer-to-Peer Systems (IPTPS), Cambridge, MA, USA, pp.251-260, (2002)
12. M. Castro, P. Druschel, A. Ganesh, A. Rowstron, D.S. Wallach, *Secure Routing for Structured Peer-to-Peer Overlay Networks*, Symposium on Operating Systems Design and Implementation (OSDI), Boston, MA, USA, (2002)
13. E. Adar and B.A. Huberman, *Free Riding on Gnutella*, Technical Report, Xerox PARC August, (2000)
14. H. Daniel, C. Geoff and W. James, *Free riding on Gnutella revisited: The bell tolls?*, IEEE Distributed Systems Online, 6(6), (2005)
15. C. Courcoubetis and R. Weber, *Incentives for Large Peer-to-Peer Systems*, IEEE Journal on selected areas in communications, 24(5), pp.1034-1050, May (2006)
16. M. Feldman, C. Papadimitriou, J. Chuang, and I. Stoica, *Free-Riding and Whitewashing in Peer-to-Peer Systems*, IEEE Journal on selected areas in communications, 24(5), pp.1010-1019, May (2006)
17. D. Purandare and R. Guha, *Preferential and Strata based P2P Model: Selfishness to Altruism and Fairness*, Proceedings of the 12th International Conference on Parallel and Distributed Systems (ICPADS'06), (2006)
18. R. Guha, R. Kumar, P. Raghavan, A. Tomkins, *Propagation of Trust and Distrust*, In: Proceedings International WWW Conference, New York, USA, pp.403-412, (2004)
19. B. Cohen, Incentives build robustness in BitTorrent, Workshop on Economics of Peer-to-Peer Systems, (2003)
20. K.G. Anagnostakis and M.B. Greenwald, *Exchange-based incentive mechanisms for peer-to-peer file sharing*, In: International Conference on Distributed Computing Systems (ICDCS), pp.524-533, (2004)
21. D. Qiu and R. SRIKANT, M*odeling and performance analysis of bittorrent-like peer-to-peer networks*, In: SIGCOMM, R. Yavatkar, E. W. Zegura, and J. Rexford, Eds. ACM, pp.367-378, (2004)
22. C. Aperjis and R. Johari, *A peer-to-peer system as an exchange economy*, Workshop on Game Theory for communications and networks,Pisa, Italy, October (2006)
23. E.J. Friedman, J.Y. Halpern and I. Kash, *Efficiency and nash equilibria in a scrip system for P2P networks*, Proceedings of the 7th ACM conference on Electronic commerce, pp.140-149, (2006)
24. V. Srivastava, et al., *Using Game Theory to Analyze Wireless Ad Hoc Networks*, IEEE Communications Surveys & Tutorials, 7(4), pp.46-56, (2005)
25. R. Mahajan, *Practical and Efficient Internet Routing with Competing Interests*, Ph.D. Dissertation (also UW-CSE TR #2005-12-02), December (2005)
26. B. Wong, Y. Vigfusson, and E. Sirer, *Hyperspaces for object clustering and approximate matching in peer-to-peer overlays*, In: Proceedings of the Workshop on Hot Topics in Operating Systems, (2007)
27. I. Brunkhorst, H. Dhraief, A. Kemper, W. Nejdl, C. Wiesner, *Distributed Queries and Query Optimization in Schema-Based P2P-Systems*, In: International Workshop on Databases, Information Systems and Peer-to-Peer Computing, Berlin, Germany, (2003)
28. D.A. Tran, K.A. Hua and T.T. Do, *A peer-to-peer architecture for media streaming*, IEEE Journal on Selected Areas in Communications, 22(1), pp.1-14, January (2004)
29. C.M. Huang, T.H. Hsu, M.F. Hsu, *Network-aware P2P file sharing over the wireless mobile networks*, IEEE Journal on Selected Areas in Communications, vol.25, pp.204-210, January (2007)
30. J. Kleinberg, *The small-world phenomenon: an algorithmic perspective*, In: Proceedings of the 32nd ACM Symposium on Theory of Computing, pp.163-170, (2000)

31. Y.J. Joung and J.C. Wang, *Reducing Maintenance Overhead in Chord via Heterogeneity*, In: Proc. 5th International Workshop on Global and P2P Computing (GP2PC), Cardiff, UK, pp.221-228, May (2006)
32. D. Li, Z.G. Chen, H. Liu, A.V. Vasilakos, *An Adaptive and Self-Supervised Structured P2P Overlay for Autonomic Communication*, In: ICWN, Las Vegas, Nevada, USA, pp. 412-417, (2007)
33. M.A. Jovanovic, *Modeling large-scale peer-to-peer networks and a case study of Gnutella*, [MS. Thesis], University of Cincinnati, Cincinnati, Ohio, USA, (2001)
34. B.F. Cooper, *An optimal overlay topology for routing peer-to-peer searches*, In: Middleware, Berlin, Heidelberg: Springer-Verlag, pp. 82-101, (2005)
35. B. Yang and H. Garcia-Molina, *Efficient Search in Peer-to-Peer Networks*, In: ICDCS, Vienna, Austria, pp.5-14, July (2002)
36. B. Yang, H. Garcia-Molina, *Improving search in peer-to-peer networks*, In: ICDCS, pp.5-14, (2002)
37. C. Gkantsidis, M. Mihail, A. Saberi, *Random walks in peer-to-peer networks*, In: INFOCOM, New York, IEEE Press, pp.120-130, (2004)
38. K.A. Hua, C. Lee, C.M. Hua, *Dynamic load balancing in Multicomputer database systems using partition tuning*, IEEE Transactions on Knowledge and Data Engineering, 7(6), pp.968-983,(1995)
39. E.W. Zegura, K.L. Calvert and S. Bhattacharjee, *How to model an internetwork*, In: INFOCOM, San Francisco, CA, USA, pp.594-602,(1996)

Autonomic and Coevolutionary Sensor Networking

Pruet Boonma and Junichi Suzuki

Abstract (WSNs) applications are often required to balance the tradeoffs among conflicting operational objectives (e.g., latency and power consumption) and operate at an optimal tradeoff. This chapter proposes and evaluates a architecture, called BiSNET/e, which allows WSN applications to overcome this issue. BiSNET/e is designed to support three major types of WSN applications: , and hybrid applications. Each application is implemented as a decentralized group of , which is analogous to a bee colony (application) consisting of bees (agents). Agents collect sensor data or detect an event (a significant change in sensor reading) on individual nodes, and carry sensor data to base stations. They perform these data collection and event detection functionalities by sensing their surrounding network conditions and adaptively invoking behaviors such as pheromone emission, reproduction, migration, swarming and death. Each agent has its own behavior policy, as a set of genes, which defines how to invoke its behaviors. BiSNET/e allows agents to evolve their behavior policies (genes) across generations and autonomously adapt their performance to given objectives. Simulation results demonstrate that, in all three types of applications, agents evolve to find optimal tradeoffs among conflicting objectives and adapt to dynamic network conditions such as traffic fluctuations and node failures/additions. Simulation results also illustrate that, in hybrid applications, data collection agents and event detection agents coevolve to augment their adaptability and performance .

1 Introduction

Autonomous adaptability is a key challenge in wireless sensor networks (WSNs) [1, 2, 6, 28, 33]. With minimal intervention to/from human operators, WSN applications are required to adapt their operations to dynamic changes in network conditions such as traffic fluctuations and node failures/additions. A critical issue in this challenge is that WSN applications have inherent tradeoffs among conflicting operational objectives [22]. For example, success rate of data transmission from individual nodes to base stations is an important objective because higher success rate ensures that base stations receive more sensor data for operators to better understand the current situation in an observation area and make better informed decisions. At

Department of Computer Science
University of Massachusetts, Boston
e-mail: {pruet,jxs}@cs.umb.edu

A.V. Vasilakos et al. (eds.), *Autonomic Communication*, DOI: 10.1007/978-0-387-09753-4_14,

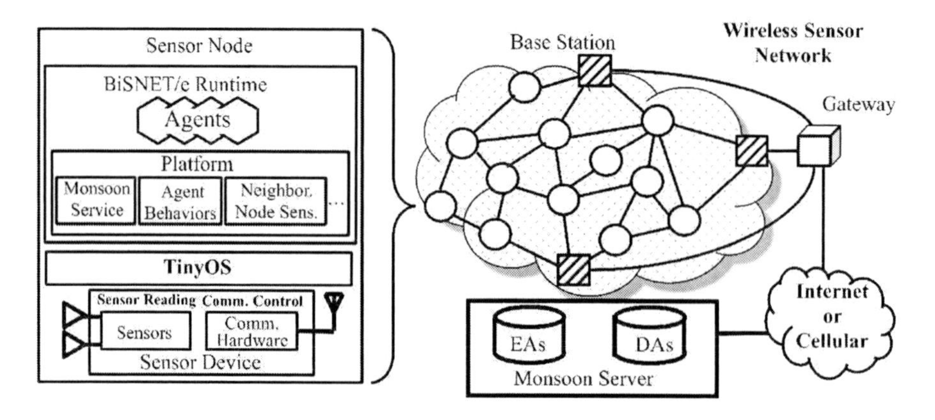

Fig. 1 BiSNET/e Runtime Architecture

the same time, latency of data transmission from individual nodes to base stations is another important objective. Lower latency ensures that base stations can collect sensor data for operators to understand the situation of an observation area more quickly and make more timely decisions. Success rate and latency conflict with each other. For improving success rate, hop-by-hop recovery is often applied; however, this can degrade latency. For improving latency, nodes may transmit data to base stations with the shortest paths; however, success rate can degrade because of traffic congestion on the paths.

In order to address this issue, the authors of the chapter envision autonomic WSN applications that understand their operational objectives, sense dynamic network conditions and act autonomously to satisfy conflicting objectives simultaneously. As inspiration for this vision, the authors observe that various biological systems have developed the mechanisms to overcome the above adaptability issue. For example, each bee colony autonomously satisfies conflicting objectives to maintain its well-being [34]. Those objectives include maximizing the amount of collected honey, maintaining temperature inside a nest and minimizing the number of dead drones. If bees focus only on foraging, they fail to ventilate their nest and remove dead drones. Based on this observation, the proposed application architecture, called BiSNET/e (Biologically-inspired architecture for Sensor NETworks, evolutionary edition), applies key biological mechanisms to design adaptive WSN applications.

Figure 1 shows the BiSNET/e runtime architecture. The BiSNET/e runtime operates atop TinyOS on each node. It consists of two software components: *agents* and *middleware platforms*, which are modeled after bees and flowers, respectively. Each WSN application is designed as a decentralized group of agents. This is analogous to a bee colony (application) consisting of bees (agents). Agents collect sensor data and/or detect an event (a significant change in sensor reading) on platforms (flowers) atop individual nodes. Then, they carry sensor data to base stations, in turn, to the MONSOON server (Figure 1). The server is modeled after a nest of bees. Agents perform these data collection and event detection functionalities by autonomously sensing their surrounding network conditions and adaptively performing biological behaviors such as pheromone emission, reproduction, migration, swarming and death. A middleware platform runs on each node, and hosts an arbitrary number of agents (Figure 1). It provides a series of runtime services that agents use to perform their functionalities and behaviors.

This chapter describes a key component in BiSNET/e, called MONSOON[1], which is an for agents. Each agent has its own behavior policy, as a set of *genes*, which defines when to and

[1] Multiobjective Optimization for Network of Sensors using a cO-evOlutionary mechaNism

how to invoke its behaviors. MONSOON allows agents to evolve their behavior policies via genetic operations (mutation and crossover) across generations and simultaneously adapt their performance to given objectives. Currently, MONSOON considers four objectives: success rate, latency, power consumption and the degree of data aggregation.

The evolution process in MONSOON frees application developers from anticipating all possible network conditions and tuning their agents' behavior policies to the conditions at design time. Instead, agents are designed to autonomously evolve and tune their behavior policies at runtime. This can significantly simplify the implementation and maintenance of agents (i.e., WSN applications).

BiSNET/e supports three major types of WSN applications: *data collection*, *event detection* and *hybrid* applications. Hybrid applications perform both data collection and event detection in order to fulfill complex sensing requirements such as target tracking [25], contour/edge detection [11] and spatiotemporal event detection/monitoring [39]. Different types of applications are implemented with different types of agents. Data collection and event detection applications use *data collection agents* (DAs) and *event detection agents* (EAs), respectively. Hybrid applications use both DAs and EAs. DAs and EAs are designed as different biological species. In hybrid applications, the two types of agents are intended to coevolve and adapt their behavior policies in a symbiotic manner. EAs help DAs improve their behavior policies, and vice versa.

This chapter is organized as follows. Section 2 overviews the structure and behaviors of agents in BiSNET/e. Section 3 describes the evolution and coevolution processes in MONSOON. Section 4 evaluates MONSOON with a series of simulation results. Simulation results demonstrate that, in all three types of applications, agents are robust and adaptive against various dynamic network conditions such as traffic fluctuations, node failures/additions and base station failures. Agents successfully evolve their behavior policies to find optimal tradeoffs among conflicting objectives. Simulation results also illustrate that, in hybrid applications, DAs and EAs coevolve to augment their adaptability and performance with each other. Sections 5 and 6 conclude with some discussion on related work and future work.

2 BiSNET/e Agents

At the beginning of a WSN's operation, one DA and one EA are deployed on each node. They have randomly-generated behavior policies. A DA collects sensor data on each node periodically (i.e., at each duty cycle) and carry the data to a base station on a hop-by-hop basis. An EA collects sensor data on each node periodically, and if it detects an event—a significant change in its sensor reading, carries the data to a base station on a hop-by-hop basis. If an event is not detected, the EA discards the data. (It is not transmitted to a base station.)

Agents are decentralized in a WSN. There are no centralized entities to control and coordinate agents. Decentralization allows agents to be scalable and survivable by avoiding a single point of performance bottlenecks and failures [3, 24].

2.1 Agent Structure and Behaviors

Each agent consists of *attributes*, *body* and *behaviors*. *Attributes* carry descriptive information on an agent. They include agent type (DA or EA), behavior policy (genes), sensor data to be reported to a base station, the data's time stamp, and the ID of a node where the data is collected.

Body implements the functionalities of an agent: collecting, processing, discarding and processing sensor data.
Behaviors implement actions inherent to all agents. Inspired by biological entities such as bees, agents sense their surrounding network conditions and behave according to the sensed conditions without any intervention from/to other agents, platforms, base stations and human operators. This chapter focuses on the following seven behaviors.

1. **Food gathering and consumption:** Biological entities strive to seek food for living. For example, bees gather nectar to produce honey. Similarly, in BiSNET/e, each agent periodically reads sensor data (as nectar) to gain *energy* (as honey)[2] and expends a constant amount of energy for living.
2. **Pheromone emission:** Agents may emit different types of pheromones: *migration* and *alert pheromones*. They emit migration pheromones on their local nodes when they migrate to neighboring nodes. Each migration pheromone references the destination node an agent has migrated to. Agents also emit alert pheromones when they fail migrations within a timeout period. Migration failures may occur because of node failures due to depleted battery and physical damages as well as link failures due to interference and congestion. Each alert pheromone references the node that an agent could not migrate to. Each of migration and alert pheromones has its own concentration, which decays by half at every duty cycle. A pheromone disappears when its concentration becomes zero.
3. **Replication:** EAs may make a copy of themselves in response to the abundance of stored energy, while DAs make a copy of themselves at each duty cycle. A replicated (child) agent is placed on the node that its parent resides on, and it inherits the parent's agent type and behavior policy (a set of genes). Replicated agents are intended to move toward base stations to report collected sensor data.
4. **Migration:** Agents may move from one node to another. Migration is used to transmit agents (sensor data) toward base stations. On an intermediate node, each agent chooses the next-hop node by sensing three types of available pheromones: base station, migration and alert pheromones.
   Each base station periodically propagates base station pheromones to individual nodes. Their concentration decays on a hop-by-hop basis. Using base station pheromones, agents can sense where base stations exist approximately, and they can move toward the base stations by climbing a concentration gradient of base station pheromones[3].
   An agent may move to a base station by following a migration pheromone trace on which many other agents have traveled. The trace can be the shortest path to a base station. Conversely, an agent may go off a migration pheromone trace and follows another path to a base station when the concentration of migration pheromones is too high on the trace (i.e., when too many agents have followed the trace). This avoids separating the network into islands. The network can be separated with the migration paths that too many agents follow, because the nodes on the paths run out of their battery earlier than the others[4].
   An agent may also avoid moving to a node referenced by an alert pheromone. This allows agents to reach base stations by bypassing failed nodes/links.
5. **Swarming:** Agents may swarm (or merge) with others at the nodes on their ways to base stations. With this behavior, multiple agents become a single agent. (A DA can merge with both DAs and EAs, and an EA can merge with both EAs and DAs.) The resulting agent (swarm) aggregates sensor data contained in other agents, and uses the behavioral policy of the best agent in the swarm in terms of given operational objectives.

---

[2] In BiSNET/e, the concept of energy does not represent the amount of physical battery in a node. It is a logical concept to affect agent behaviors.

[3] Base station pheromones are designed after the Nasonov gland pheromone, which guides bees to move toward their nest [14].

[4] Data transmission imposes the highest power consumption among all the operations that each node performs [26].

In order to increase the chances of swarming, at each intermediate node toward a base station, an agent may wait for other agents. If an agent(s) arrives at the node during a waiting period, the waiting agent merges with the arriving agent(s). The swarming behavior saves power consumption of nodes because in-node data aggregation requires much less power consumption than data transmission does [26].

6. **Reproduction:** Once agents arrive at the MONSOON server (Figure 1), they are evaluated according to their four objectives. Then, MONSOON selects best-performing (or elite) agents, and propagates them to individual nodes. An agent running on each node performs reproduction with one of the propagated agents. A reproduced agent inherits a behavior policy (gene) from its parents via crossover, and mutation may occur on the inherited behavior policy. Reproduced agents trigger a generation change by taking over existing agents running on individual nodes.
   Reproduction is intended to evolve agents so that the agents that fit better to the environment become more abundant. It retains the agents whose fitness to the current network conditions is high (i.e., the agents that have effective behavior policies, such as moving toward a base station in a short latency), and eliminates the agents whose fitness is low (i.e., the agents that have ineffective behavior policies, such as consuming too much power to reach a base station). Through successive generations, effective behavior policies become abundant in agent population while ineffective ones become dormant or extinct. This allows agents to adapt to dynamic network conditions.
7. **Death:** Agents periodically consume energy for living and expend energy to invoke their behaviors. The energy costs to invoke behaviors are constant for all agents. Agents die due to lack of energy when they cannot balance energy gain and expenditure. The death behavior is intended to eliminate the agents that have ineffective behavior policies. For example, an agent would die before arriving at a base station if it follows a too long migration path. When an agent dies, the local platform removes the agent and releases all resources allocated to it.

2.2 Behavior Sequence for DAs

Figures 2 shows the sequence of behaviors that each DA performs on a node at each duty cycle. A DA reads sensor data with the underlying sensor device and gains a constant amount of energy. Given the energy intake (E_F), each agent updates its energy level as follows.

$$E(t) = E(t-1) + E_F \tag{1}$$

$E(t)$ and $E(t-1)$ denote a DA's energy level at the current and previous duty cycle. t is incremented by one at each duty cycle.

If a DA's energy level ($E(t)$) goes below the death threshold (T_D), the DA dies due to starvation[5].

A DA replicates itself at each duty cycle. A replicating (parent) agent splits its energy units to halves ($\frac{E(t)-E_R}{2}$), gives a half to its child agent, and keeps the other half. E_R is the energy cost for an agent to perform the replication behavior. A child agent contains the sensor data that its parent collected, and carries it to a base station.

Each replicated DA migrates toward a base station on a hop by hop basis. On each intermediate node, it decides whether it migrates to a next-hop node or wait for other agents to swarm (or merge) with them. This decision is made based on a migration probability (p_m). If the agent decides to migrate, it examines Equation 2 to determine which next-hop node it migrates to.

[5] If all agents are dying on a node at the same time, a randomly selected agent for each type (i.e., EA and DA) will survive. At least one agent of each type runs on each node.

for each *duty cycle*

do
- Read sensor data and gain energy (E_F).
- Update energy level ($E(t)$).
- **if** $E(t) <$ *the death threshold* (T_D)
  - **then** Invoke the death behavior.
- Invoke the replication behavior to make a child agent.
- Give the half of the current energy level to a replicated (child) agent.
- **for each** *migrating agent*
  - **do**
    - **if not** *waiting*
      - **then**
        - Determine the destination node of migration.
        - Emit a migration pheromone on the local node.
        - Migrate to a neighboring node.
        - **if** *Migration fails*
          - **then**
            - Emit an alert pheromone on the local node.
            - Propagate it to neighboring nodes.

Fig. 2 Sequence of DA Behaviors

$$WS_j = \sum_{t=1}^{3} w_t \frac{P_{t,j} - P_{t_{min}}}{P_{t_{max}} - P_{t_{min}}} \tag{2}$$

A DA calculates this weighted sum (WS_j) for each neighboring node j, and moves to a node that generates the highest weighted sum. t denotes pheromone type; P_{1j}, P_{2j} and P_{3j} represent the concentrations of base station, migration and alert pheromones on the node j. $P_{t_{max}}$ and $P_{t_{min}}$ denote the maximum and minimum concentrations of P_t among all neighboring nodes. When a DA is migrating to a neighboring node, it emits a migration pheromone on the local node. If the DA's migration fails, it emits an alert pheromone, and it spreads to one-hop away neighboring nodes.

2.3 Behavior Sequence for EAs

Figures 3 shows the sequence of behaviors that each EA performs on a node at each duty cycle. When an EA reads sensor data (as nectar) with the underlying sensor device and gains energy (as honey), its current energy level ($E(t)$) is updated with Equation 3.

$$E(t) = E(t-1) + S \cdot M \tag{3}$$

S denotes the absolute difference between the current and previous sensor data. M is metabolic rate, which is a constant between 0 and 1.

Each EA replicates itself if its energy level exceeds the replication threshold: $T_R(t)$. The replication threshold is continuously adjusted as an EWMA (Exponentially Weighted Moving Average) of energy level:

$$T_R(t) = (1-\alpha)T_R(t-1) + \alpha E(t) \tag{4}$$

$T_R(t)$ and $T_R(t-1)$ denote the replication thresholds at the current and previous duty cycle, respectively. EWMA is used to smooth out short-term minor oscillations in the data series of E. It places more emphasis on the long-term transition trend of E; only significant changes in E have the effects to change T_R. The α value is a constant to control the responsiveness of EWMA against the changes of E.

A parent EA splits its energy units to halves, gives a half to its child agent, and keeps the other half. The parent EA keeps replicating itself until its energy level becomes less than its

T_R. Each child agent contains the sensor data that its parent collected, and carries it to a base station.

As DAs do, each migrating EA decides whether it performs the migration behavior or the swarming behavior using its migration probability (p_m). It performs the migration behavior with Equation 2, followed by the pheromone emission behavior, in the same way as DAs do.

for each *duty cycle*
do
- Read sensor data and gain energy (E_F).
- Update energy level ($E(t)$).
- **if** $E(t)$ < *the death threshold* (T_D)
  - **then** Invoke the death behavior.
- **while** $E(t)$ > *the replication threshold* ($T_R(t)$)
  - **do**
    - Invoke the replication behavior to make a child agent.
    - Give the half of the current energy level to the child agent.
- **for each** *migrating agent*
  - **do**
    - **if not** *waiting*
      - **then**
        - Determine the destination node of migration.
        - Emit a migration pheromone on the local node.
        - Migrate to a neighboring node.
        - **if** *Migration fails*
          - **then**
            - Emit an alert pheromone on the local node.
            - Propagate it to neighboring nodes.

Fig. 3 Sequence of EA Behaviors

2.4 Agent Behavior Policy

EAs and DAs have the same structure of behavior policies (genes). Each behavior policy consists of two distinctive information: migration probability (p_m) and a set of weight values in Equation 2 ($w_t, 1 \leq t \leq 3$). Migration probability is a non-negative value between zero and one. With higher migration probability, an agent has a higher chance to perform the migration behavior instead of the swarming behavior. With a lower migration probability, an agent has a higher chance to perform the swarming behavior. Weight values govern how agents perform the migration behavior. For example, if an agent has zero for w_2 and w_3, the agent ignores migration and alert pheromones, and moves toward the base stations by climbing the concentration gradient of base station pheromones. If an agent has a positive value for w_2, it follows a migration pheromone trace on which many other agents have traveled. A negative w_2 value allows an agent to go off a migration pheromone trace and follow another path toward a base station. If an agent has a negative value for w_3, it moves to a base station by bypassing failed nodes/links.

3 MONSOON

In order to drive agent evolution and coevolution, MONSOON performs *elite selection* and *genetic operations*. The elite selection process evaluates each type of agents (DAs and EAs) that arrive at base stations, based on given objectives, and chooses the best (or elite) ones.

Elite agents are propagated to individual nodes in the network. Through genetic operations (crossover and mutation), an agent running on each node performs the reproduction behavior with one of elite agents. A reproduced agent inherits a behavior policy (a set of genes) from its parents via crossover, and mutation may occur on the inherited behavior policy. Reproduced agents trigger a generation change by taking over parent agents. Elite selection is performed in the MONSOON server (Figure 1), and genetic operations are performed in each node. Reproduction is intended to evolve agents so that the agents that fit better to the current network conditions become more abundant. It retains the agents that have effective behavior policies, such as moving toward a base station in a short latency, and eliminates the agents that have ineffective behavior policies, such as consuming too much power to reach a base station. Through successive generations, effective behavior policies become abundant in agent population while ineffective ones become dormant or extinct. This allows agents to adapt to dynamic network conditions.

3.1 Operational Objectives

Each agent (DA or EA) considers four conflicting objectives: *latency*, *cost*, *success rate* and *data yield*. MONSOON strives to minimize latency and cost and maximize success rate and data yield.

1. **Latency** represents the time required for an agent (DA or EA) to travel to a base station from a node where the agent is born (replicated). As depicted below, latency (L) is measured as a ratio of this agent travel time (t) to the physical distance (d) between a base station and a node where the agent is born. The MONSOON server knows the location of each node with a certain localization mechanism.

$$L = \frac{t}{d} \tag{5}$$

2. **Cost** represents power consumption required for an agent (DA or EA) to travel to a base station from a node where the agent is born. Cost (C) is measured with the total number of node-to-node data transmissions required for an agent to arrive at a base station (n_{tran}), each node's radio transmission range (r_{tran}), and physical distance (d).

$$C = \frac{n_{tran}}{d/r_{tran}} \tag{6}$$

The total number of data transmissions include successful and unsuccessful (failed) agent migrations as well as the transmissions of migration or alert pheromones.

3. **Success rate** is measured differently for DAs and EAs. For DAs, it is measured as follows.

$$S_{DA} = \frac{n_{arrive}}{N} \tag{7}$$

n_{arrive} indicates the number of agents that arrive at base stations, and N indicates the total number of nodes in the network.

For EAs, success rate is measured as follows.

$$S_{EA} = \frac{m_{success}}{m_{total}} \tag{8}$$

$m_{success}$ indicates the number of successful migrations that an EA performs until it arrives at a base station. m_{total} indicates the total number of migration attempts that an EA makes. This includes the number of successful migrations (i.e., $m_{success}$) and the number of failed migrations.

4. **Data yield** is measured as the number of sensor data that an agent (DA or EA) aggregates and carries to a base station. Its initial value is one; however, it increases as the an agent swarms with other agents.

3.2 Elite Selection

Figure 4 shows how elite selection occurs at the MONSOON server in each duty cycle. The MONSOON server performs the same selection process for EAs and DAs separately. The first step is to measure four objective values (i.e., latency, cost, success rate and data yield) of each agent that reaches the MONSOON server via base stations. Then, each agent is evaluated whether it is *dominated* by another one. MONSOON determines that agent A dominates agent B iif:

- A's objective values are better than, or equal to, B's in all objectives, and
- A's objective values are better than B's in at least one objective.

```
Empty the archive
for each duty cycle
     ⎧Empty the population pool.
     ⎪Collect agents from the network.
     ⎪Add collected agents to the population pool.
     ⎪Move agents from the archive to the population pool.
     ⎪Empty the archive
  do ⎨for each agent of the ones in the population pool
     ⎪        ⎧if not dominated by all other agents in
     ⎪     do ⎨ the population pool
     ⎪        ⎩   then Add the agent to the archive.
     ⎪Select elite agents from the archive.
     ⎩Propagate elite agents to the network.
```

Fig. 4 Elite Selection in MONSOON

In the next step, a subset of non-dominated agents are selected as elite agents. This is performed with a four dimensional hypercube space whose axes represent four objectives. Each axis of the hypercube space is divided so that the space contains small cubes. Non-nominated agents are plotted in this hypercube space based on their objective values. If multiple non-dominated agents are plotted in a cube, one of them is randomly selected as an elite agent. If no non-dominated agents are plotted in a cube, no elite agent is selected from the cube. This elite selection is designed to maintain the diversity of elite agents. Diversity of agents can improve their adaptability to unanticipated network conditions.

Figure 5 shows an example hypercube space. For simplicity, it shows only three of four objectives (i.e., cost, latency and data yield). Each axis is divided into two ranges; therefore, eight cubes exist in total. In this example, six non-dominated agents (A to F) are plotted in the hypercube space. Three agents (B, C, and D) are plotted in a lower left cube, while the other three agents (A, E, and F) are plotted in three different cubes. From the lower left cube, only one agent is randomly selected as an elite agent. A, E, and F are selected as elite agents because they exist in different cubes.

In addition to select elite agents, the MONSOON server adjusts the mutation rate of agents based on performance improvement of non-dominated agents. The smaller improvement they

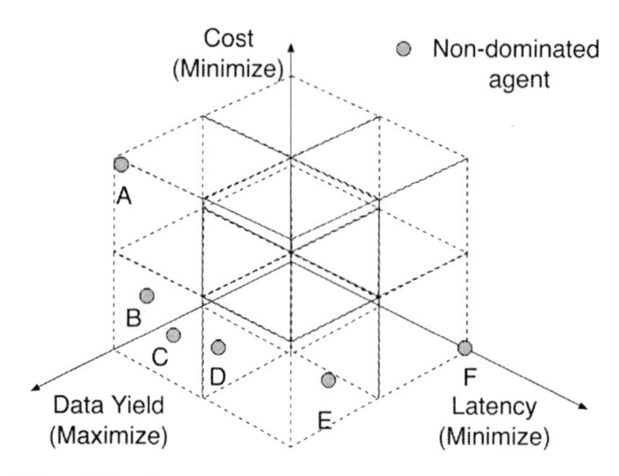

Fig. 5 An Example Elite Selection

make in objective values, the higher mutation rage the MONSOON server assigns to agents, thereby accelerating agent evolution/coevolution.

The performance of non-dominated agents is measured as a set of performance representative points in different objectives. Equation 9 shows how to obtain a performance representative point ($\bar{o}_i$) in each objective i.

$$\bar{o}_i = \frac{\sum_{a \in A} o_i(a)}{|A|} \tag{9}$$

A denotes the set of non-dominated agents. $o_i(a)$ denotes the objective value that agent a yields in objective i. o_i is a value that a performance representative point is projected on objective i. It is normalized between 0 and 1.

The improvement of performance is measured as the Euclidean distance (d) between the performance representative points at the current and previous duty cycles:

$$d = \sqrt{\frac{\sum_{i \in O} (\bar{o}_i(t) - \bar{o}_i(t-1))^2}{|O|}} \tag{10}$$

O denotes the set of all objectives. $\bar{o}_i(t)$ and $\bar{o}_i(t-1)$ denote the performance representative points projected on objective i in the current and previous duty cycles, respectively.

Mutation rate (m) is adjusted with Equation 11 where k is a constant and less than one.

$$m = k(1-d) \tag{11}$$

3.3 Genetic Operations

Once elite DAs and EAs are selected, the MONSOON server propagates them and adjusted mutation rate to each node in the network. They are propagated with a base station pheromone. Upon receiving a base station pheromone, an agent running on each node performs the reproduction behavior with a certain reproduction rate through genetic operations (crossover and mutation). It selects one of propagated elite agents, as a mating partner, which has the most similar behavior policy (genes). This similarity is measured with the Euclidean distance between the values of behavior policies. If two or more elite agents have the same similarity to the local agent, one of them is randomly selected. During reproduction, a child agent performs one-point half-and-half crossover; it randomly inherits the half of its gene from its parent agent and the other half from the parent's mating partner.

DAs can mate with elite EAs, and EAs can mate with elite DAs. This cross-mating allows DAs and EAs to coevolve their behavior policies; DAs can improve EAs' genes, and vice versa. This is particularly important when no events occur in a WSN. In this case, EAs have no chance to evolve their genes because they do not migrate toward the MONSOON server. Through cross-mating with DAs, EAs can reproduce offspring and coevolve their genes even if no events occur.

Mutation occurs on a child agent's gene with a certain mutation rate. Mutation randomly changes gene values within a predefined value range. As discussed in Section 3.2, the MONSOON server periodically adjusts mutation rate. After reproduction, a child agent takes over the local parent agent as the next generation agent.

4 Simulation Results

This section shows a set of simulation results to evaluate BiSNET/e and MONSOON. Sections 4.1, 4.2 and 4.3 discuss the simulation results obtained with a data collection application, event detection application and hybrid application. Each application is used to monitor an oil spill at the sea. The spill is simulated as 100 barrels (approximately 3,100 gallons) of crude oil spreads at the middle of the Dorchester Bay of Massachusetts. Simulation data of this spill is generated with an oil spill trajectory model implemented in the General NOAA Oil Modeling Environment [5].

A simulated WSN consists of 100 nodes uniformly deployed in an observation area of 300x300 square meters. An oil spill starts at the middle of this observation area. Each node's communication range is 30 meters and equips a surface roughness sensor to detect spilled oil. A base station is deployed on the observation area's northwestern corner. The base station links the MONSOON server via emulated serial port connection. All software components in the BiSNET/e runtime are implemented in nesC, and the MONSOON server is implemented in Java.

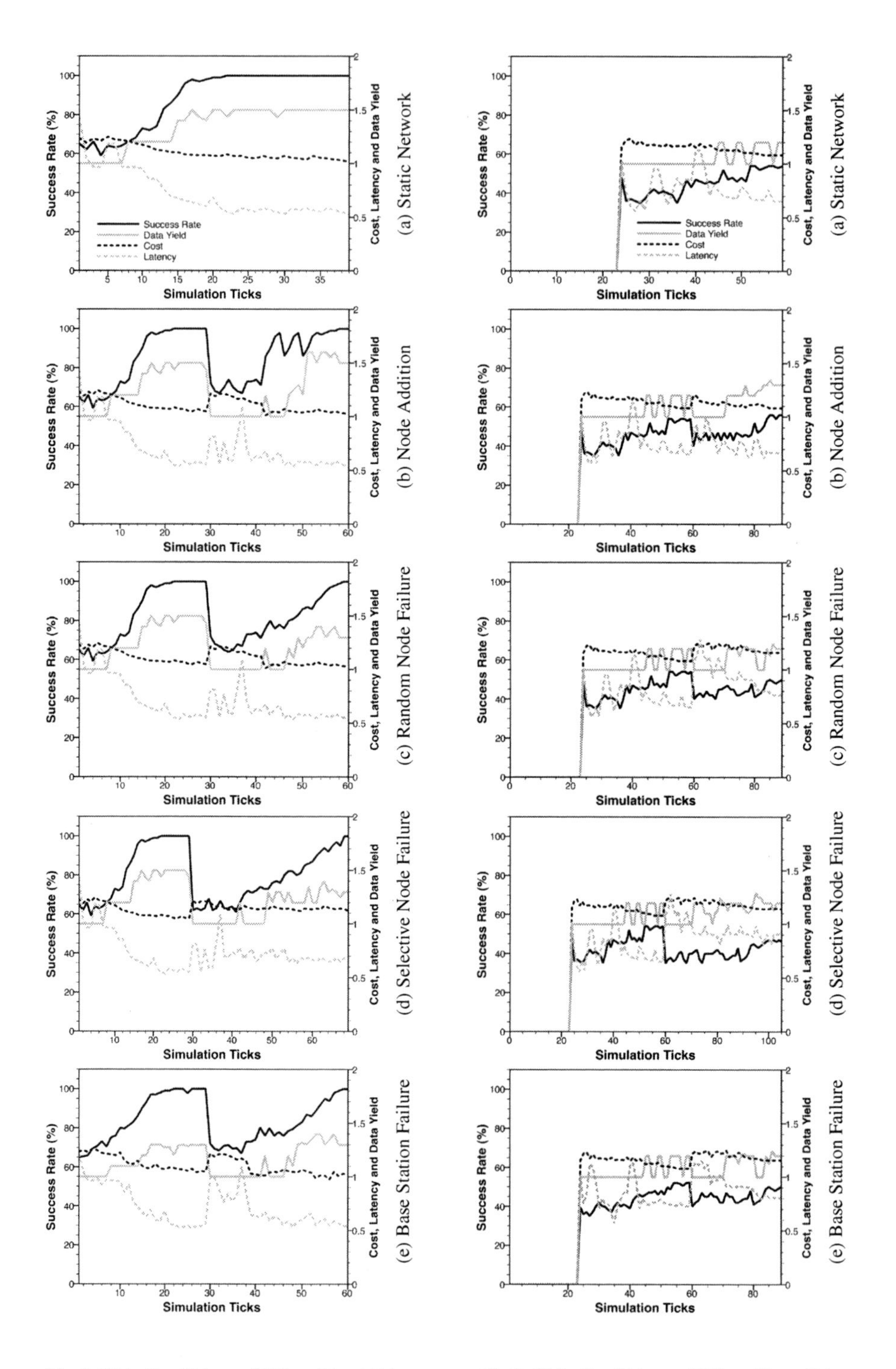

Fig.6. Objective Values of DAs without EAs

Fig.7. Objective Values of EAs without DAs

Simulation time is counted with ticks. Each tick represents five minutes. For genetic operations in MONSOON, reproduction probability and the maximum mutation rate are configured as 0.75 and 0.2, respectively.

4.1 Data Collection Application

A data collection application is implemented with DAs that perform the sequence of behaviors shown in Figure 2. No EAs are used in this application. The duty cycle corresponds to a simulation tick (five minutes).

Figure 6 (a) shows the average objective values produced by DAs at each simulation tick. Each objective value gradually improves and converges at the 22nd tick. This simulation result shows that MONSOON allows DAs to simultaneously satisfy conflicting objectives by evolving their behavior policies.

Figure 6 (b) shows how the performance of DAs changes against a dynamic node addition. 25 nodes are added at random locations at the 30th tick. Upon this change in the network environment, objective values degrade dramatically because DAs have randomly-generated behavior policies on the new nodes. Those DAs cannot migrate efficiently toward the base station. Also, enough pheromones are not available on new nodes; DAs cannot make proper migration decisions when they move to the new nodes. However, DAs gradually improve their performance again, and objective values converge again at the 56th tick. Interestingly, after 50th tick, average data yield is greater than that before 30th tick. Because there are more DAs from the additional nodes, so DAs have higher chance to swarm. MONSOON allows DAs to autonomously recover application performance despite dynamic node addition by evolving their behavior policies.

Figure 6 (c) shows how the performance of DAs changes against dynamic node failures. 25 nodes randomly fail at the 30th tick. Objective values degrade because some DAs try to migrate to failed nodes referenced by migration pheromones. This increases the number of unsuccessful agent migrations. However, DAs gradually improve their performance again, and objective values converge again at the 56th tick. MONSOON allows DAs to autonomously recover application performance despite dynamic node failures by evolving their behavior policies.

Figure 6 (d) shows how the performance of DAs changes when nodes selectively fail in a specific area. At the 30th tick, 20 nodes fail in the middle of WSN observation area. Hence, a WSN has a hole in its middle area. Compared with Figure 6 (c), it takes longer time for DAs to recover their performance. Objective values converge at 66th tick again. The converged cost and latency are worse than the ones at the 30th tick because DAs have to detour a hole (i.e., a set of failed nodes) and take longer migration paths to the base station. This simulation results shows that MONSOON allows DAs to survive selective node failures through evolution.

Figure 6 (e) shows how the performance of DAs changes against base station failures. In this simulation scenario, two base stations are deployed at the northwestern and southeastern corners of WSN observation area. At the 30th tick, a base station at the southeastern corner fails. Objective values degrade because some DAs try to migrate toward the failed base station referenced by base station pheromones. This increases the number of unsuccessful agent migrations. However, DAs gradually improve their performance again, and objective values converge again at the 56th tick. MONSOON allows DAs to autonomously evolve and recover application performance despite dynamic base station failures.

4.2 Event Detection Application

An event detection application is implemented with EAs that perform the sequence of behaviors shown in Figure 3. No DAs are used in this application. This simulation study simulates an oil spill, which occurs in the middle of WSN observation area at the 24th tick and radially spreads over time.

Figure 7 (a) shows the average objective values at each simulation tick. Upon an event detection, objective values are low because EAs use random behavior policies at first. However, each objective value gradually improves and converges at the 52nd tick. This simulation result shows that MONSOON allows EAs to simultaneously satisfy conflicting objectives by evolving their behavior policies.

Figure 7 (b) shows how the performance of EAs changes against a dynamic node addition. 25 nodes are added at random locations at the 60th tick. Upon this environmental change, objective values degrade slightly because EAs have randomly-generated behavior policies on the new nodes. Those EAs cannot migrate efficiently toward the base station. However, EAs gradually improve their performance immediately, and objective values converge again at the 85th tick. MONSOON allows EAs to autonomously recover application performance despite dynamic node addition by evolving their behavior policies.

Figure 7 (c) shows how the performance of EAs changes against dynamic node failures. 25 nodes randomly fail at the 60th tick. Objective values degrade slightly because some EAs try to migrate to failed nodes referenced by migration pheromones. This increases the number of unsuccessful agent migrations. However, EAs gradually improve their performance again, and objective values converge again at the 85th tick. MONSOON allows EAs to autonomously recover application performance despite dynamic node failures by evolving their behavior policies.

Figure 7 (d) shows the result of a simulation when 20 sensor nodes are selected in selective fashion, i.e. create a hole in the middle of network, to be deactivated at the 60th tick. Compared with the result in Figure 7 (c), MONSOON takes longer time to improve the performance of the WSN. The success rate converges at about the 100th tick to approximately 48%. The cost and latency also show the similar trend. Particularly, after the 52nd tick, the average value of cost and latency are higher than the values just before the 20th tick because agents have to detour in a longer path to avoid the hole in the middle of the network.

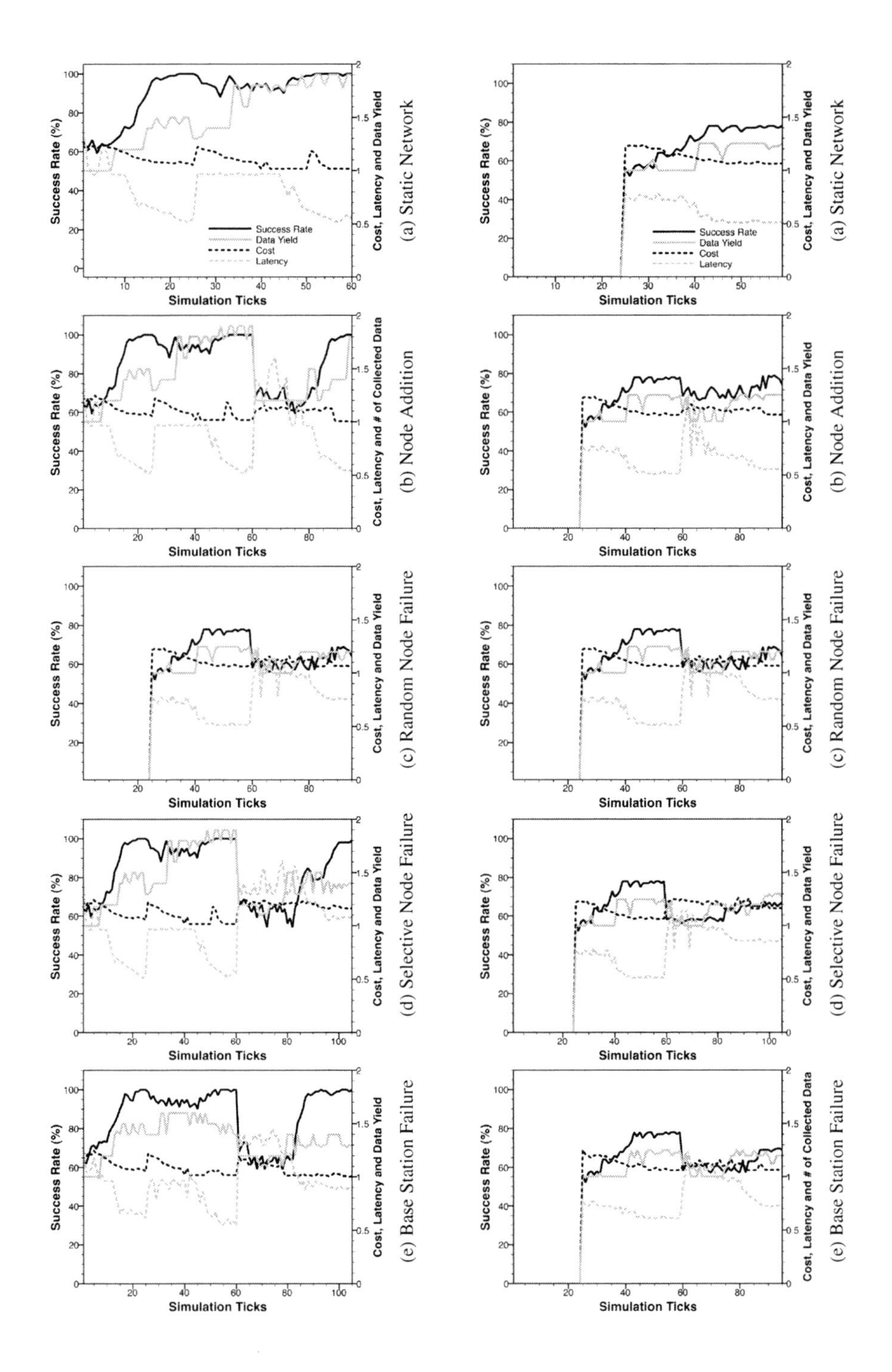

Fig.8. Objective Values of DAs with EAs

Fig.9. Objective Values of EAs with DAs

The simulation results shows that MONSOON allows WSN to survives a selective sensor nodes failure by adjusting the operational parameters of WSN to be suitable to the changes in network condition.
Figure 7 (e) shows the result of a simulation which initially has two base stations deployed at the northwestern and southeastern corner of the observation area. Then, at the 60th tick, the base station at the southeastern corner is deactivated. In this figure, at the 61st tick, the success rate drops to about 40% from around 50%. However, the success rate is improved successively and reach the same level as before the base station is deactivated at the 85th tick. Cost and latency show the same trend. MOSOON allows WSN to survives a base station failure by autonomously directing all agents to the remaining base station.

4.3 Hybrid Application

This section represents simulation results from a sensor network with two applications deployed simultaneously. Figure 8 shows the average objective values from collected DAs, i.e. for data collection application, in each simulation ticks. On the other hand, Figure 9 shows the average objective values from collected EAs, i.e. for event collection application, in each simulation ticks.
In Figure 9 (a), at the 24th simulation tick, oil spill happens and EAs start detecting and moving to the base station. The impact of EAs on DAs can be observed from Figure 8 (a) with the drop in success rate and the increase of cost and latency around 24th tick. However, within thirty simulation ticks, MONOON allows DAs to adapt to the EAs and retain their performance. The simulation results shows that MONSOON allows a WSN application to adapt to the other application such that they can co-exist tranquilly in a same sensor network.
Figure 9 (b), (c), (d) and (e) show the similar scenario as in Figure 7 (b), (c), (d) and (e), respectively. The simulation result in the former set of the figures also show the similar trend as in the later set of the figures; therefore, MONSOON allows a WSN application to adapt to network changes, i.e. partial node failure or the base station failure, even when it has to work simultaneously with another application on the same network.
Figure 9 (a) portraits the same scenario as in Figure 7 (a). In Figure 9 (a), sensor network hosts two applications, data collection and event detection. However, the objective values of event detection application, i.e. EAs, in Figure 9 (a) are improved faster than in Figure 7 (a). For example, the latency is reduced to lower than 0.5 at around the 44th tick in Figure 9 (a) but it takes about the 58th tick in Figure 7 (a) to reduce to about 0.6. Thanks to cross-mating (see section 3.3) , MONSOON allows event detection application, i.e., EAs, to improve its objective values by using information from the other application. Figure 9 (b), (c), (d) and (e) also show the similar results.

4.4 Adaptive Mutation

In the current implementation of BiSNET/e, mutation rate of EAs and DAs is adaptively adjusted by MONSOON server. Figure 10 and 11 show simulation result from the same simulation setup as in Figure 6 (a) and 7 (a) respectively; however, in Figure 10 and 11, the BiSNET/e does not use , a fix mutation rate of 0.05 is used instead. It is clear that, without adaptive mutation, MONSOON has to take about two times longer to archive the same optimized objective values. The simulation results shows that adaptive mutation in BiSNET/e allow MONSOON to quickly adjust the WSN applications to suit to environment condition.

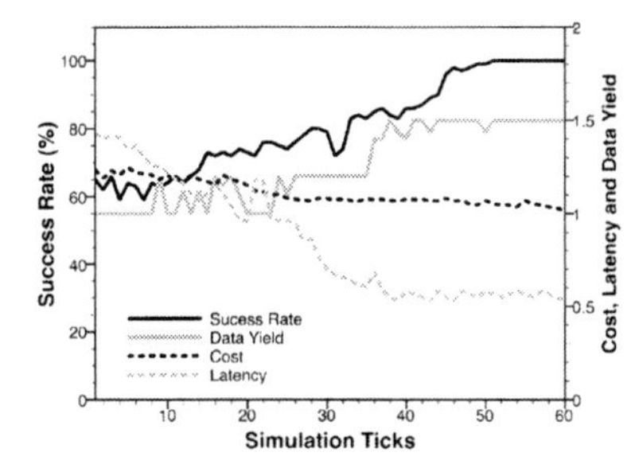

Fig. 10 Objective Values of DAs without EAs

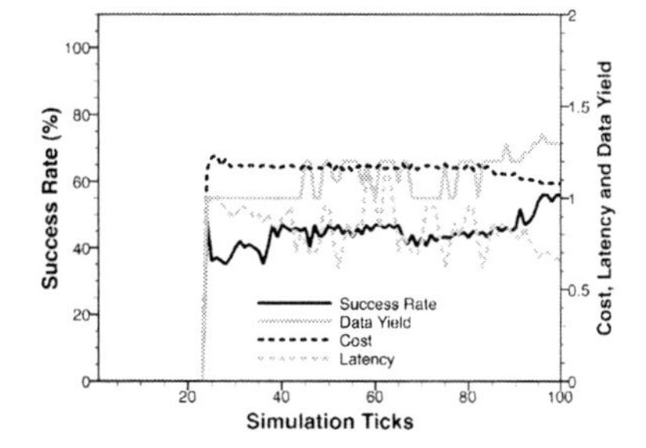

Fig. 11 Objective Values of EAs without DAs

4.5 Power Consumption

Figure 12 shows the impact of MONSOON and BiSNET/e on power consumption, and compare between hybrid application and individual applications. The figure represents the power consumption on each simulation tick for the sensor network with node addition scenario, e.g. as in Figure 6 (b). In this figure, individual applications represents summation of the power consumption of data collection and event detection application when they are implemented separately, i.e. the summation of power consumption from sensor network in Figures 6 (b) and 7 (b) in each simulation tick. On the other hand, hybrid application represents the power consumption of a sensor network which implements both data collection and event detection on the same application, i.e. from Figure 8 (b). In this figure, MONSOON and BiSNET/e can reduce the power consumption of WSN by optimizing the agent's behavior policy. Moreover, by implementing hybrid application on a same framework, the power consumption can be further reduced which can be seen when compare the power consumption of hybrid application and individual applications.

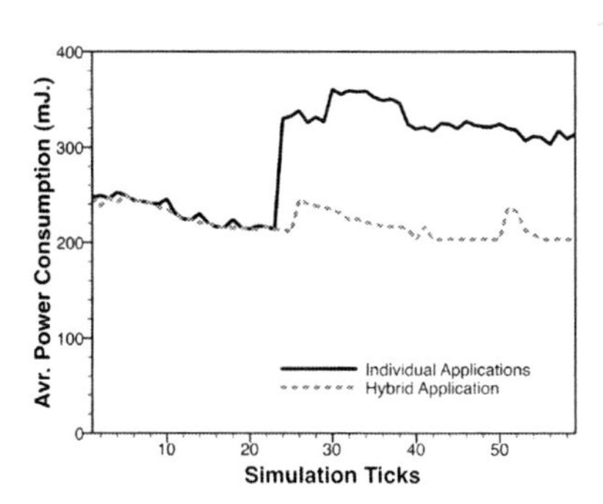

Fig. 12 Average Power Consumption

4.6 Memory Footprint

Table 1 shows the memory footprint of the BiSNET/e runtime in a MICA2 mote, and compares it with the footprint of Blink (an example program in TinyOS), which periodically turns on and off an LED, and Agilla, which is a mobile agent platform for WSNs [13]. The BiSNET/e runtime is lightweight in its footprint thanks to the simplicity of the biologically-inspired mechanisms in BiSNET/e. BiSNET/e can even run on a smaller-scale nodes, for example, TelosB, which has 48KB ROM.

Table 1 Memory Footprint in a MICA2 Node

	RAM (KB)	ROM (KB)
BiSNET	2.8	31.2
Blink	0.04	1.6
Agilla	3.59	41.6

5 Related Work

This chapter extends the authors' prior work [7–9]. In [7], the authors proposed a biologically-inspired WSN architecture, called BiSNET. BiSNET does not investigate evolutionary adaptation. Thus, agent behavior policies are manually configured through trial-and-errors and fixed at runtime. Unlike BiSNET, BiSNET/e allows agents to dynamically adapt their behavior policies to unanticipated network conditions. In [8], MONSOON was proposed and studied with data collection applications. This chapter considers event detection applications and hybrid applications as well as data collection applications. Moreover, this chapter evaluates how coevolution between DAs and EAs augments agent adaptability. This is beyond the scope of [8]. Compared with [9], this chapter investigates new operational objective (the degree of data aggregation) and new mechanisms in MONSOON (e.g., swarming behavior, migration probability and adaptive mutation).

Agilla proposes a programming language to implement mobile agents for WSNs, and provides an interpreter to operate agents on TinyOS [13]. Similarly, BiSNET/e exploits mobile agents (DAs and EAs); however, this chapter does not focus on investigating a new programming language for those agents. While BiSNET/e and Agilla implement a similar set of agent behaviors such as migration and replication, BiSNET/e studies a wider range of agent behaviors. For example, Agilla does not consider energy gain/expenditure, swarming and pheromone emission. Moreover, Agilla does not consider evolutionary and coevolutionary adaptation of agents to seek optimal tradeoffs among conflicting objectives. As shown in Table 1, BiSNET/e is implemented more lightweight than Agilla.

Virtual pheromone (VP) is a biologically-inspired node-to-node communication primitive in TinyOS-based WSNs [37]. It has a generic set of properties such as pheromone type, strength, source and payload. Therefore, VP can be used to implement base station, migration and alert pheromones in BiSNET/e. However, VP does not address a research issue that BiSNET/e does: autonomous adaptability of WSN applications.

Quasar is similar to BiSNET/e in that it proposes a data collection protocol that balances the tradeoff between data accuracy and power efficiency [16]. Although BiSNET/e does not focus on data accuracy as its operational objective, it studies extra objectives in data transmission such as success rate and latency. Also, it considers not only data collection applications but also event detection and hybrid applications in dynamic WSNs. (Quasar is considered and evaluated for static WSNs.) Quasar and BiSNET/e employ different optimization/adaptation processes; BiSNET performs a population-based evolutionary algorithm while Quasar employs time series data analysis.

[4] proposes a cost function (or fitness function) that comprises conflicting objectives regarding data transmission cost, power consumption, latency, reliability (the time between node/link failures) and link interference. These objectives are similar to the ones BiSNET/e considers. However, in [4], the total cost (or fitness) is calculated as a weighted sum of objective values. This means that application designers need to manually configure every weight value in a fitness function through trial-and-errors. In BiSNET/e, no manually-configured parameters exist for elite selection because of a domination ranking mechanism. BiSNET/e minimizes the number of manually-configured parameters to minimize configuration costs for application designers. Moreover, BiSNET/e does not require each node to have global network information as [4] does.

Genetic algorithms (GAs) have been investigated in various aspects in WSNs; for example, routing [10, 12, 18, 20, 23], data processing [17], localization [38, 42], node placement [15, 43] and object tracking [10]. All of these work use fitness functions, each of which combines multiple objective values as a weighted sum and rank agents/genes in elite selection. As discussed above, it is always non-trivial to manually configure weight values in a fitness function through. In contrast, BiSNET/e eliminates parameters in elite selection by design. Moreover, [10, 12, 15, 18, 23, 38, 43] do not assume dynamic WSNs, but static WSNs.

Beyond these classical GAs, multiobjective GAs (MOGAs) have also been investigated in WSNs; for example, for routing [30, 31, 35, 40], node placement [19, 21, 27, 29, 32] and duty cycle management [41]. In all of these work except [35], a central server performs an evolutionary optimization process. This can lead to scalability issue as network size increases. In contrast, MONSOON is carefully designed to perform its optimization process in both the MONSOON server and individual nodes. Moreover, all of these work do not assume dynamic WSNs, but static WSNs.

[30, 31, 40] investigates MOGAs that optimize migration routes for mobile agents to travel from a base station to cluster head nodes and collect sensor data from individual clusters. In BiSNET/e, agents make their migration and other behavior decisions by themselves. MONSOON optimizes their behavior policies, not agents' migration routes.

[35] is similar to BiSNET/e in that both follow the agent designs proposed in BiSNET and exploit MOGAs to adapt agent behavior policies. Unlike [35], BiSNET/e studies coevolution between DAs and EAs as well as their regular evolution (i.e., single-species evolution). [35] considers data collection applications only in static WSNs. Also, BiSNET/e performs adaptive mutation and crossover, which [35] does not consider.

Adaptive mutation was initially proposed in [36], and it has been used in WSNs [10, 23]. In [10, 23, 36], mutation rate is dynamically adjusted based on the current fitness that is a weighted sum of objective values. In MONSOON, mutation rate is adjusted based on the progress of performance improvement by the non-dominated individuals.

6 Conclusion

This chapter describes a coevolutionary multiobjective adaptation framework for WSNs, called MONSOON. MONSOON allows WSN applications to simultaneously satisfy conflicting operational objectives by adapting to dynamic network conditions (e.g., network traffic and node/link failures) through evolution. Thanks to a set of simple biologically-inspired mechanisms, the BiSNET/e runtime is implemented lightweight.

Some extensions to MONSOON and BiSNET/e are planed. The extensions include associating a constraint(s) with each operational objective. A constraint is defined as an upper or lower bound for each objective. For example, a tolerable (upper) bound may be defined for the latency objective. Constraints allow agent designers to flexibly specify their specific requirements (or priorities) on objectives. They can also improve evolution speed by dedicating agents to satisfy those constraints.

References

1. Akkaya, K., Younis, M.: A survey of routing protocols in wireless sensor networks. Elsevier Ad Hoc Networks **3**(3), 325–349 (2005)
2. Akyildiz, I.F., Su, W., Sankarasubramaniam, Y., Cayirci, E.: Wireless sensor networks: A survey. Elsevier J. of Computer Networks **38**(4), 393–422 (2002)
3. Albert, R., Jeong, H., Barabasi, A.: Error and attack tolerance of complex networks. Nature **406**(6794), 378–382 (2000)
4. Baldi, P., Nardis, L.D., Benedetto, M.G.D.: Modeling and optimization of uwb communication networks through a flexible cost function. IEEE J. on Sel. Areas in Comm. **20**(9), 1733–1744 (2002)
5. Beegle-Krause, C.: General NOAA oil modeling environment (GNOME): A new spill trajectory model. In: Proc. of Int'l Oil Spill Conf. (2001)
6. Blumenthal, J., Handy, M., Golatowski, F., Haase, M., Timmermann, D.: Wireless sensor networks - new challenges in software engineering. In: Proc. of IEEE Emerging Technologies and Factory Automation (2003)
7. Boonma, P., Suzuki, J.: BiSNET: A biologically-inspired middleware architecture for self-managing wireless sensor networks. Elsevier J. of Computer Networks **51** (2007)
8. Boonma, P., Suzuki, J.: Evolutionary constraint-based multiobjective adaptation for self-organizing wireless sensor networks. In: Proc. of ACM/IEEE/Create-Net/ICST Int'l Conf. Bio-Inspired Models of Network, Info. and Comp. Sys. (2007)
9. Boonma, P., Suzuki, J.: Monsoon: A coevolutionary multiobjective adaptation framework for dynamic wireless sensor networks. In: Proc. of IEEE Hawaii Int'l Conf on System Sciences (2008)
10. Buczaka, A.L., Wangb, H.: Optimization of fitness functions with non-ordered parameters by genetic algorithms. In: Proc. of IEEE Congress on Evolutionary Comp. (2001)
11. Chintalapudi, K.K., Govindan, R.: Localized edge detection in sensor fields. Elsevier Ad-hoc Networks **1**, 59–70 (2003)
12. Ferentinos, K.P., Tsiligiridis, T.A.: Adaptive design optimization of wireless sensor networks using genetic algorithms. Elsevier J. of Computer Nets. **51**(4), 1031–1051 (2007)
13. Fok, C.L., Roman, G.C., Lu, C.: Rapid development and flexible deployment of adaptive wireless sensor network applications. In: Proc. of IEEE Int'l Conf. on Distributed Computing Systems (2005)
14. Free, J.B., Williams, I.H.: The role of the nasonov gland pheromone in crop communication by honey bees. Brill Int'l J. of Behavioural Biology **41**(3–4), 314–318 (1972)
15. Guo, H.Y., Zhang, L., Zhang, L.L., Zhou, J.X.: Optimal placement of sensors for structural health monitoring using improved genetic algorithms. IOP Smart Materials and Structures **13**(3), 528–534 (2004)
16. Han, Q., Mehrotra, S., Venkatasubramanian, N.: Energy efficient data collection in distributed sensor environments. In: Proc. of IEEE Int'l Conf. on Distributed Computing Systems (2004)
17. Hauser, J., Purdy, C.: Sensor data processing using genetic algorithms. In: Proc. of IEEE Midwest Symp. on Circuits and Systems (2000)
18. Hussain, S., Matin, A.W.: Hierarchical cluster-based routing in wireless sensor networks. In: Proc. of IEEE/ACM Conf. on Info. Processing in Sensor Nets (2006)
19. Jia, J., Chen, J., Chang, G., Tan, Z.: Energy efficient coverage control in wireless sensor networks based on multi-objective genetic algorithm. Elsevier Computers & Mathematics with Applications **10** (2008)

20. Jin, S., Zhou, M., Wu, A.S.: Sensor network optimization using a genetic algorithm. In: Proc. of IIIS World Multiconf. on Systemics, Cybernetics and Informatics (2003)
21. Jourdan, D.B., de Weck, O.L.: Multi-objective genetic algorithm for the automated planning of a wireless sensor network to monitor a critical facility. In: Proc. of SPIE Defense and Security Symp. (2004)
22. Karl, H., Willig, A.: Protocols and Architectures for Wireless Sensor Networks. Wiley-Interscience (2007)
23. Khanna, R., Liu, H., Chen, H.: Self-organisation of sensor networks using genetic algorithms. Inderscience Int'l J. of Sensor Networks **1**(3), 241–252 (2006)
24. Leibnitz, K., Wakamiya, N., Murata, M.: Biologically inspired networking. In: Q. Mahmoud (ed.) Cognitive Networks: Towards Self-Aware Networks. Wiley (2007)
25. Li, D., Wong, K.D., Hu, Y., Sayeed, A.M.: Detection, classification, and tracking of targets. IEEE Signal Processing Magazine **19**(2), 17–20 (2002)
26. Mathur, G., Desnoyers, P., Genesan, D., Shenoy, P.: Ultra-low power data storage for sensor networks. In: Proc. of IEEE/ACM Conf. on Info. Processing in Sensor Nets (2006)
27. Molina, G., Alba, E., Talbi, E.G.: Optimal sensor network layout using multi-objective metaheuristics. J. of Universal Computer Science **14**(15), 2549–2565 (2008)
28. Phoha, S., La Porta, T.F., Griffin, C.: Sensor Network Operations. Wiley-IEEE Press (2006)
29. Raich, A.M., Liszkai, T.R.: Multi-objective genetic algorithm methodology for optimizing sensor layouts to enhance structural damage identification. In: Proc. of Int'l Workshop on Structural Health Monitoring (2003)
30. Rajagopalan, R., Mohan, C., Varshney, P., Mehrotra, K.: Multi-objective mobile agent routing in wireless sensor networks. In: Proc. of IEEE Congress on Evolutionary Comp. (2005)
31. Rajagopalan, R., Varshney, P.K., Mehrotra, K.G., Mohan, C.K.: Fault tolerant mobile agent routing in sensor networks: A multi-objective optimization approach. In: Proc. of IEEE Upstate New York Workshop on Communication and Networking (2005)
32. Rajagopalan, R., Varshney, P.K., Mohan, C.K., Mehrotra, K.G.: Sensor placement for energy efficient target detection in wireless sensor networks: A multi-objective optimization approach. In: Proc. of IEEE Annual Conf. on Information Sciences and Systems (2005)
33. Rentala, P., Musunuri, R., Gandham, S., Sexena, U.: Survey on sensor networks. In: Proc. of ACM Int'l Conf. on Mobile Computing and Networking (2001)
34. Seeley, T.: The Wisdom of the Hive. Harvard University Press (2005)
35. Sin, H., Lee, J., Lee, S., Yoo, S., Lee, S., Lee, J., Lee, Y., , Kim, S.: Agent-based framework for energy efficiency in wireless sensor networks. World Academy of Science, Engineering and Technology **35**, 305–309 (2008)
36. Srinivas, M., Patnaik, L.: Adaptive probabilities of crossover and mutation in genetic algorithms. IEEE Tran. on Systems, Man and Cybernetics **24**(4), 656–667 (1994)
37. Szumel, L., Owens, J.D.: The virtual pheromone communication primitive. In: Proc. of IEEE Int'l Conf. on Distributed Computing in Sensor Systems (2006)
38. Tam, V., Cheng, K.Y., Lui, K.S.: Using micro-genetic algorithms to improve localization in wireless sensor networks. Academy J. of Comm. **1**(4), 1–10 (2006)
39. Wada, H., Boonma, P., Suzuki, J.: Macroprogramming spatio-temporal event detection and data collection in wireless sensor networks: An implementation and evaluation study. In: Proc. of IEEE Hawaii Int'l Conf on System Sciences (2008)
40. Xuea, F., Sanderson, A., Graves, R.: Multi-objective routing in wireless sensor networks with a differential evolution algorithm. In: Proc. of IEEE Int'l Conf. on Networking, Sensing and Control (2006)
41. Yang, E., Erdogan, A.T., Arslan, T., Barton, N.: Multi-objective evolutionary optimizations of a space-based reconfigurable sensor network under hard constraints. In: Proc. of ECSIS Symp. on Bio-inspired, Learning, and Intelligent Sys. for Security (2007)
42. Zhang, Q., Wang, J., Jin, C., Ye, J., Ma, C., Zhang, W.: Genetic algorithm based wireless sensor network localization. In: Proc. of IEEE Int'l Conf. on Natural Computation (2008)

43. Zhao, J., Wen, Y., Shang, R., Wang, G.: Optimizing sensor node distribution with genetic algorithm in wireless sensor network. In: Proc. of IEEE Int'l Symp. on Neural Nets. (2004)

Index

abstract computer, 248
abstract regions, 309
adaptive mutation, 356
alternative modality, 258
automatic parallelization, 227
Autonomic, 67–88
  Computing (Initiative), 69
  Element, 67, 69
  Manager, 71, 78
  Networking, 69, 75
autonomic communication, 245
autonomous flight systems, 223

biologically-inspired, 341

caching policies
  generation-based, 127
  location-based, 125
case studies and applications
  connectivity to rural villages, 35
  Haggle project, 35
  ZebraNet, 35
case-based reasoning, 267
choropleth, 305
classification of modalities, 255
code transformation, 231
complex interactions, 224
constraint propagation, 227
context, 244, 247
context storage, 254
contour map, 308

data collection, 341
deadlock, 225
determining context, 251
DiffServ, 68, 77–78
dynamically-resizable array, 225

energy map, 302
energy scan (eScan), 308
environment context, 247
event, 306
event detection, 341
event service, 306
evolutionary adaptation framework, 342
exemplar, 265

gMAP, 309
goal networks, 224

hotspot model, 314
hovering information, 114
  accessibility of, 118
  anchor area, 114
  anchor location, 114
  anchor radius, 114
  availability of, 117
  survivability of, 117
  system, 115
human-machine interface, 249

incremental context, 244
Inductive Logic Programming, 74
input modality, 246
interaction context, 245–247
iso-map, 309
isoline, 308
isomap, 305

Knowledge
  Plane, 67–88
  Plane (Collaborative), 71
  Sharing
    Algorithm, 84
    Processes, 75
  Sharing (Situated), 84

layered virtual machine, 244
linearizability, 234
livelock, 225
lock-free synchronization, 225, 233

Machine Learning, 67–88
  Algorithm, 77
  Module, 78
machine learning, 244
map, 302, 305
map-based prediction, 310
Map-based World Model, 302, 305
Map-based WSN design, 302
Mars Pathfinder, 225
media, 255
media devices priority table, 259
media priority, 258
Mission Data System, 223
mission-critical space software, 224
mobility models
  CMM, 37
  HCMM, 37
modality, 255
modality's suitability, 256
multimodal multimedia, 244
multimodality, 244
mutual exclusion locks, 225
MWM, 302, 305

network maps, 305
network partitioning, 311
network world, 303
nonblocking object, 225
nonblocking synchronization, 233
nonblocking vector, 226

Opportunistic networks, 32
  against DTNs, 33
optimal input modality, 257
optimal output modality, 257
output modality, 246

parallel propagation, 236
pervasive computing, 244
physical world, 303
pop_back, 235
post-condition scenario, 263
pre-condition scenario, 263
priority inversion, 225
push_back, 235

query service, 306

real-time C++, 223
real-world situation, 306
region, 302, 305
replication algorithms
  attractor point, 123
  broadcast-based, 124
  irrelevant area, 122
  relevant area, 121
  risk area, 121
  safe area, 121
routing protocols
  BubbleRap, 45
  CAR, 46
  Epidemic, 44
  HiBOp, 47
  MobySpace, 45
  PROPHET, 45

scenario, 263
scenario number, 265
scenario table, 265
Self-adaptation, 67–88
self-healing, 247
Self-organization, 67–88
semantic parallelization, 223, 230
similarity relevance score, 268
similarity scoring algorithm, 266
Situated View, 81
software agents, 341
software certification, 223
sources of complexity, 224
specimen interaction context, 260
STL vector, 225
supervised learning, 267
system context, 248

temporal constraint, 226
Temporal Constraint Network, 226
tight coupling, 224
time point, 226
total scenarios, 265

ubiquitous computing, 244
user context, 247
user handicap, 263
user maps, 305
user profile, 261

virtual machine, 244
virtualization, 248

Wireless sensor networks, 341
world model, 303
write descriptor, 235

Breinigsville, PA USA
08 October 2009
225398BV00004B/35/P

9 780387 097527